Dmitry Panchenko

Lecture Notes on Probability Theory

ISBN-13: 978-1-9994190-4-2
ISBN-10: 1-9994190-4-9

Contents

Chapter 1
Introduction

1.1 Probability spaces

$(\Omega, \mathscr{A}, \mathbb{P})$ is a probability space if $(\Omega, \mathscr{A})$ is a measurable space and $\mathbb{P}$ is a measure on $\mathscr{A}$ such that $\mathbb{P}(\Omega) = 1$. Let us recall some definitions from measure theory. A pair $(\Omega, \mathscr{A})$ is a *measurable space* if $\mathscr{A}$ is a σ-algebra of subsets of Ω. A collection A of subsets of Ω is called *an algebra* if:

(i) $\Omega \in A$,
(ii) $C, B \in A \Longrightarrow C \cap B, C \cup B \in A$,
(iii) $B \in A \Longrightarrow \Omega \setminus B \in A$.

A collection $\mathscr{A}$ of subsets of Ω is called a *σ-algebra* if it is an algebra and

(iv) $C_i \in \mathscr{A}$ for all $i \geq 1 \Longrightarrow \cup_{i \geq 1} C_i \in \mathscr{A}$.

Elements of the σ-algebra $\mathscr{A}$ are often called *events*. $\mathbb{P}$ is a *probability measure* on the σ-algebra $\mathscr{A}$ if

(1) $\mathbb{P}(\Omega) = 1$,
(2) $\mathbb{P}(A) \geq 0$ for all events $A \in \mathscr{A}$,
(3) $\mathbb{P}$ is *countably additive*: given any disjoint $A_i \in \mathscr{A}$ for $i \geq 1$, i.e. $A_i \cap A_j = \emptyset$ for all $i \neq j$,

$$\mathbb{P}\Big(\bigcup_{i=1}^{\infty} A_i\Big) = \sum_{i=1}^{\infty} \mathbb{P}(A_i).$$

It is sometimes convenient to use an equivalent formulation of property (3):

(3′) $\mathbb{P}$ is finitely additive and *continuous*, i.e. for any decreasing sequence of events $B_n \in \mathscr{A}$, $B_n \supseteq B_{n+1}$,

$$B = \bigcap_{n \geq 1} B_n \Longrightarrow \mathbb{P}(B) = \lim_{n \to \infty} \mathbb{P}(B_n).$$

Lemma 1.1. *Properties (3) and (3′) are equivalent.*

Proof. First, let us show that (3) implies (3′). If we denote $C_n = B_n \setminus B_{n+1}$ then B_n is the disjoint union $\big(\cup_{k\geq n} C_k\big) \cup B$ and, by (3),

$$\mathbb{P}(B_n) = \mathbb{P}(B) + \sum_{k\geq n} \mathbb{P}(C_k).$$

Since the last sum is the tail of convergent series, $\lim_{n\to\infty} \mathbb{P}(B_n) = \mathbb{P}(B)$. Next, let us show that (3′) implies (3). If, given disjoint sets (A_n), we define $B_n = \cup_{i\geq n+1} A_i$, then

$$\bigcup_{i\geq 1} A_i = A_1 \bigcup A_2 \bigcup \cdots \bigcup A_n \bigcup B_n,$$

and, by finite additivity,

$$\mathbb{P}\Big(\bigcup_{i\geq 1} A_i\Big) = \sum_{i=1}^{n} \mathbb{P}(A_i) + \mathbb{P}(B_n).$$

Clearly, $B_n \supseteq B_{n+1}$ and, since (A_n) are disjoint, $\cap_{n\geq 1} B_n = \emptyset$. Therefore, by (3′), $\lim_{n\to\infty} \mathbb{P}(B_n) = 0$ and (3) follows.

Let us give several examples of probability spaces. One most basic example of a probability space is $([0,1], \mathscr{B}([0,1]), \lambda)$, where $\mathscr{B}([0,1])$ is the Borel σ-algebra on $[0,1]$ and λ is the Lebesgue measure. Let us quickly recall how this measure is constructed. More generally, let us consider the construction of the Lebesgue-Stieltjes measure on $\mathbb{R}$ corresponding to a non-decreasing right-continuous function $F(x)$. One considers an algebra of finite unions of disjoint intervals

$$A = \Big\{ \bigcup_{i\leq n} (a_i, b_i] \ : n \geq 1, \text{ all } (a_i, b_i] \text{ are disjoint} \Big\}$$

and defines the measure F on the sets in this algebra by (we slightly abuse the notations here)

$$F\Big(\bigcup_{i\leq n} (a_i, b_i]\Big) = \sum_{i=1}^{n} \big(F(b_i) - F(a_i)\big).$$

It is not difficult to show that F is countably additive on the algebra A, i.e.,

$$F\Big(\bigcup_{i=1}^{\infty} A_i\Big) = \sum_{i=1}^{\infty} F(A_i)$$

whenever all A_i and $\cup_{i\geq 1} A_i$ are finite unions of disjoint intervals. The proof is exactly the same as in the case of $F(x) = x$ corresponding to the Lebesgue measure case. Once countable additivity is proved on the algebra, it remains to appeal to the following key result. Recall that, given an algebra A, the σ-algebra $\mathscr{A} = \sigma(A)$ generated by A is the smallest σ-algebra that contains A.

Theorem 1.1 (Caratheodory's extension theorem). *If A is an algebra of sets and $\mu : A \to \mathbb{R}$ is a non-negative countably additive function on A, then μ can be ex-*

tended to a measure on the σ-algebra $\sigma(A)$. If μ is σ-finite, then this extension is unique.

Therefore, F above can be uniquely extended to the measure on the σ-algebra $\sigma(A)$ generated by the algebra of finite unions of disjoint intervals. This is the σ-algebra $\mathcal{B}(\mathbb{R})$ of Borel sets on $\mathbb{R}$. Clearly, $(\mathbb{R}, \mathcal{B}(\mathbb{R}), F)$ will be a probability space if

(1) $F(x) = F((-\infty, x])$ is non-decreasing and right-continuous,
(2) $\lim_{x \to -\infty} F(x) = 0$ and $\lim_{x \to \infty} F(x) = 1$.

The reason we required F to be right-continuous corresponds to our choice that intervals $(a,b]$ in the algebra are closed on the right, so the two conventions agree and the measure F is continuous, as it should be, e.g.

$$F((a,b]) = F\Big(\bigcap_{n \ge 1} (a, b+n^{-1}]\Big) = \lim_{n \to \infty} F((a, b+n^{-1}]).$$

In Probability Theory, functions satisfying properties (1) and (2) above are called *cumulative distribution functions*, or c.d.f. for short, and we will give an alternative construction of the probability space $(\mathbb{R}, \mathcal{B}(\mathbb{R}), F)$ in the next section.

Other basic ways to define a probability is through a probability function or density function. If the measurable space is such that all singletons are measurable, we can simply assign some weights $p_i = \mathbb{P}(\omega_i)$ to a sequence of distinct points $\omega_i \in \Omega$, such that $\sum_{i \ge 1} p_i = 1$, and let

$$\mathbb{P}(A) = \mathbb{P}\big(A \cap \{\omega_i\}_{i \ge 1}\big).$$

The function $i \to p_i = \mathbb{P}(\{\omega_i\})$ is called a *probability function*. Now, suppose that we already have a σ-finite measure $\mathbb{Q}$ on $(\Omega, \mathcal{A})$, and consider any measurable function $f \colon \Omega \to \mathbb{R}^+$ such that

$$\int_\Omega f(\omega)\, d\mathbb{Q}(\omega) = 1.$$

Then we can define a probability measure $\mathbb{P}$ on $(\Omega, \mathcal{A})$ by

$$\mathbb{P}(A) = \int_A f(\omega)\, d\mathbb{Q}(\omega).$$

The function f is called the *density function* of $\mathbb{P}$ with respect to $\mathbb{Q}$ and, in a typical setting when $\Omega = \mathbb{R}^k$ and $\mathbb{Q}$ is the Lebesgue measure λ, f is simply called the density function of $\mathbb{P}$.

Example 1.1.1. A probability measure on $\mathbb{R}$ corresponding to the probability function

$$p_i = \mathbb{P}(\{i\}) = \frac{\lambda^i}{i!} e^{-\lambda}$$

for integer $i \geq 0$ is called the Poisson distribution with the parameter $\lambda > 0$. (*Notation*: given a set A, we will denote by $\mathrm{I}(x \in A)$ or $\mathrm{I}_A(x)$ the indicator that x belongs to A.) □

Example 1.1.2. A probability measure on $\mathbb{R}$ corresponding to the density function $f(x) = \lambda e^{-\lambda x}\mathrm{I}(x \geq 0)$ is called the exponential distribution with the parameter $\lambda > 0$. □

Example 1.1.3. A probability measure on $\mathbb{R}$ corresponding to the density function

$$f(x) = \frac{1}{\sqrt{2\pi}} e^{-x^2/2}$$

is called the standard normal, or standard Gaussian, distribution on $\mathbb{R}$. □

Recall that a measure $\mathbb{P}$ is called *absolutely continuous* with respect to another measure $\mathbb{Q}$, $\mathbb{P} \prec \mathbb{Q}$, if for all $A \in \mathscr{A}$,

$$\mathbb{Q}(A) = 0 \Longrightarrow \mathbb{P}(A) = 0,$$

in which case the existence of the density is guaranteed by the following classical result from measure theory.

Theorem 1.2 (Radon-Nikodym). *On a measurable space $(\Omega, \mathscr{A})$ let μ be a σ-finite measure and ν be a finite measure absolutely continuous with respect to μ, $\nu \prec \mu$. Then there exists the* Radon-Nikodym derivative $h \in \mathscr{L}^1(\Omega, \mathscr{A}, \mu)$ *such that*

$$\nu(A) = \int_A h(\omega)\, d\mu(\omega)$$

for all $A \in \mathscr{A}$. Such h is unique modulo μ-a.e. equivalence.

Of course, the Radon-Nikodym theorem also applies to finite signed measures ν, which can be decomposed into $\nu = \nu^+ - \nu^-$ for some finite measures ν^+, ν^- – the so-called Hahn-Jordan decomposition. Let us recall a proof of the Radon-Nikodym theorem for convenience.

Proof. Clearly, we can assume that μ is a finite measure. Consider the Hilbert space $H = \mathscr{L}^2(\Omega, \mathscr{A}, \mu + \nu)$ and the linear functional $T : H \to \mathbb{R}$ given by $T(f) = \int f\, d\nu$. Since

$$\Big|\int f\, d\nu\Big| \leq \int |f|\, d(\mu + \nu) \leq C\|f\|_H,$$

T is a continuous linear functional and, by the Riesz-Fréchet theorem, $\int f\, d\nu = \int f g\, d(\mu + \nu)$ for some $g \in H$. This implies $\int f\, d\mu = \int f(1 - g)\, d(\mu + \nu)$. Now $g(\omega) \geq 0$ for $(\mu + \nu)$-almost all ω, which can be seen by taking $f(\omega) = \mathrm{I}(g(\omega) < 0)$, and similarly $g(\omega) \leq 1$ for $(\mu + \nu)$-almost all ω. Therefore, we can take $0 \leq g \leq 1$. Let $E = \{\omega : g(\omega) = 1\}$. Then

$$\mu(E) = \int \mathrm{I}(\omega \in E)\, d\mu(\omega) = \int \mathrm{I}(\omega \in E)(1 - g(\omega))\, d(\mu + \nu)(\omega) = 0,$$

and since $\nu \prec \mu$, $\nu(E) = 0$. Since $\int fg\,d\mu = \int f(1-g)\,d\nu$, we can restrict both integral to E^c, and replacing f by $f/(1-g)$ and denoting $h := g/(1-g)$ (defined on E^c) we get $\int_{E^c} f\,d\nu = \int_{E^c} fh\,d\mu$ (more carefully, one can truncate $f/(1-g)$ first and then use the monotone convergence theorem). Therefore, $\nu(A\cap E^c) = \int_{A\cap E^c} h\,d\mu$ and this finishes the proof if we set $h = 0$ on E. To prove uniqueness, consider two such h and h' and let $A = \{\omega : h(\omega) > h'(\omega)\}$. Then $0 = \int_A (h-h')d\mu$ and, therefore, $\mu(A) = 0$. □

Let us now write down some important properties of σ-algebras and probability measures.

Lemma 1.2 (Approximation lemma). *If A is an algebra of sets then for any $B \in \sigma(A)$ there exists a sequence $B_n \in A$ such that* $\lim_{n\to\infty} \mathbb{P}(B \triangle B_n) = 0$.

Proof. Here $B \triangle B_n$ denotes the symmetric difference $(B\cup B_n)\setminus(B\cap B_n)$. Let

$$\mathscr{D} = \big\{B \in \sigma(A) : \lim_{n\to\infty} \mathbb{P}(B \triangle B_n) = 0 \text{ for some } B_n \in A\big\}.$$

We will prove that $\mathscr{D}$ is a σ-algebra and, since $A \subseteq \mathscr{D}$, this will imply that $\sigma(A) \subseteq \mathscr{D}$. One can easily check that

$$d(B,C) := \mathbb{P}(B \triangle C) = \int_\Omega |I_B(\omega) - I_C(\omega)|\,d\mathbb{P}(\omega)$$

is a semi-metric, which satisfies

(a) $d(B\cap C, D\cap E) \le d(B,D) + d(C,E)$,
(b) $|\mathbb{P}(B) - \mathbb{P}(C)| \le d(B,C)$,
(c) $d(B^c,C^c) = d(B,C)$.

Now, consider $D_1,\ldots,D_N \in \mathscr{D}$. If a sequence $C_{in} \in A$ for $n \ge 1$ approximates D_i, i.e.

$$\lim_{n\to\infty} \mathbb{P}(C_{in} \triangle D_i) = 0,$$

then, by the properties (a) – (c), $C_n^N = \bigcup_{i\le N} C_{in}$ approximates $D^N = \bigcup_{i\le N} D_i$, which means that $D^N \in \mathscr{D}$. Let $D = \bigcup_{i\ge 1} D_i$. Since $\mathbb{P}(D\setminus D^N) \to 0$ as $N \to \infty$, it is clear that $D \in \mathscr{D}$, so $\mathscr{D}$ is a σ-algebra. □

Dynkin's theorem. We will now describe a tool, the so-called Dynkin's theorem, or π–λ theorem, which is often quite useful in checking various properties of probabilities.

π-systems: A collection of sets $\mathscr{P}$ is called a π-system if it is closed under taking intersections, i.e.

1. if $A,B \in \mathscr{P}$ then $A\cap B \in \mathscr{P}$.

λ-systems: A collection of sets $\mathscr{L}$ is called a λ-system if

1. $\Omega \in \mathscr{L}$,
2. if $A \in \mathscr{L}$ then $A^c \in \mathscr{L}$,
3. if $A_n \in \mathscr{L}$ are disjoint for $n \geq 1$ then $\cup_{n\geq 1} A_n \in \mathscr{L}$.

Given any collection of sets $\mathscr{C}$, by analogy with the σ-algebra $\sigma(\mathscr{C})$ generated by $\mathscr{C}$, we will denote by $\mathscr{L}(\mathscr{C})$ the smallest λ-system that contains $\mathscr{C}$. If is easy to see that the intersection of all λ-systems that contain $\mathscr{C}$ is again a λ-system that contains $\mathscr{C}$, so this intersection is precisely $\mathscr{L}(\mathscr{C})$.

Theorem 1.3 (Dynkin's theorem). *If $\mathscr{P}$ is a π-system, $\mathscr{L}$ is a λ-system and $\mathscr{P} \subseteq \mathscr{L}$, then $\sigma(\mathscr{P}) \subseteq \mathscr{L}$.*

We will give typical examples of application of this result below.

Proof. First of all, it should be obvious that the collection of sets which is both a π-system and a λ-system is a σ-algebra. Therefore, if we can show that $\mathscr{L}(\mathscr{P})$ is a π-system then it is a σ-algebra and

$$\mathscr{P} \subseteq \sigma(\mathscr{P}) \subseteq \mathscr{L}(\mathscr{P}) \subseteq \mathscr{L},$$

which proves the result. Let us prove that $\mathscr{L}(\mathscr{P})$ is a π-system. For a fixed set $A \subseteq \Omega$, let us define

$$\mathscr{G}_A = \big\{ B \subseteq \Omega \bigm| B \cap A \in \mathscr{L}(\mathscr{P}) \big\}.$$

Step 1. Let us show that if $A \in \mathscr{L}(\mathscr{P})$ then $\mathscr{G}_A$ is a λ-system. Obviously, $\Omega \in \mathscr{G}_A$. If $B \in \mathscr{G}_A$ then $B \cap A \in \mathscr{L}(\mathscr{P})$ and, since $A^c \in \mathscr{L}(\mathscr{P})$ is disjoint from $B \cap A$,

$$B^c \cap A = \big((B \cap A) \cup A^c \big)^c \in \mathscr{L}(\mathscr{P}).$$

This means that $B^c \in \mathscr{G}_A$. Finally, if $B_n \in \mathscr{G}_A$ are disjoint then $B_n \cap A \in \mathscr{L}(\mathscr{P})$ are disjoint and

$$(\cup_{n\geq 1} B_n) \cap A = \cup_{n\geq 1} (B_n \cap A) \in \mathscr{L}(\mathscr{P}),$$

so $\cup_{n\geq 1} B_n \in \mathscr{G}_A$. We showed that $\mathscr{G}_A$ is a λ-system.

Step 2. Next, let us show that if $A \in \mathscr{P}$ then $\mathscr{L}(\mathscr{P}) \subseteq \mathscr{G}_A$. Since $\mathscr{P} \subseteq \mathscr{L}(\mathscr{P})$, by Step 1, $\mathscr{G}_A$ is a λ-system. Also, since $\mathscr{P}$ is a π-system, closed under taking intersections, $\mathscr{P} \subseteq \mathscr{G}_A$. This implies that $\mathscr{L}(\mathscr{P}) \subseteq \mathscr{G}_A$. In other words, we showed that if $A \in \mathscr{P}$ and $B \in \mathscr{L}(\mathscr{P})$ then $A \cap B \in \mathscr{L}(\mathscr{P})$.

Step 3. Finally, let us show that if $B \in \mathscr{L}(\mathscr{P})$ then $\mathscr{L}(\mathscr{P}) \subseteq \mathscr{G}_B$. By step 2, $\mathscr{G}_B$ contains $\mathscr{P}$ and, by Step 1, $\mathscr{G}_B$ is a λ-system. Therefore, $\mathscr{L}(\mathscr{P}) \subseteq \mathscr{G}_B$. We showed that if $B \in \mathscr{L}(\mathscr{P})$ and $A \in \mathscr{L}(\mathscr{P})$ then $A \cap B \in \mathscr{L}(\mathscr{P})$, so $\mathscr{L}(\mathscr{P})$ is a π-system.
□

Example 1.1.4. Suppose that Ω is a topological space with the Borel σ-algebra $\mathscr{B}$ generated by open sets. Given two probability measures $\mathbb{P}_1$ and $\mathbb{P}_2$ on $(\Omega, \mathscr{B})$, the collection of sets

$$\mathscr{L} = \{B \in \mathscr{B} : \mathbb{P}_1(B) = \mathbb{P}_2(B)\}$$

is trivially a λ-system, by the properties of probability measures. On the other hand, the collection $\mathscr{P}$ of all open sets is a π-system and, therefore, if we know that $\mathbb{P}_1(B) = \mathbb{P}_2(B)$ for all open sets then, by Dynkin's theorem, this holds for all Borel sets $B \in \mathscr{B}$. Similarly, one can see that a probability on the Borel σ-algebra on the real line is determined by probabilities of the sets $(-\infty, t]$ for all $t \in \mathbb{R}$. □

Regularity of measures. Let us now consider the case of $\Omega = S$ where (S,d) is a metric space, and let $\mathscr{A}$ be the Borel σ-algebra generated by open (or closed) sets. A probability measure $\mathbb{P}$ on this space is called *closed regular* if

$$\mathbb{P}(A) = \sup\{\mathbb{P}(F) : F \subseteq A, F \text{ - closed}\} \tag{1.1}$$

for all $A \in \mathscr{A}$. Similarly, probability measure $\mathbb{P}$ is called *regular* if

$$\mathbb{P}(A) = \sup\{\mathbb{P}(K) : K \subseteq A, K \text{ - compact}\} \tag{1.2}$$

for all $A \in \mathscr{A}$. It is a standard result in measure theory that every finite measure on $(\mathbb{R}^k, \mathscr{B}(\mathbb{R}^k))$ is regular. In the setting of complete separable metric spaces, this is known as Ulam's theorem, which we will prove below.

Theorem 1.4. *Every probability measure $\mathbb{P}$ on a metric space (S,d) is closed regular.*

Proof. Let us consider a collection of sets

$$\mathscr{L} = \{A \in \mathscr{A} : \text{both } A \text{ and } A^c \text{ satisfy } (1.1)\}. \tag{1.3}$$

First of all, let us show that each closed set $F \in \mathscr{L}$; we only need to show that an open set $U = F^c$ satisfies (1.1). Let us consider sets

$$F_n = \{s \in S : d(s,F) \geq 1/n\}.$$

It is obvious that all F_n are closed, and $F_n \subseteq F_{n+1}$. One can also easily check that, since F is closed, $\cup_{n\geq 1} F_n = U$ and, by the continuity of measure,

$$\mathbb{P}(U) = \lim_{n\to\infty} \mathbb{P}(F_n) = \sup_{n\geq 1} \mathbb{P}(F_n).$$

This proves that $U = F^c$ satisfies (1.1) and $F \in \mathscr{L}$. Next, one can easily check that $\mathscr{L}$ is a λ-system, which we will leave as an exercise below. Since the collection of all closed sets is a π-system, and closed sets generate the Borel σ-algebra $\mathscr{A}$, by Dynkin's theorem, all measurable sets are in $\mathscr{L}$. This proves that $\mathbb{P}$ is closed regular. □

Theorem 1.5 (Ulam's theorem). *If (S,d) is a complete separable metric space then every probability measure $\mathbb{P}$ is regular.*

Proof. First, let us show that there exists a compact set $K \subseteq S$ such that $\mathbb{P}(S \setminus K) \le \varepsilon$. Consider a sequence $\{s_1, s_2, \ldots\}$ that is dense in S. For any $m \ge 1$, $S = \bigcup_{i=1}^{\infty} \bar{B}(s_i, \frac{1}{m})$, where $\bar{B}(s_i, \frac{1}{m})$ is the closed ball of radius $1/m$ centered at s_i. By the continuity of measure, for large enough $n(m)$,

$$\mathbb{P}\Bigl(S \setminus \bigcup_{i=1}^{n(m)} \bar{B}\Bigl(s_i, \frac{1}{m}\Bigr)\Bigr) \le \frac{\varepsilon}{2^m}.$$

If we take

$$K = \bigcap_{m \ge 1} \bigcup_{i=1}^{n(m)} \bar{B}\Bigl(s_i, \frac{1}{m}\Bigr)$$

then

$$\mathbb{P}(S \setminus K) \le \sum_{m \ge 1} \frac{\varepsilon}{2^m} = \varepsilon.$$

Obviously, by construction, K is closed and totally bounded. Since S is complete, K is compact. By the previous theorem, given $A \in \mathscr{A}$, we can find a closed subset $F \subseteq A$ such that $\mathbb{P}(A \setminus F) \le \varepsilon$. Therefore, $\mathbb{P}(A \setminus (F \cap K)) \le 2\varepsilon$, and since $F \cap K$ is compact, this finishes the proof. □

Exercise 1.1.1. Let $\mathscr{F} = \{F \subseteq \mathbb{N} : \mathbb{N} \setminus F \text{ is finite}\}$ be the collection of all sets in $\mathbb{N}$ with finite complements. $\mathscr{F}$ is a filter, which means that (a) $\emptyset \notin \mathscr{F}$, (b) if $F_1, F_2 \in \mathscr{F}$ then $F_1 \cap F_2 \in \mathscr{F}$, and (c) if $F \in \mathscr{F}$ and $F \subseteq G$ then $G \in \mathscr{F}$. It is well known (by Zorn's lemma) that $\mathscr{F} \subseteq \mathscr{U}$ for some ultrafilter $\mathscr{U}$, which in addition to (a) – (c) also satisfies: (d) for any set $A \subseteq \mathbb{N}$, either A or $\mathbb{N} \setminus A$ is in $\mathscr{U}$. If we define $\mathbb{P}(A) = 1$ for $A \in \mathscr{U}$ and $\mathbb{P}(A) = 0$ for $A \notin \mathscr{U}$, show that $\mathbb{P}$ is finitely additive, but not countably additive, on $2^{\mathbb{N}}$. If $\mathbb{P}$ could be interpreted as a "probability", suppose that two people pick a number in $\mathbb{N}$ according to $\mathbb{P}$, and whoever picks the bigger number wins. If one shows his number first, what is the probability that the other one wins?

Exercise 1.1.2. Suppose that $\mathscr{C}$ is a class of subsets of Ω and $B \in \sigma(\mathscr{C})$. Show that there exists a countable class $\mathscr{C}_B \subseteq \mathscr{C}$ such that $B \in \sigma(\mathscr{C}_B)$.

Exercise 1.1.3. Check that $\mathscr{L}$ in (1.3) is a λ-system. (In fact, one can check that this is a σ-algebra.)

1.2 Random variables

Let $(\Omega, \mathscr{A}, \mathbb{P})$ be a probability space and $(S, \mathscr{B})$ be a measurable space where $\mathscr{B}$ is a σ-algebra of subsets of S. Recall that a function $X : \Omega \to S$ is called *measurable* if for all $B \in \mathscr{B}$,

$$X^{-1}(B) = \{\omega \in \Omega : X(\omega) \in B\} \in \mathscr{A}.$$

In Probability Theory, such functions are called *random variables*, especially, when $(S, \mathscr{B}) = (\mathbb{R}, \mathscr{B}(\mathbb{R}))$. Depending on the target space S, X may be called a random vector, sequence, or, more generally, a random element in S. Recall that measurability can be checked on the sets that generate the σ-algebra $\mathscr{B}$ and, in particular, the following holds.

Lemma 1.3. *$X : \Omega \to \mathbb{R}$ is a random variable if and only if, for all $t \in \mathbb{R}$,*

$$\{X \le t\} := \{\omega \in \Omega : X(\omega) \in (-\infty, t]\} \in \mathscr{A}.$$

Proof. Only "if" direction requires proof. We will prove that

$$\mathscr{D} = \{D \subseteq \mathbb{R} : X^{-1}(D) \in \mathscr{A}\}$$

is a σ-algebra. Since sets $(-\infty, t] \in \mathscr{D}$, this will imply that $\mathscr{B}(\mathbb{R}) \subseteq \mathscr{D}$. The fact that $\mathscr{D}$ is a σ-algebra follows simply because taking pre-image preserves set operations. For example, if we consider a sequence $D_i \in \mathscr{D}$ for $i \ge 1$ then

$$X^{-1}\Big(\bigcup_{i\ge 1} D_i\Big) = \bigcup_{i \ge 1} X^{-1}(D_i) \in \mathscr{A},$$

because $X^{-1}(D_i) \in \mathscr{A}$ and $\mathscr{A}$ is a σ-algebra. Therefore, $\bigcup_{i\ge 1} D_i \in \mathscr{D}$. Other properties can be checked similarly, so $\mathscr{D}$ is a σ-algebra. □

Given a random element X on $(\Omega, \mathscr{A}, \mathbb{P})$ with values in $(S, \mathscr{B})$, let us denote the image measure on $\mathscr{B}$ by $\mathbb{P}_X = \mathbb{P} \circ X^{-1}$, which means that for $B \in \mathscr{B}$,

$$\mathbb{P}_X(B) = \mathbb{P}(X \in B) = \mathbb{P}(X^{-1}(B)) = \mathbb{P} \circ X^{-1}(B).$$

$(S, \mathscr{B}, \mathbb{P}_X)$ is called the *sample space* of a random element X and $\mathbb{P}_X$ is called the *law* of X, or the *distribution* of X. Clearly, on this space a random variable $\xi : S \to S$ defined by the identity $\xi(s) = s$ has the same law as X.When $S = \mathbb{R}$, the function $F(t) = \mathbb{P}(X \le t)$ is called the *cumulative distribution function* (c.d.f.) of X. Clearly, this function satisfies the following properties that already appeared in the previous section:

(1) $F(x) = F((-\infty, x])$ is non-decreasing and right-continuous,
(2) $\lim_{x\to-\infty} F(x) = 0$ and $\lim_{x\to\infty} F(1) = 1$.

On the other hand, any such function is a c.d.f. of some random variable, for example, the random variables $X(x) = x$ on the space $(\mathbb{R}, \mathscr{B}(\mathbb{R}), F)$ constructed in the previous section, since

$$\mathbb{P}\big(x : X(x) \le t\big) = F\big((-\infty,t]\big) = F(t).$$

Another construction can be given on the probability space $([0,1],\mathscr{B}([0,1]),\lambda)$ with the Lebesgue measure λ, using the *quantile transformation*. Given a c.d.f. F, let us define a random variable $X: [0,1] \to \mathbb{R}$ by the quantile transformation

$$X(x) = \inf\big\{s \in \mathbb{R} : F(s) \ge x\big\}.$$

What is the c.d.f. of X? Notice that, since F is right-continuous,

$$X(x) \le t \Longleftrightarrow \inf\{s : F(s) \ge x\} \le t \Longleftrightarrow \lim_{s \downarrow t} F(s) \ge x \Longleftrightarrow F(t) \ge x.$$

This implies that F is the c.d.f. of X, since

$$\lambda(x : X(x) \le t) = \lambda(x : F(t) \ge x) = F(t).$$

This means that, to define the probability space $(\mathbb{R},\mathscr{B}(\mathbb{R}),F)$, we can start with $([0,1],\mathscr{B}([0,1]),\lambda)$ and let $F = \lambda \circ X^{-1}$ be the image of the Lebesgue measure by the quantile transformation, or the law of X on $\mathbb{R}$. A related "inverse" property is left as an exercise below.

Given random element $X: (\Omega,\mathscr{A}) \to (S,\mathscr{B})$, the σ-algebra

$$\sigma(X) = \big\{X^{-1}(B) : B \in \mathscr{B}\big\}$$

is called a *σ-algebra generated by X*. It is obvious that this collection of sets is, indeed, as σ-algebra.

Example 1.2.1. Consider a random variable X on $([0,1],\mathscr{B}([0,1]),\lambda)$ defined by

$$X(x) = \begin{cases} 0, \ 0 \le x \le 1/2, \\ 1, \ 1/2 < x \le 1. \end{cases}$$

Then, the σ-algebra generated by X consists of the sets

$$\sigma(X) = \Big\{\emptyset, \Big[0,\frac{1}{2}\Big], \Big(\frac{1}{2},1\Big], [0,1]\Big\},$$

and $\mathbb{P}(X=0) = \mathbb{P}(X=1) = 1/2$. □

Lemma 1.4. *Consider a probability space $(\Omega,\mathscr{A},\mathbb{P})$, a measurable space $(S,\mathscr{B})$ and random elements $X: \Omega \to S$ and $Y: \Omega \to \mathbb{R}$. Then the following are equivalent:*

1. *$Y = g(X)$ for some (Borel) measurable function $g: S \to \mathbb{R}$,*
2. *$Y: \Omega \to \mathbb{R}$ is $\sigma(X)$-measurable.*

It should be obvious from the proof that $\mathbb{R}$ can be replaced by any separable metric space.

Proof. The fact that 1 implies 2 is obvious, since for any Borel set $B \subseteq \mathbb{R}$ the set $B' = g^{-1}(B) \in \mathscr{B}$ and, therefore,

$$\{Y = g(X) \in B\} = \{X \in g^{-1}(B) = B'\} = X^{-1}(B') \in \sigma(X).$$

Let us show that 2 implies 1. For all integer n and k, consider sets

$$A_{n,k} = \Big\{\omega : Y(\omega) \in \Big[\frac{k}{2^n}, \frac{k+1}{2^n}\Big)\Big\} = Y^{-1}\Big(\Big[\frac{k}{2^n}, \frac{k+1}{2^n}\Big)\Big).$$

By 2, $A_{n,k} \in \sigma(X) = \{X^{-1}(B) : B \in \mathscr{B}\}$ and, therefore, $A_{n,k} = X^{-1}(B_{n,k})$ for some $B_{n,k} \in \mathscr{B}$. Let us consider a function

$$g_n(x) = \sum_{k\in\mathbb{Z}} \frac{k}{2^n} \mathrm{I}(x \in B_{n,k}).$$

By construction, $|Y - g_n(X)| \leq 2^{-n}$, since

$$Y(\omega) \in \Big[\frac{k}{2^n}, \frac{k+1}{2^n}\Big) \Longleftrightarrow X(\omega) \in B_{n,k} \Longleftrightarrow g_n(X(\omega)) = \frac{k}{2^n}.$$

It is easy to see that $g_n(x) \leq g_{n+1}(x)$ and, therefore, the limit $g(x) = \lim_{n\to\infty} g_n(x)$ exists and is measurable as a limit of measurable functions. Clearly, $Y = g(X)$. □

Independence. Consider a probability space $(\Omega, \mathscr{A}, \mathbb{P})$. Then, σ-algebras $\mathscr{A}_i \subseteq \mathscr{A}$, $i \leq n$, are *independent* if

$$\mathbb{P}(A_1 \cap \cdots \cap A_n) = \prod_{i\leq n} \mathbb{P}(A_i)$$

for all $A_i \in \mathscr{A}_i$. Similarly, σ-algebras $\mathscr{A}_i \subseteq \mathscr{A}$ for $i \leq n$ are *pairwise independent* if

$$\mathbb{P}(A_i \cap A_j) = \mathbb{P}(A_i)\mathbb{P}(A_j)$$

for all $A_i \in \mathscr{A}_i, A_j \in \mathscr{A}_j, i \neq j$. Random variables $X_i : (\Omega, \mathscr{A}) \to (S, \mathscr{B})$ for $i \leq n$ are independent if the σ-algebras $\sigma(X_i)$ are independent, which is just another convenient way to state that

$$\mathbb{P}(X_1 \in B_1, \ldots, X_n \in B_n) = \mathbb{P}(X_1 \in B_1) \times \ldots \times \mathbb{P}(X_n \in B_n)$$

for any events $B_1, \ldots, B_n \in \mathscr{B}$. Pairwise independence is defined similarly.

Example 1.2.2. Consider a regular tetrahedron die with red, green and blue sides and a red-green-blue base. If we roll this die then the colors provide an example of pairwise independent random variables that are not independent, since

$$\mathbb{P}(r) = \mathbb{P}(b) = \mathbb{P}(g) = \frac{1}{2} \text{ and } \mathbb{P}(rb) = \mathbb{P}(rg) = \mathbb{P}(bg) = \frac{1}{4},$$

while $\mathbb{P}(rbg) = \frac{1}{4} \neq \mathbb{P}(r)\mathbb{P}(b)\mathbb{P}(g) = \left(\frac{1}{2}\right)^3$. □

First of all, independence can be checked on generating algebras.

Lemma 1.5. *If algebras $A_i, i \le n$ are independent then σ-algebras $\sigma(A_i)$ are independent.*

Proof. Obvious by the Approximation Lemma 1.2. □

A more flexible criterion follows from Dynkin's theorem.

Lemma 1.6. *If collections of sets $\mathscr{C}_i$ for $i \le n$ are π-systems (closed under finite intersections) then their independence implies the independence of the σ-algebras $\sigma(\mathscr{C}_i)$ they generate.*

Proof. Le us consider the collection $\mathscr{C}$ of sets $C \in \mathscr{A}$ such that

$$\mathbb{P}(C \cap C_2 \cap \cdots \cap C_n) = \mathbb{P}(C) \prod_{i=2}^{n} \mathbb{P}(C_i)$$

for all $C_i \in \mathscr{C}_i$ for $2 \le i \le n$. It is obvious that $\mathscr{C}$ is a λ-system, and it contains $\mathscr{C}_1$ by assumption. Since $\mathscr{C}_1$ is a π-system, by Dynkin's theorem, $\mathscr{C}$ contains $\sigma(\mathscr{C}_1)$. This means that we can replace $\mathscr{C}_1$ by $\sigma(\mathscr{C}_1)$ in the statement of the theorem and, similarly, we can continue to replace each $\mathscr{C}_i$ by $\sigma(\mathscr{C}_i)$. □

Lemma 1.7. *Let us consider random variables $X_i \colon \Omega \to \mathbb{R}$ on a probability space $(\Omega, \mathscr{A}, \mathbb{P})$. The following holds.*

(a) Random variables (X_i) are independent if and only if, for all $t_i \in \mathbb{R}$,

$$\mathbb{P}(X_1 \le t_1, \ldots, X_n \le t_n) = \prod_{i=1}^{n} \mathbb{P}(X_i \le t_i). \tag{1.4}$$

(b) If the laws of X_i have densities f_i on $\mathbb{R}$ then these random variables are independent if and only if a joint density f on $\mathbb{R}^n$ of the vector $(X_i)_{1 \le i \le n}$ exists and

$$f(x_1, \ldots, x_n) = \prod_{i=1}^{n} f_i(x_i).$$

Proof. (a) This is obvious by Lemma 1.6, because the collection of sets $(-\infty, t]$ for $t \in \mathbb{R}$ is a π-system that generates the Borel σ-algebra on $\mathbb{R}$.

(b) Let us start with the "if" part. If we denote $X = (X_1, \ldots, X_n)$ then, for any $A_i \in \mathscr{B}(\mathbb{R})$,

$$
\begin{aligned}
\mathbb{P}\Big(\bigcap_{i=1}^{n}\{X_i \in A_i\}\Big) &= \mathbb{P}(X \in A_1 \times \cdots \times A_n) \\
&= \int_{A_1\times\cdots\times A_n} \prod_{i=1}^{n} f_i(x_i)\,dx_1\cdots dx_n \\
&= \prod_{i=1}^{n}\int_{A_i} f_i(x_i)\,dx_i = \prod_{i=1}^{n}\mathbb{P}(X_i \in A_i).
\end{aligned}
$$

Next, we prove the "only if" part. First of all, by independence,

$$
\mathbb{P}(X \in A_1 \times \cdots \times A_n) = \prod_{i=1}^{n}\mathbb{P}(X_i \in A_i) = \int_{A_1\times\cdots\times A_n} \prod_{i=1}^{n} f_i(x_i)\,dx_1\cdots dx_n.
$$

We would like to show that this implies that

$$
\mathbb{P}(X \in A) = \int_{A}\prod_{i=1}^{n} f_i(x_i)\,dx_1\cdots dx_n.
$$

for all A in the Borel σ-algebra on $\mathbb{R}^n$, which would means that the joint density exists and is equal to the product of individual densities. One can prove the above equality for all $A \in \mathscr{B}(\mathbb{R}^n)$ by appealing to the Monotone Class Theorem from measure theory, or the Caratheodory Extension Theorem 1.1, since the above equality, obviously, can be extended from the semi-algebra of measurable rectangles $A_1 \times \cdots \times A_n$ to the algebra of disjoint unions of measurable rectangles, which generates the Borel σ-algebra. However, we can also appeal to the Dynkin's theorem, since the family $\mathscr{L}$ of sets A that satisfy the above equality is a λ-system by properties of measures and integrals, and it contains the π-system $\mathscr{P}$ of measurable rectangles $A_1 \times \cdots \times A_n$ that generates the Borel σ-algebra, $\mathscr{B}(\mathbb{R}^n) = \sigma(\mathscr{P})$. □

More generally, a collection of σ-algebras $\mathscr{A}_t \subseteq \mathscr{A}$ indexed by $t \in T$ for some set T is called independent if any finite subset of these σ-algebras are independent. Let $T = T_1 \cup \cdots \cup T_n$ be a partition of T into disjoint sets. In this case, the following holds.

Lemma 1.8 (Grouping lemma). *The σ-algebras*

$$
\mathscr{B}_i = \bigvee_{t\in T_i} \mathscr{A}_t = \sigma\Big(\bigcup_{t\in T_i} \mathscr{A}_t\Big)
$$

generated by the subsets of σ-algebras $(\mathscr{A}_t)_{t\in T_i}$ are independent.

Proof. For each $i \le n$, consider a collection of sets

$$
\mathscr{C}_i = \Big\{\bigcap_{t\in F} A_t \;:\; \text{for all finite } F \subseteq T_i \text{ and } A_t \in \mathscr{A}_t\Big\}.
$$

It is obvious that $\mathscr{B}_i = \sigma(\mathscr{C}_i)$ since $\mathscr{A}_t \subseteq \mathscr{C}_i$ for all $t \in T_i$, each $\mathscr{C}_i$ is a π-system, and $\mathscr{C}_1, \ldots, \mathscr{C}_n$ are independent by the definition of independence of the σ-algebras $\mathscr{A}_t$ for $t \in T$. Using Lemma 1.6 finishes the proof. (Of course, one should recognize from measure theory that $\mathscr{C}_i$ is a semi-algebra that generates $\mathscr{B}_i$.) □

If we would like to construct finitely many independent random variables $(X_i)_{i \le n}$ with arbitrary distributions $(\mathbb{P}_i)_{i \le n}$ on $\mathbb{R}$, we can simply consider the space $\Omega = \mathbb{R}^n$ with the product measure

$$\mathbb{P}_1 \times \ldots \times \mathbb{P}_n$$

and define a random variable X_i by $X_i(x_1, \ldots, x_n) = x_i$. The main result in the next section will imply that one can construct an infinite sequence of independent random variables with arbitrary distributions on the same probability space, and here we will give a sketch of another construction on the space $([0,1], \mathscr{B}([0,1]), \lambda)$. We will write $\mathbb{P} = \lambda$ to emphasize that we think of the Lebesgue measure as a probability.

Step 1. If we write the dyadic decomposition of $x \in [0,1]$,

$$x = \sum_{n \ge 1} 2^{-n} \varepsilon_n(x),$$

then it is easy to see that $(\varepsilon_n)_{n \ge 1}$ are independent random variables with the distribution $\mathbb{P}(\varepsilon_n = 0) = \mathbb{P}(\varepsilon_n = 1) = 1/2$, since for any $n \ge 1$ and any $a_i \in \{0,1\}$,

$$\mathbb{P}(X \colon \varepsilon_1(x) = a_1, \ldots, \varepsilon_n(x) = a_n) = 2^{-n},$$

since fixing the first n coefficients in the dyadic expansion places x into an interval of length 2^{-n}.

Step 2. Let us consider injections $k_m : \mathbb{N} \to \mathbb{N}$ for $m \ge 1$ such that their ranges $k_m(\mathbb{N})$ are all disjoint and let us define

$$X_m = X_m(x) = \sum_{n \ge 1} 2^{-n} \varepsilon_{k_m(n)}(x).$$

It is an easy exercise to check that each X_m is well defined and has the uniform distribution on $[0,1]$, which can be seen by looking at the dyadic intervals first. Moreover, by the Grouping Lemma above, the random variables $(X_m)_{m \ge 1}$ are all independent since they are defined in terms groups of independent random variables.

Step 3. Given a sequence of probability distributions $(\mathbb{P}_m)_{m \ge 1}$ on $\mathbb{R}$, let $(F_m)_{m \ge 1}$ be the sequence of the corresponding c.d.f.s and let $(Q_m)_{m \ge 1}$ be their quantile transforms. We have seen above that each $Y_m = Q_m(X_m)$ has the distribution $\mathbb{P}_m$ on $\mathbb{R}$, and they are obviously independent of each other. Therefore, we constructed a sequence of independent random variables Y_m on the space $([0,1], \mathscr{B}([0,1]), \lambda)$ with arbitrary distributions $\mathbb{P}_m$.

Expectation. If $X : \Omega \to \mathbb{R}$ is a random variable on $(\Omega, \mathscr{A}, \mathbb{P})$ then the *expectation* of X is defined as

$$\mathbb{E}X = \int_\Omega X(\omega)\, d\mathbb{P}(\omega).$$

In other words, expectation is just another term for the integral with respect to a probability measure and, as a result, expectation has all the usual properties of the integrals in measure theory: convergence theorems, change of variables formula, Fubini's theorem, etc. Let us write down some special cases of the change of variables formula.

Lemma 1.9. *(1) If F is the c.d.f. (and the law) of X on $\mathbb{R}$ then, for any measurable function $g : \mathbb{R} \to \mathbb{R}$,*

$$\mathbb{E}g(X) = \int_{\mathbb{R}} g(x)\, dF(x).$$

(2) If the distribution of X is discrete, i.e. $\mathbb{P}(X \in \{x_i\}_{i\geq 1}) = 1$, then

$$\mathbb{E}g(X) = \sum_{i\geq 1} g(x_i)\mathbb{P}(X = x_i).$$

(3) If the distribution of $X : \Omega \to \mathbb{R}^n$ on $\mathbb{R}^n$ has the density function $f(x)$ then, for any measurable function $g : \mathbb{R}^n \to \mathbb{R}$,

$$\mathbb{E}g(X) = \int_{\mathbb{R}^n} g(x) f(x)\, dx.$$

Proof. All these properties follow by making the change of variables $x = X(\omega)$,

$$\mathbb{E}g(X) = \int_\Omega g(X(\omega))\, d\mathbb{P}(\omega) = \int g(x)\, d\mathbb{P} \circ X^{-1}(x) = \int g(x)\, d\mathbb{P}_X(x),$$

where $\mathbb{P}_X = \mathbb{P} \circ X^{-1}$ is the law of X on $\mathbb{R}$ or $\mathbb{R}^n$. □

Another simple fact is the following.

Lemma 1.10. *If $X, Y : \Omega \to \mathbb{R}$ are independent and $\mathbb{E}|X|, \mathbb{E}|Y| < \infty$ then $\mathbb{E}XY = \mathbb{E}X\mathbb{E}Y$.*

Proof. Independence implies that the distribution of (X,Y) on $\mathbb{R}^2$ is the product measure $\mathbb{P} \times \mathbb{Q}$, where $\mathbb{P}$ and $\mathbb{Q}$ are the distributions of X and Y on $\mathbb{R}$, and, therefore,

$$\mathbb{E}XY = \int_{\mathbb{R}^2} xy\, d(\mathbb{P} \times \mathbb{Q})(x,y) = \int_{\mathbb{R}} x\, d\mathbb{P}(x) \int_{\mathbb{R}} y\, d\mathbb{Q}(y) = \mathbb{E}X\mathbb{E}Y,$$

by the change of variables and Fubini theorems. □

Exercise 1.2.1. If a random variable X has continuous c.d.f. $F(t)$, show that $F(X)$ is uniform on $[0,1]$, i.e. the law of $F(X)$ is the Lebesgue measure on $[0,1]$.

Exercise 1.2.2. If F is a continuous distribution function, show that $\int F(x)\,dF(x) = 1/2$.

Exercise 1.2.3. $\mathrm{ch}(\lambda)$ is a moment generating function of a random variable X with distribution $\mathbb{P}(X = \pm 1) = 1/2$, since

$$\mathbb{E}e^{\lambda X} = \frac{e^{\lambda} + e^{-\lambda}}{2} = \mathrm{ch}(\lambda).$$

Does there exist a (bounded) random variable X such that $\mathbb{E}e^{\lambda X} = \mathrm{ch}^m(\lambda)$ for $0 < m < 1$? (Hint: compute several derivatives at zero.)

Exercise 1.2.4. Consider a measurable function $f : X \times Y \to \mathbb{R}$ and a product probability measure $\mathbb{P} \times \mathbb{Q}$ on $X \times Y$. For $0 < p \leq q$, prove that

$$\big\|\|f\|_{L^p(\mathbb{P})}\big\|_{L^q(\mathbb{Q})} \leq \big\|\|f\|_{L^q(\mathbb{Q})}\big\|_{L^p(\mathbb{P})},$$

i.e.

$$\Big(\int\Big(\int |f(x,y)|^p d\mathbb{P}(x)\Big)^{q/p} d\mathbb{Q}(y)\Big)^{1/q} \leq \Big(\int\Big(\int |f(x,y)|^q d\mathbb{Q}(y)\Big)^{p/q} d\mathbb{P}(x)\Big)^{1/p}.$$

Assume that both sides are well-defined, for example, that f is bounded.

Exercise 1.2.5. If the event $A \in \sigma(\mathscr{P})$ is independent of the π-system $\mathscr{P}$ then $\mathbb{P}(A) = 0$ or 1.

Exercise 1.2.6. Suppose X is a random variable and $g : \mathbb{R} \to \mathbb{R}$ is measurable. Prove that if X and $g(X)$ are independent then $\mathbb{P}(g(X) = c) = 1$ for some constant c.

Exercise 1.2.7. Suppose that $(e_m)_{1 \leq m \leq n}$ are i.i.d. exponential random variables with the parameter $\lambda > 0$, and let

$$e_{(1)} \leq e_{(2)} \leq \ldots \leq e_{(n)}$$

be the order statistics (the random variables arranged in the increasing order). Prove that the spacings

$$e_{(1)}, e_{(2)} - e_{(1)}, \ldots, e_{(n)} - e_{(n-1)}$$

are independent exponential random variables, and $e_{(k+1)} - e_{(k)}$ has the parameter $(n-k)\lambda$.

Exercise 1.2.8. Suppose that $(e_n)_{n \geq 1}$ are i.i.d. exponential random variables with the parameter $\lambda = 1$. Let $S_n = e_1 + \ldots + e_n$ and $R_n = S_{n+1}/S_n$ for $n \geq 1$. Prove that $(R_n)_{n \geq 1}$ are independent and R_n has density $nx^{-(n+1)}\mathrm{I}(x \geq 1)$. *Hint:* Let $R_0 = e_1$ and compute the joint density of $(R_0, R_1, \ldots, R_n)$ first.

Exercise 1.2.9. Let N be a Poisson random variable with the mean λ, i.e. $\mathbb{P}(N = j) = \lambda^j e^{-\lambda}/j!$ for integer $j \geq 0$. Then, consider N i.i.d. random variables, independent of N, taking values $1, \ldots, k$ with probabilities $p_1, \ldots, p_k$. Let N_j be the number of these random variables taking value j, so that $N_1 + \ldots + N_k = N$. Prove that $N_1, \ldots, N_k$ are independent Poisson random variables with means $\lambda p_1, \ldots, \lambda p_k$.

Exercise 1.2.10. Suppose that a measurable subset $P \subseteq [0,1]$ and the interval $I = [a,b] \subseteq [0,1]$ are such that $\lambda(P) = \lambda(I)$, where λ is the Lebesgue measure on $[0,1]$. Show that there exists a measure-preserving transformation $T : [0,1] \to [0,1]$, i.e. $\lambda \circ T^{-1} = \lambda$, such that $T(I) \subseteq P$ and T is one-to-one (injective) outside a set of measure zero.

1.3 Kolmogorov's consistency theorem

In this section we will describe a typical way to construct an infinite family of random variables $(X_t)_{t\in T}$ on the same probability space, namely, when we are given all their finite dimensional marginals. This means that for any finite subset $N \subseteq T$, we are given

$$\mathbb{P}_N(B) = \mathbb{P}\big((X_t)_{t\in N} \in B\big)$$

for all B in the Borel σ-algebra $\mathscr{B}_N = \mathscr{B}(\mathbb{R}^N)$. Clearly, these laws must satisfy a natural *consistency condition*,

$$\mathbb{P}_N(B) = \mathbb{P}_M(B \times \mathbb{R}^{M\setminus N}), \tag{1.5}$$

for any finite subsets $N \subseteq M$ and any Borel set $B \in \mathscr{B}_N$. (Of course, to be careful, we should define these probabilities for ordered subsets and also make sure they are consistent under rearrangements, but the notations for unordered sets is clear and should not cause any confusion.)

Our goal is to construct a probability space $(\Omega, \mathscr{A}, \mathbb{P})$ and random variables $X_t : \Omega \to \mathbb{R}$ that have $(\mathbb{P}_N)$ as their finite dimensional distributions. We take

$$\Omega = \mathbb{R}^T = \big\{\omega : T \to \mathbb{R}\big\}$$

to be the space of all real-valued functions on T, and let X_t be the coordinate projection $X_t = X_t(\omega) = \omega(t)$. For the coordinate projections to be measurable, the following collection of events,

$$A = \big\{B \times \mathbb{R}^{T\setminus N} : B \in \mathscr{B}_N\big\},$$

must be contained in the σ-algebra $\mathscr{A}$. It is easy to see that A is, in fact, an algebra of sets, and it is called the *cylindrical algebra* on $\mathbb{R}^T$. We will then take $\mathscr{A} = \sigma(A)$ to be the smallest σ-algebra on which all coordinate projections are measurable. This is the so-called *cylindrical σ-algebra* on $\mathbb{R}^T$. A set $B \times \mathbb{R}^{T\setminus N}$ is called a *cylinder*. As we already agreed, the probability $\mathbb{P}$ on the sets in algebra A is given by

$$\mathbb{P}\big(B \times \mathbb{R}^{T\setminus N}\big) = \mathbb{P}_N(B).$$

Given two finite subsets $N \subseteq M \subset T$ and $B \in \mathscr{B}_N$, the same set can be represented as two different cylinders,

$$B \times \mathbb{R}^{T\setminus N} = \big(B \times \mathbb{R}^{M\setminus N}\big) \times \mathbb{R}^{T\setminus M}.$$

However, by the consistency condition, the definition of $\mathbb{P}$ will not depend on the choice of the representation. To finish the construction, we need to show that $\mathbb{P}$ can be extended from algebra A to a probability measure on the σ-algebra $\mathscr{A}$. By the Caratheodory Extension Theorem 1.1, we only need to show that the following holds.

Theorem 1.6. $\mathbb{P}$ *is countably additive on the cylindrical algebra A.*

Proof. Equivalently, it is enough to show that $\mathbb{P}$ satisfies continuity of measure property on A, namely, given a sequence $B_n \in A$,

$$B_n \supseteq B_{n+1}, \bigcap_{n\geq 1} B_n = \emptyset \Longrightarrow \lim_{n\to\infty} \mathbb{P}(B_n) = 0.$$

We will prove that if there exists $\varepsilon > 0$ such that $\mathbb{P}(B_n) > \varepsilon$ for all n then $\bigcap_{n\geq 1} B_n \neq \emptyset$. We can represent the cylinders B_n as $B_n = C_n \times \mathbb{R}^{T\setminus N_n}$ for some finite subset $N_n \subseteq T$ and $C_n \in \mathscr{B}_{N_n}$. Since $B_n \supseteq B_{n+1}$, we can assume that $N_n \subseteq N_{n+1}$. First of all, by the regularity of probability on a finite dimensional space, there exists a compact set $K_n \subseteq C_n$ such that

$$\mathbb{P}_{N_n}(C_n \setminus K_n) \leq \frac{\varepsilon}{2^{n+1}}.$$

It is easy to see that

$$\bigcap_{i\leq n} C_i \times \mathbb{R}^{T\setminus N_i} \setminus \bigcap_{i\leq n} K_i \times \mathbb{R}^{T\setminus N_i} \subseteq \bigcup_{i\leq n} (C_i \setminus K_i) \times \mathbb{R}^{T\setminus N_i}$$

and, therefore,

$$\begin{aligned} &\mathbb{P}\Big(\bigcap_{i\leq n} C_i \times \mathbb{R}^{T\setminus N_i} \setminus \bigcap_{i\leq n} K_i \times \mathbb{R}^{T\setminus N_i}\Big) \leq \mathbb{P}\Big(\bigcup_{i\leq n} (C_i \setminus K_i) \times \mathbb{R}^{T\setminus N_i}\Big) \\ &\leq \sum_{i\leq n} \mathbb{P}\Big((C_i \setminus K_i) \times \mathbb{R}^{T\setminus N_i}\Big) \leq \sum_{i\leq n} \frac{\varepsilon}{2^{i+1}} \leq \frac{\varepsilon}{2}. \end{aligned}$$

Since we assumed that

$$\mathbb{P}(B_n) = \mathbb{P}\Big(\bigcap_{i\leq n} C_i \times \mathbb{R}^{T\setminus N_i}\Big) > \varepsilon,$$

this implies that

$$\mathbb{P}\Big(\bigcap_{i\leq n} K_i \times \mathbb{R}^{T\setminus N_i}\Big) \geq \frac{\varepsilon}{2} > 0.$$

Let us rewrite this intersection as

$$\bigcap_{i\leq n} K_i \times \mathbb{R}^{T\setminus N_i} = \bigcap_{i\leq n} (K_i \times \mathbb{R}^{N_n\setminus N_i}) \times \mathbb{R}^{T\setminus N_n} = K^n \times \mathbb{R}^{T\setminus N_n},$$

where

$$K^n = \bigcap_{i\leq n} (K_i \times \mathbb{R}^{N_n\setminus N_i})$$

is a compact in $\mathbb{R}^{N_n}$, since K_n is a compact in $\mathbb{R}^{N_n}$. We proved that

$$\mathbb{P}_{N_n}(K^n) = \mathbb{P}(K^n \times \mathbb{R}^{T\setminus N_n}) = \mathbb{P}\Big(\bigcap_{i\leq n} K_i \times \mathbb{R}^{T\setminus N_i}\Big) > 0$$

and, therefore, there exists a point $\omega^n = (\omega^n(t))_{t\in N_n} \in K^n$. By construction, we also have the following inclusion property. For $m > n$,

$$\omega^m \in K^m \subseteq K^n \times \mathbb{R}^{N_m\setminus N_n}$$

and, therefore, $(\omega^m(t))_{t\in N_n} \in K^n$. Any sequence on a compact has a converging subsequence. Let $(n_k^1)_{k\geq 1}$ be such that

$$(\omega^{n_k^1}(t))_{t\in N_1} \to (\omega(t))_{t\in N_1} \in K^1$$

as $k \to \infty$. Then we can take a subsequence $(n_k^2)_{k\geq 1}$ of the sequence $(n_k^1)_{k\geq 1}$ such that

$$(\omega^{n_k^2}(t))_{t\in N_2} \to (\omega(t))_{t\in N_2} \in K^2.$$

Notice that the values of $\omega(t)$ must agree on N_1. Iteratively, we can find a subsequence $(n_k^m)_{k\geq 1}$ of $(n_k^{m-1})_{k\geq 1}$ such that

$$(\omega^{n_k^m}(t))_{t\in N_m} \to (\omega(t))_{t\in N_m} \in K^m.$$

This proves the existence of a point

$$\omega \in \bigcap_{n\geq 1} K^n \times \mathbb{R}^{T\setminus N_n} \subseteq \bigcap_{n\geq 1} B_n,$$

so this last set is not empty. This finishes the proof. □

This results gives us another way to construct a sequence $(X_n)_{n\geq 1}$ of independent random variables with given distributions $(\mathbb{P}_n)_{n\geq 1}$ as the coordinate projections on the infinite product space $\mathbb{R}^{\mathbb{N}}$ with the cylindrical (product) σ-algebra $\mathscr{A} = \mathscr{B}^{\mathbb{N}}$ and infinite product measure $\mathbb{P} = \prod_{n\geq 1} \mathbb{P}_n$.

Exercise 1.3.1. Does the set $C([0,1],\mathbb{R})$ of continuous functions on $[0,1]$ belong to the cylindrical σ-algebra $\mathscr{A}$ on $\mathbb{R}^{[0,1]}$? *Hint*: An exercise in Section 1 might be helpful.

Exercise 1.3.2. Let $\mathscr{B}$ be the cylindrical σ-algebra on $\mathbb{R}^{[0,1]}$ and let Ω be the set of all continuous functions on $[0,1]$, i.e. $\Omega = C[0,1] \subseteq \mathbb{R}^{[0,1]}$. Consider the σ-algebra $\mathscr{A} := \{B\cap\Omega : B \in \mathscr{B}\}$ on $\Omega = C[0,1]$, which consists of intersections of sets in $\mathscr{B}$ with the set of continuous functions $\Omega = C[0,1]$. Show that the set

$$A := \Big\{\omega \in C[0,1] : \int_0^1 \omega(t)\,dt < 1\Big\}$$

belongs to the σ-algebra $\mathscr{A}$.

1.4 Conditional expectations and distributions

This section can be studied anytime before Chapter 4. The topic of this section can be viewed as a vast abstract generalization of the Fubini theorem for product spaces. For example, given a pair of random variables (X,Y), how can we average a function $f(X,Y)$ by fixing $X=x$ and averaging with respect to Y first? Can we define some distribution of Y given $X=x$ that can be used to compute this average? We will begin by defining these conditional averages, or conditional expectations, and then we will use them to define the corresponding conditional distributions.

Let $(\Omega,\mathscr{A},\mathbb{P})$ be a probability space and $X:\Omega\to\mathbb{R}$ be a random variable such that $\mathbb{E}|X|<\infty$. Let $\mathscr{B}$ be a σ-subalgebra of $\mathscr{A}$, $\mathscr{B}\subseteq\mathscr{A}$. Random variable $Y:\Omega\to\mathbb{R}$ is called the *conditional expectation* of X given $\mathscr{B}$ if

1. Y is measurable on $\mathscr{B}$, i.e. for every Borel set B on $\mathbb{R}$, $Y^{-1}(B)\in\mathscr{B}$.
2. For any set $B\in\mathscr{B}$, we have $\mathbb{E}X\mathrm{I}_B=\mathbb{E}Y\mathrm{I}_B$.

This conditional expectation is denoted by $Y=\mathbb{E}(X|\mathscr{B})$. If X,Z are random variables then the conditional expectation of X given Z is defined by

$$Y=\mathbb{E}(X|Z)=\mathbb{E}(X|\sigma(Z)).$$

Since Y is measurable on $\sigma(Z)$, $Y=f(Z)$ for some measurable function f. Let us go over a number of properties of conditional expectation, most of which follow easily from the definition.

1. *(Existence)* Let us define

$$\mu(B)=\int_B X\,d\mathbb{P}\ \text{ for }\ B\in\mathscr{B}.$$

Since $\mathbb{E}|X|<\infty$, $\mu(B)$ is a σ-additive signed measure on $\mathscr{B}$. Moreover, if $\mathbb{P}(B)=0$ then $\mu(B)=0$, which means that μ is absolutely continuous with respect to $\mathbb{P}$. By the Radon-Nikodym theorem, there exists a function $Y=d\mu/d\mathbb{P}$ measurable on $\mathscr{B}$ such that, for $B\in\mathscr{B}$,

$$\mu(B)=\int_B X\,d\mathbb{P}=\int_B Y\,d\mathbb{P}.$$

By definition, $Y=\mathbb{E}(X|\mathscr{B})$. Another way to define at conditional expectations is as follows. Notice that $\mathscr{L}^2(\Omega,\mathscr{B},\mathbb{P})$ is a linear subspace of $\mathscr{L}^2(\Omega,\mathscr{A},\mathbb{P})$ for $\mathscr{B}\subseteq\mathscr{A}$. If $f\in\mathscr{L}^2(\Omega,\mathscr{A},\mathbb{P})$ then the conditional expectation $g=\mathbb{E}(f|\mathscr{B})$ is simply an orthogonal projection of f onto that subspace, since such projection is $\mathscr{B}$-measurable and the orthogonal projection is defined by $\mathbb{E}fh=\mathbb{E}gh$ for all $h\in\mathscr{L}^2(\Omega,\mathscr{B},\mathbb{P})$. In particular, one can take $h=I_B$ for any $B\in\mathscr{B}$. When $f\in\mathscr{L}^1(\Omega,\mathscr{A},\mathbb{P})$, the conditional expectation can be defined by approximating f by functions in $\mathscr{L}^2(\Omega,\mathscr{A},\mathbb{P})$.

2. *(Uniqueness)* Suppose there exists another $Y' = \mathbb{E}(X|\mathscr{B})$ such that $\mathbb{P}(Y \neq Y') > 0$, i.e. either $\mathbb{P}(Y > Y') > 0$ or $\mathbb{P}(Y < Y') > 0$. Since both Y, Y' are measurable on $\mathscr{B}$, the set $B = \{Y > Y'\} \in \mathscr{B}$. If $\mathbb{P}(B) > 0$ then $\mathbb{E}(Y - Y')\mathrm{I}_B > 0$. On the other hand,

$$\mathbb{E}(Y - Y')\mathrm{I}_B = \mathbb{E}X\mathrm{I}_B - \mathbb{E}X\mathrm{I}_B = 0,$$

a contradiction. Similarly, $\mathbb{P}(Y < Y') = 0$. Of course, the conditional expectation can be redefined on a set of measure zero, so uniqueness is understood in this sense.

3. *(Linearity)* Conditional expectation is linear,

$$\mathbb{E}(cX + Y|\mathscr{B}) = c\mathbb{E}(X|\mathscr{B}) + \mathbb{E}(Y|\mathscr{B}),$$

just like the usual integral. It is obvious that the right hand side satisfies the definition of the conditional expectation $\mathbb{E}(cX + Y|\mathscr{B})$ and the equality follows by uniqueness. Again, the equality holds almost surely.

4. *(Smoothing)* If σ-algebras $\mathscr{C} \subseteq \mathscr{B} \subseteq \mathscr{A}$ then

$$\mathbb{E}(X|\mathscr{C}) = \mathbb{E}(\mathbb{E}(X|\mathscr{B})|\mathscr{C}).$$

To see this, consider a set $C \in \mathscr{C}$. Since C is also in $\mathscr{B}$,

$$\mathbb{E}\mathrm{I}_C(\mathbb{E}(\mathbb{E}(X|\mathscr{B})|\mathscr{C})) = \mathbb{E}\mathrm{I}_C\mathbb{E}(X|\mathscr{B}) = \mathbb{E}\mathrm{I}_C X,$$

so the right hand side satisfies the definition of $\mathbb{E}(X|\mathscr{C})$.

5. *(Extreme cases)* When $\mathscr{B}$ is either the entire σ-algebra $\mathscr{A}$ or the trivial σ-algebra $\{\emptyset, \Omega\}$,

$$\mathbb{E}(X|\mathscr{A}) = X, \ \mathbb{E}(X|\{\emptyset, \Omega\}) = \mathbb{E}X.$$

This is obvious by definition.

6. *(Independent σ-algebra)* If X is independent of $\mathscr{B}$ then

$$\mathbb{E}(X|\mathscr{B}) = \mathbb{E}X.$$

This is also obvious by independence, since for $B \in \mathscr{B}$,

$$\mathbb{E}X\mathrm{I}_B = \mathbb{E}X\mathbb{P}(B) = \mathbb{E}(\mathbb{E}X)\mathrm{I}_B$$

and $\mathbb{E}X$ is, of course, $\mathscr{B}$-measurable.

7. *(Monotonicity)* If $X \leq Z$ then

$$\mathbb{E}(X|\mathscr{B}) \leq \mathbb{E}(Z|\mathscr{B})$$

almost surely. The proof is by contradiction, similar to the proof of uniqueness above.

8. *(Monotone convergence)* If $\mathbb{E}|X_n| < \infty, \mathbb{E}|X| < \infty$ and $X_n \uparrow X$ then

$$\mathbb{E}(X_n|\mathscr{B}) \uparrow \mathbb{E}(X|\mathscr{B}).$$

To prove this, let us first note that, since $\mathbb{E}(X_n|\mathscr{B}) \leq \mathbb{E}(X_{n+1}|\mathscr{B}) \leq \mathbb{E}(X|\mathscr{B})$, there exists a limit

$$g = \lim_{n\to\infty} \mathbb{E}(X_n|\mathscr{B}) \leq \mathbb{E}(X|\mathscr{B}).$$

Since $\mathbb{E}(X_n|\mathscr{B})$ are measurable on $\mathscr{B}$, so is the limit g. It remains to check that for any set $B \in \mathscr{B}$, $\mathbb{E}g\mathrm{I}_B = \mathbb{E}X\mathrm{I}_B$. Since $X_n\mathrm{I}_B \uparrow X\mathrm{I}_B$ and $\mathbb{E}(X_n|\mathscr{B})\mathrm{I}_B \uparrow g\mathrm{I}_B$, the usual monotone convergence theorem implies that

$$\mathbb{E}X_n\mathrm{I}_B \uparrow \mathbb{E}X\mathrm{I}_B \text{ and } \mathbb{E}\mathrm{I}_B\mathbb{E}(X_n|\mathscr{B}) \uparrow \mathbb{E}g\mathrm{I}_B.$$

Since $\mathbb{E}\mathrm{I}_B\mathbb{E}(X_n|\mathscr{B}) = \mathbb{E}X_n\mathrm{I}_B$, this implies that $\mathbb{E}g\mathrm{I}_B = \mathbb{E}X\mathrm{I}_B$ and, therefore, $g = \mathbb{E}(X|\mathscr{B})$ almost surely.

9. *(Dominated convergence)* If $|X_n| \leq Y, \mathbb{E}Y < \infty$, and $X_n \to X$ then

$$\lim_{n\to\infty} \mathbb{E}(X_n|\mathscr{B}) = \mathbb{E}(X|\mathscr{B}).$$

The proof of the same as for usual integrals. We write

$$-Y \leq g_n = \inf_{m\geq n} X_m \leq X_n \leq h_n = \sup_{m\geq n} X_m \leq Y$$

and observe that $g_n \uparrow X$, $h_n \downarrow X$, $|g_n| \leq Y$ and $|h_n| \leq Y$. Therefore, by the monotone convergence theorem,

$$\mathbb{E}(g_n|\mathscr{B}) \uparrow \mathbb{E}(X|\mathscr{B}),\ \mathbb{E}(h_n|\mathscr{B}) \downarrow \mathbb{E}(X|\mathscr{B})$$

and the monotonicity of conditional expectation implies the claim.

10. *(Measurable factor)* If $\mathbb{E}|X| < \infty, \mathbb{E}|XY| < \infty$ and Y is measurable on $\mathscr{B}$ then

$$\mathbb{E}(XY|\mathscr{B}) = Y\mathbb{E}(X|\mathscr{B}).$$

First of all, it is enough to prove this for non-negative $X, Y \geq 0$ by decomposing $X = X^+ - X^-$ and $Y = Y^+ - Y^-$. Since Y is $\mathscr{B}$-measurable, we can find a sequence of simple functions Y_n of the form $\sum_k w_k\mathrm{I}_{C_k}$, where $C_k \in \mathscr{B}$, such that $0 \leq Y_n \uparrow Y$. By the monotone convergence theorem, it is enough to prove that

$$\mathbb{E}(X\mathrm{I}_{C_k}|\mathscr{B}) = \mathrm{I}_{C_k}\mathbb{E}(X|\mathscr{B}).$$

Take any $B \in \mathscr{B}$. Since $B \cap C_k \in \mathscr{B}$,

$$\mathbb{E}\mathrm{I}_B\mathrm{I}_{C_k}\mathbb{E}(X|\mathscr{B}) = \mathbb{E}\mathrm{I}_{B\cap C_k}\mathbb{E}(X|\mathscr{B}) = \mathbb{E}X\mathrm{I}_{B\cap C_k} = \mathbb{E}(X\mathrm{I}_{C_k})\mathrm{I}_B$$

and this finishes the proof.

11. *(Jensen's inequality)* If $f\colon \mathbb{R} \to \mathbb{R}$ is convex and $\mathbb{E}|f(X)| < \infty$ then

$$f(\mathbb{E}(X|\mathscr{B})) \le \mathbb{E}(f(X)|\mathscr{B}).$$

Let $h(x)$ be some monotone (thus, measurable) function such that $h(x)$ is in the subgradient $\partial f(x)$ of the convex function f for all x. Then, by convexity,

$$f(X) - f(\mathbb{E}(X|\mathscr{B})) \ge h(\mathbb{E}(X|\mathscr{B}))(X - \mathbb{E}(X|\mathscr{B})).$$

If we ignore any integrability issues then, taking conditional expectations of both sides,

$$\mathbb{E}(f(X)|\mathscr{B}) - f(\mathbb{E}(X|\mathscr{B})) \ge h(\mathbb{E}(X|\mathscr{B}))(\mathbb{E}(X|\mathscr{B}) - \mathbb{E}(X|\mathscr{B})) = 0,$$

where we used the previous property, since $h(\mathbb{E}(X|\mathscr{B}))$ is $\mathscr{B}$-measurable. Let us now make this simple idea rigorous. Let $B_n = \{\omega : |\mathbb{E}(X|\mathscr{B})| \le n\}$ and $X_n = XI_{B_n}$. As above,

$$f(X_n) - f(\mathbb{E}(X_n|\mathscr{B})) \ge h(\mathbb{E}(X_n|\mathscr{B}))(X_n - \mathbb{E}(X_n|\mathscr{B})).$$

Notice that $B_n \in \mathscr{B}$ and, by property 9, $\mathbb{E}(X_n|\mathscr{B}) = I_{B_n}\mathbb{E}(X|\mathscr{B}) \in [-n,n]$. Since both f and h are bounded on $[-n,n]$, now we don't have any integrability issues and, taking conditional expectations of both sides, we get

$$f(\mathbb{E}(X_n|\mathscr{B})) \le \mathbb{E}(f(X_n)|\mathscr{B}).$$

Now, let $n \to \infty$. Since $\mathbb{E}(X_n|\mathscr{B}) = I_{B_n}\mathbb{E}(X|\mathscr{B}) \to \mathbb{E}(X|\mathscr{B})$ and f is continuous, $f(\mathbb{E}(X_n|\mathscr{B}))$ converges to $f(\mathbb{E}(X|\mathscr{B}))$. On the other hand,

$$\begin{aligned}\mathbb{E}(f(X_n)|\mathscr{B})7 &= \mathbb{E}(f(X)I_{B_n} + f(0)I_{B_n^c}|\mathscr{B})\\ &= \mathbb{E}(f(X)|\mathscr{B})I_{B_n} + f(0)I_{B_n^c} \to \mathbb{E}(f(X)|\mathscr{B}),\end{aligned}$$

and this finishes the proof.

Conditional distributions. Again, let $(\Omega,\mathscr{A},\mathbb{P})$ be a probability space and let $\mathscr{B} \subseteq \mathscr{A}$ be a sub-σ-algebra of $\mathscr{A}$. Let $(Y,\mathscr{Y})$ be a measurable space and f be a measurable function from $(\Omega,\mathscr{A})$ into $(Y,\mathscr{Y})$. A *conditional distribution* of f given $\mathscr{B}$ is a function $\mathbb{P}_{f|\mathscr{B}}\colon \mathscr{Y} \times \Omega \to [0,1]$ such that

(i) for all ω, $\mathbb{P}_{f|\mathscr{B}}(\,\cdot\,,\omega)$ is a probability measure on $\mathscr{Y}$;
(ii) for each $C \in \mathscr{Y}$, $\mathbb{P}_{f|\mathscr{B}}(C,\cdot\,) = \mathbb{E}(\mathrm{I}(f \in C)|\mathscr{B})(\,\cdot\,)$ almost surely and $\mathbb{P}_{f|\mathscr{B}}(C,\cdot)$ is $\mathscr{B}$-measurable.

The meaning of this definition will become more transparent in the examples below, but we will study the existence of conditional distribution in this abstract setting first. For any given C, the conditional expectation in (ii) is defined up to a

$\mathbb{P}_{|\mathscr{B}}$-null set so there is some freedom in defining $\mathbb{P}_{f|\mathscr{B}}(C,\cdot)$ for a given C, but it is not obvious that we can define $\mathbb{P}_{f|\mathscr{B}}(C,\cdot)$ *simultaneously for all* C in such a way that (i) is also satisfied. However, when the space $(Y,\mathscr{Y})$ is *regular*, i.e. (Y,d) is a complete separable metric space and $\mathscr{Y}$ is its Borel σ-algebra, then a conditional distribution always exists.

Theorem 1.7. *If $(Y,\mathscr{Y})$ is regular then a conditional distribution $\mathbb{P}_{f|\mathscr{B}}$ exists. Moreover, if $\mathbb{P}'_{f|\mathscr{B}}$ is another conditional distribution then*

$$\mathbb{P}_{f|\mathscr{B}}(\cdot,\omega) = \mathbb{P}'_{f|\mathscr{B}}(\cdot,\omega)$$

for $\mathbb{P}_{|\mathscr{B}}$-almost all ω, i.e. the conditional distribution is unique modulo $\mathbb{P}_{|\mathscr{B}}$-almost everywhere equivalence.

Proof. The Borel σ-algebra $\mathscr{Y}$ is generated by a countable collection $\mathscr{Y}'$ of sets in $\mathscr{Y}$, for example, by open balls with rational radii and centers in a countable dense subset of (Y,d). The algebra $\mathscr{C}$ generated by $\mathscr{Y}'$ is countable since it can be written as a countable union of increasing finite algebras corresponding to finite subsets of $\mathscr{Y}'$. Since (Y,d) is complete and separable, by Ulam's theorem (Theorem 1.5), for any set $C \in \mathscr{Y}$ one can find a sequence of compact subsets $K_1 \subset K_2 \subset \ldots \subset C$ such that $\mathbb{P}(f \in K_n) \uparrow \mathbb{P}(f \in C)$. If $K = \cup_{n\geq 1} K_n$ then, by conditional monotone convergence theorem,

$$\mathbb{E}(\mathrm{I}(f \in K_n)|\mathscr{B}) \uparrow \mathbb{E}(\mathrm{I}(f \in K)|\mathscr{B}) = \mathbb{E}(\mathrm{I}(f \in C)|\mathscr{B}) \tag{1.6}$$

almost surely, where the last equality holds because $\mathbb{P}(f \in C \setminus K) = 0$. Let $\mathscr{D}$ be a countable algebra generated by the union of $\mathscr{C}$ and the set of all K_n for all $C \in \mathscr{C}$. For each $D \in \mathscr{D}$, let $\mathbb{P}_{f|\mathscr{B}}(D,\cdot)$ be any fixed version of the conditional expectation $\mathbb{E}(\mathrm{I}(f \in D)|\mathscr{B})(\cdot)$. We have:

(a) for each $D \in \mathscr{D}$, $\mathbb{P}_{f|\mathscr{B}}(D,\cdot) \geq 0$ a.s.;
(b) $\mathbb{P}_{f|\mathscr{B}}(Y,\cdot) = 1$ and $\mathbb{P}_{f|\mathscr{B}}(\emptyset,\cdot) = 0$ a.s.;
(c) for any $k \geq 1$ and any disjoint $D_1,\ldots,D_k \in \mathscr{D}$,

$$\mathbb{P}_{f|\mathscr{B}}\Big(\bigcup_{i\leq k} D_i,\cdot\Big) = \sum_{i\leq k} \mathbb{P}_{f|\mathscr{B}}(D_i,\cdot) \text{ a.s.;}$$

(d) for any $C \in \mathscr{C}$ and the specific sequence K_n as chosen, (1.6) holds a.s..

Since there are countably many conditions in (a) - (d), there exists a set $\mathscr{N} \in \mathscr{B}$ such that $\mathbb{P}(\mathscr{N}) = 0$ and conditions (a) - (d) hold for all $\omega \notin \mathscr{N}$. We will now show that for all such ω, $\mathbb{P}_{f|\mathscr{B}}(\cdot,\omega)$ can be extended to a probability measure on $\mathscr{Y}$ and this extension is such that for any $C \in \mathscr{Y}$, $\mathbb{P}_{f|\mathscr{B}}(C,\cdot)$ is a version of the conditional expectation, i.e. (ii) holds.

First, we need to show that for all $\omega \notin \mathscr{N}$, $\mathbb{P}_{f|\mathscr{B}}(\cdot,\omega)$ is countably additive on $\mathscr{C}$ or, equivalently, for any sequence $C_n \in \mathscr{C}$ such that $C_n \downarrow \emptyset$ we have $\mathbb{P}_{f|\mathscr{B}}(C_n,\omega) \downarrow 0$.

If not, then there exists a sequence $C_n \downarrow \emptyset$ such that $\mathbb{P}_{f|\mathscr{B}}(C_n,\omega) \geq \delta > 0$. Using (d), take compact $K_n \in \mathscr{D}$ such that $\mathbb{P}_{f|\mathscr{B}}(C_n \setminus K_n,\omega) \leq \delta/3^n$. Then

$$\mathbb{P}_{f|\mathscr{B}}(K_1 \cap \ldots \cap K_n,\omega) \geq \mathbb{P}_{f|\mathscr{B}}(C_n,\omega) - \sum_{1\leq i\leq n} \frac{\delta}{3^i} > \frac{\delta}{2}$$

so that each set $K^n = K_1 \cap \ldots \cap K_n$ is not empty. Since $C_n \downarrow \emptyset$ and, therefore, $K^n \downarrow \emptyset$, the complements of K^n form an open cover of the compact set K_1 (and the entire space) and there exists a finite open sub-cover of K_1. Since K^n are decreasing, it is easy to see that this can only happen if K^n is empty for some n, which is a contradiction.

By Caratheodory's extension theorem, for all $\omega \notin \mathscr{N}$, $\mathbb{P}_{f|\mathscr{B}}(\cdot,\omega)$ can be uniquely extended to a probability measure on $\mathscr{Y}$ so that (i) holds (for $\omega \in \mathscr{N}$, we can define $\mathbb{P}_{f|\mathscr{B}}(\cdot,\omega)$ to be any fixed probability measure on $\mathscr{Y}$). To prove (ii), notice that the collection of sets $C \in \mathscr{Y}$ that satisfy (ii) is, obviously, a λ-system that contains algebra $\mathscr{C}$ (which is a π-system) and, by Dynkin's theorem, Theorem 1.3, it contains the σ-algebra $\mathscr{Y}$.

To prove uniqueness, by the property (ii), we have

$$\mathbb{P}_{f|\mathscr{B}}(C,\omega) = \mathbb{P}'_{f|\mathscr{B}}(C,\omega)$$

for all $C \in \mathscr{C}$ on the set of ω of measure one, since $\mathscr{C}$ is countable, and we know that the probability measure on the σ-algebra $\mathscr{Y}$ is uniquely determined by its values on the generating algebra $\mathscr{C}$. □

When conditional distribution exists, it can be used to compute the conditional expectations with respect to $\mathscr{B}$.

Theorem 1.8. *Let $g : Y \to \mathbb{R}$ be a measurable functions such that $\mathbb{E}|g(f)| < \infty$. Then for $\mathbb{P}_{|\mathscr{B}}$-almost all ω, g is integrable with respect to $\mathbb{P}_{f|\mathscr{B}}(\cdot,\omega)$ and*

$$\mathbb{E}(g(f)|\mathscr{B})(\omega) = \int g(y)\,\mathbb{P}_{f|\mathscr{B}}(dy,\omega) \ \text{ almost surely.} \tag{1.7}$$

Proof. If $C \in \mathscr{Y}$ and $g(y) = \mathrm{I}(y \in C)$ then, by (ii),

$$\mathbb{E}(\mathrm{I}(f \in C)|\mathscr{B})(\omega) = \mathbb{P}_{f|\mathscr{B}}(C,\omega) = \int \mathrm{I}(y \in C)\,\mathbb{P}_{f|\mathscr{B}}(dy,\omega)$$

for $\mathbb{P}_{|\mathscr{B}}$-almost all ω. Therefore, (1.7) holds for all simple functions. For $g \geq 0$, take a sequence of simple functions $0 \leq g_n \uparrow g$. Since $\mathbb{E}g(f) < \infty$, by monotone convergence theorem for conditional expectations,

$$\mathbb{E}(g_n(f)|\mathscr{B})(\omega) \uparrow \mathbb{E}(g(f)|\mathscr{B})(\omega) < \infty$$

for all $\omega \notin \mathscr{N}$ for some $\mathscr{N} \in \mathscr{B}$ with $\mathbb{P}(\mathscr{N}) = 0$. Assume also that for $\omega \notin \mathscr{N}$, (1.7) holds for all functions in the sequence (g_n). By the usual monotone convergence theorem,

$$\int g_n(y)\,\mathbb{P}_{f|\mathscr{B}}(dy,\omega) \uparrow \int g(y)\,\mathbb{P}_{f|\mathscr{B}}(dy,\omega)$$

for all ω and, therefore, (1.7) holds for all $\omega \notin \mathscr{N}$. This prove the claim for $g \geq 0$ and the general case follows. □

Product space case. Consider measurable spaces $(X,\mathscr{X})$ and $(Y,\mathscr{Y})$ and let $(\Omega,\mathscr{A})$ be the product space, $\Omega = X\times Y$ and $\mathscr{A} = \mathscr{X}\otimes\mathscr{Y}$, with some probability measure $\mathbb{P}$ on it. Let

$$h(\omega) = h(x,y) = x : X\times Y \to X \text{ and } f(\omega) = f(x,y) = y : X\times Y \to Y$$

be the coordinate projections. Let $\mathscr{B}$ be the σ-algebra generated by the first coordinate projection h, i.e.

$$\mathscr{B} = h^{-1}(\mathscr{X}) = \mathscr{X}\times Y.$$

Suppose that a conditional distribution $\mathbb{P}_{f|\mathscr{B}}$ of f given $\mathscr{B}$ exists, for example, if $(Y,\mathscr{Y})$ is regular. For a fixed $C\in\mathscr{Y}$, $\mathbb{P}_{f|\mathscr{B}}(C,\cdot)$ is $\mathscr{B}$-measurable, by definition, and since $\mathscr{B} = h^{-1}(\mathscr{X})$,

$$\mathbb{P}_{f|\mathscr{B}}(C,(x,y)) = \mathbb{P}_x(C) \tag{1.8}$$

for some $\mathscr{X}$-measurable function $\mathbb{P}_x(C)$. In the product space setting, $\mathbb{P}_x$ is called a *conditional distribution of* y *given* x. Notice that for any set $D\in\mathscr{X}$, $\{\omega : h(\omega)\in D\}\in\mathscr{B}$ and

$$\begin{aligned}\mathrm{I}(x\in D)\mathbb{P}_x(C) &= \mathrm{I}(h\in D)\mathbb{E}(\mathrm{I}(f\in C)|\mathscr{B})\\ &= \mathbb{E}(\mathrm{I}(h\in D)\mathrm{I}(f\in C)|\mathscr{B}) = \mathbb{E}(\mathrm{I}(\omega\in D\times C)|\mathscr{B}) \text{ a.s.}\end{aligned}$$

Integrating both sides over Ω we get

$$\mathbb{P}(D\times C) = \int \mathrm{I}(x\in D)\mathbb{P}_x(C)\,d\mathbb{P}(x,y) = \int_D \mathbb{P}_x(C)\,d\mu(x),$$

where $\mu = \mathbb{P}\circ h^{-1}$ is the *marginal* of $\mathbb{P}$ on X. Therefore, conditional distribution of y given x satisfies:

(1) for all $x\in X$, $\mathbb{P}_x(\cdot)$ is a probability measure on $\mathscr{Y}$;
(2) for each set $C\in\mathscr{Y}$, $\mathbb{P}_x(C)$ is $\mathscr{X}$-measurable;
(3) for any $D\in\mathscr{X}$ and $C\in\mathscr{Y}$, $\mathbb{P}(D\times C) = \int_D\int_C d\mathbb{P}_x(y)\,d\mu(x)$.

Of course, (1) and (2) imply that $\mathbb{P}_x(C)$ defines a version of the conditional expectation of $\mathrm{I}(y\in C)$ given the first coordinate x. Moreover, (3) implies the following more general statement.

Theorem 1.9. *If* $g : X\times Y\to\mathbb{R}$ *is* $\mathbb{P}$*-integrable then*

$$\int g(x,y)\,d\mathbb{P}(x,y) = \iint g(x,y)\,d\mathbb{P}_x(y)d\mu(x). \tag{1.9}$$

Proof. This coincides with (3) for the indicator of the rectangle $D \times C$ and as a result holds for indicators of disjoint unions of rectangles. Then, using the monotone class theorem (or Dynkin's theorem) as in the proof of Fubini's theorem, (1.9) can be extended to indicators of all measurable sets in the product σ-algebra $\mathscr{A} = \mathscr{X} \otimes \mathscr{Y}$ and then to simple functions, positive measurable functions and all integrable functions. □

This result is a generalization of Fubini's theorem in the case when the measure is not a product measure. One can also view this as a way to construct more general measures on product spaces than product measures. One can start with any function $\mathbb{P}_x(C)$ that satisfies properties (1) and (2) and define a measure $\mathbb{P}$ on the product space as in property (3). Such functions satisfying properties (1) and (2) are called *probability kernels* or *transition functions*. In the case when $\mathbb{P}_x(C) = \nu(C)$, we recover the product measure $\mu \times \nu$.

A special version of the product space case is the so called *disintegration theorem*. Let $(Y,\mathscr{Y})$ be regular and let ν be a probability measure on $\mathscr{Y}$. Consider a measurable function $\pi : (Y,\mathscr{Y}) \to (X,\mathscr{X})$ and let $\mu = \nu \circ \pi^{-1}$ be the push-forward measure on $\mathscr{X}$. Let $\mathbb{P}$ be the push-forward measure on the product σ-algebra $\mathscr{X} \otimes \mathscr{Y}$ by the map $y \to (\pi(y), y)$, so that $\mathbb{P}$ has marginals μ and ν.

Theorem 1.10 (Disintegration Theorem). *For any bounded measurable functions* $h : X \to \mathbb{R}, f : Y \to \mathbb{R}$,

$$\int f(y)h(\pi(y))\,d\nu(y) = \int \Big(\int f(y)\,d\mathbb{P}_x(y)\Big) h(x)\,d\mu(x). \tag{1.10}$$

Moreover, for μ-almost all x, $h(\pi(y)) = h(x)$ for $\mathbb{P}_x$-almost all y.

Proof. To prove (1.10) just take $g(x,y) = h(x)f(y)$ in (1.9). If we replace $h(x)$ with $h(x)g(x)$ and use (1.10) twice we get

$$\begin{aligned}
&\int \Big(\int f(y)h(\pi(y))\,d\mathbb{P}_x(y)\Big) g(x)\,d\mu(x) = \int f(y)h(\pi(y))g(\pi(y))\,d\nu(y) \\
&= \int \Big(\int f(y)d\mathbb{P}_x(y)\Big) h(x)g(x)\,d\mu(x) = \int \Big(\int f(y)h(x)\,d\mathbb{P}_x(y)\Big) g(x)\,d\mu(x).
\end{aligned}$$

Since g is arbitrary, this means that for μ-almost all x,

$$\int f(y)h(\pi(y))\,d\mathbb{P}_x(y) = \int f(y)h(x)\,d\mathbb{P}_x(y).$$

This also holds for μ-almost all x simultaneously for countably many functions f. We can choose these f to be the indicators of sets in the algebra $\mathscr{C}$ in the proof of Theorem 1.7, and this choice obviously implies that $h(\pi(y)) = h(x)$ for $\mathbb{P}_x$-almost all y. □

Example 1.4.1. *(Same space case)* Suppose that now $X = Y$, $\mathscr{X} \subseteq \mathscr{Y}$ and $\pi : Y \to X$ is the identity map, $\pi(y) = y$. Then $\mu = \nu\lfloor_{\mathscr{X}}$ is just the restriction of ν to a smaller σ-algebra $\mathscr{X}$. In this case, (1.10) becomes

$$\int f(y)h(y)\,d\nu(y) = \int\Big(\int f(y)\,d\mathbb{P}_x(y)\Big)h(x)\,d\nu\lfloor_{\mathscr{X}}(x),$$

for any bounded $\mathscr{X}$-measurable function h and $\mathscr{Y}$-measurable function f. □

Exercise 1.4.1. Let (X,Y) be a random variable on $\mathbb{R}^2$ with the density $f(x,y)$. Show that if $\mathbb{E}|X| < \infty$ then

$$\mathbb{E}(X|Y) = \int_{\mathbb{R}} xf(x,Y)dx \Big/ \int_{\mathbb{R}} f(x,Y)dx,$$

where the right hand side is defined almost surely.

Exercise 1.4.2. Suppose that X,Y and U are random variables on some probability space such that U is independent of (X,Y) and has the uniform distribution on $[0,1]$. Prove that there exists a measurable function $f\colon \mathbb{R}\times[0,1] \to \mathbb{R}$ such that $(X, f(X,U))$ has the same distribution as (X,Y) on $\mathbb{R}^2$.

Exercise 1.4.3. Consider a measurable space $(\Omega,\mathscr{B})$ and suppose that random pairs (X,Y) and (X',Y'), defined on different probability spaces and taking values in $\mathbb{R}\times\Omega$, have the same distribution. If $\mathbb{E}|X|, \mathbb{E}|X'| < \infty$, show that the conditional expectations $\mathbb{E}(X|Y)$ and $\mathbb{E}(X'|Y')$ have the same distribution.

Exercise 1.4.4. Consider a triple (X,Y,Z) of random elements taking values on some regular spaces (i.e. complete separable metric spaces with their Borel σ-algebras). Show that the the following are equivalent:

(a) the conditional distribution of X given (Y,Z) depends only on Y (i.e. $\sigma(Y)$-measurable),
(b) X and Z are independent conditionally on Y.

Exercise 1.4.5. Consider two probability measures ν and η on $\mathbb{R}^2$.

(a) If possible, construct any probability measure $\mathbb{P}$ on $\mathbb{R}^3$ such that, for any Borel set $A \subseteq \mathbb{R}^2$,

$$\mathbb{P}\big((x,y,z) : (x,y)\in A\big) = \nu(A) \;\text{ and }\; \mathbb{P}\big((x,y,z) : (x,z)\in A\big) = \eta(A).$$

In other words, if possible, construct any probability measure $\mathbb{P}$ on $\mathbb{R}^3$ with (x,y)-marginal ν and (x,z)-marginal η.
(b) For which ν and η it is impossible to construct such $\mathbb{P}$?

Exercise 1.4.6. In the setting of the Disintegration Theorem, if $X = Y = [0,1]$, ν is the Lebesgue measure on Y, $\pi(y) = 2y\mathrm{I}(y \le 1/2) + 2(1-y)\mathrm{I}(y > 1/2)$, what is μ and $\mathbb{P}_x$?

Exercise 1.4.7. In the setting of the Disintegration Theorem, show that if X is a separable metric space then, for μ-almost every x, $\mathbb{P}_x(\pi^{-1}(x)) = 1$.

Exercise 1.4.8. Consider three random variables $X, Y, Z \colon \Omega \to \mathbb{R}$ such that

(a) X and Z are independent;
(b) Y and Z are independent conditionally on X;
(c) $Z \sim N(0,1)$.

Let $\mathscr{F} = \sigma(X+Y, X-Y)$ be the σ-algebra generated by $X+Y$ and $X-Y$. Compute the conditional expectation $\mathbb{E}(\sin(Z)|\mathscr{F})$.

Chapter 2
Laws of Large Numbers

2.1 Inequalities for sums of independent random variables

When we toss an unbiased coin many times, we expect the number of heads or tails to be close to a half — a phenomenon called the *law of large numbers*. More generally, if $(X_n)_{n\geq 1}$ is a sequence of independent identically distributed (i.i.d.) random variables, we expect that their average

$$\overline{X}_n = \frac{S_n}{n} = \frac{1}{n}\sum_{i=1}^{n} X_i$$

is, in some sense, close to the expectation $\mu = \mathbb{E}X_1$, assuming that it exists. In the next section, we will prove a general qualitative result of this type, but we begin with more quantitative statements for bounded random variables. We will need a couple of simple lemmas.

Lemma 2.1 (Chebyshev's inequality). *If a random variable $X \geq 0$ then, for $t > 0$,*

$$\mathbb{P}(X \geq t) \leq \frac{\mathbb{E}X\mathrm{I}(X \geq t)}{t} \leq \frac{\mathbb{E}X}{t}.$$

Proof. The first inequality follows from

$$\mathbb{E}X\mathrm{I}(X \geq t) \geq t\mathbb{E}\mathrm{I}(X \geq t) = t\mathbb{P}(X \geq t).$$

In the second inequality we use that $X \geq 0$. □

This inequality is often used in the exponential form,

$$\mathbb{P}(X \geq t) \leq e^{-\lambda t}\mathbb{E}e^{\lambda X} \text{ for } \lambda > 0, \tag{2.1}$$

which is obvious if we rewrite $\{X \geq t\} = \{e^{\lambda X} \geq e^{\lambda t}\}$.

Lemma 2.2. *If a random variable X satisfies $|X| \leq 1$ and $\mathbb{E}X = 0$ then, for any $\lambda \geq 0$,*

$$\mathbb{E}e^{\lambda X} \le e^{\lambda^2/2}.$$

Proof. Let us write λX as a convex combination

$$\lambda X = \frac{1+X}{2}\lambda + \frac{1-X}{2}(-\lambda),$$

where $(1+X)/2, (1-X)/2 \in [0,1]$ and their sum is equal to one. By the convexity of exponential function,

$$e^{\lambda X} \le \frac{1+X}{2}e^{\lambda} + \frac{1-X}{2}e^{-\lambda},$$

and taking expectations we get $\mathbb{E}e^{\lambda X} \le \mathrm{ch}(\lambda) \le e^{\lambda^2/2}$, where the last inequality is obvious by comparing the Taylor series on both sides. □

The first main result in this section is the following classical inequality.

Theorem 2.1 (Hoeffding's inequality). *If $\varepsilon_1, \ldots, \varepsilon_n$ are independent random variables such that $|\varepsilon_i| \le 1$ and $\mathbb{E}\varepsilon_i = 0$ then, for any $a_1, \ldots, a_n \in \mathbb{R}$ and $t \ge 0$,*

$$\mathbb{P}\Big(\sum_{i=1}^n \varepsilon_i a_i \ge t\Big) \le \exp\Big(-\frac{t^2}{2\sum_{i=1}^n a_i^2}\Big).$$

The most classical case is that of i.i.d. Rademacher random variables $(\varepsilon_n)_{n\ge 1}$ such that

$$\mathbb{P}(\varepsilon_n = 1) = \mathbb{P}(\varepsilon_n = -1) = \frac{1}{2},$$

which is equivalent to tossing a coin with heads and tails replaced by ± 1.

Proof. We begin by writing, for $\lambda > 0$,

$$\mathbb{P}\Big(\sum_{i=1}^n \varepsilon_i a_i \ge t\Big) \le e^{-\lambda t}\mathbb{E}\exp\Big(\lambda\sum_{i=1}^n \varepsilon_i a_i\Big) = e^{-\lambda t}\prod_{i=1}^n \mathbb{E}e^{\lambda\varepsilon_i a_i},$$

where in the last step we used Lemma 1.10. By Lemma 2.2, $\mathbb{E}e^{\lambda\varepsilon_i a_i} \le e^{\lambda^2 a_i^2/2}$ and, therefore,

$$\mathbb{P}\Big(\sum_{i=1}^n \varepsilon_i a_i \ge t\Big) \le \exp\Big(-\lambda t + \frac{\lambda^2}{2}\sum_{i=1}^n a_i^2\Big).$$

Optimizing over $\lambda \ge 0$ finishes the proof. □

We can also apply the same inequality to $(-\varepsilon_i)$ and, combining both cases,

$$\mathbb{P}\Big(\Big|\sum_{i=1}^n \varepsilon_i a_i\Big| \ge t\Big) \le 2\exp\Big(-\frac{t^2}{2\sum_{i=1}^n a_i^2}\Big).$$

This implies, for example, that

$$\mathbb{P}\Big(\Big|\frac{1}{n}\sum_{i=1}^{n}\varepsilon_i\Big|\geq t\Big)\leq 2\exp\Big(-\frac{nt^2}{2}\Big).$$

This show that, no matter how small $t>0$ is, the probability that the average deviates from the expectation $\mathbb{E}\varepsilon_1=0$ by more than t decreases exponentially fast with n.

Let us now consider a more general case. For $p,q\in(0,1)$, we consider the function

$$D(p,q)=p\log\frac{p}{q}+(1-p)\log\frac{1-p}{1-q},$$

which is called the *Kullback-Leibler divergence.* To see that $D(p,q)\geq 0$, with equality only if $p=q$, just use that $\log x\leq x-1$ with equality only if $x=1$.

Theorem 2.2 (Hoeffding-Chernoff inequality). *Suppose that random variables $(X_i)_{i\geq 1}$ are independent, $0\leq X_i\leq 1$, and $\mu=\mathbb{E}X$. Then*

$$\mathbb{P}\Big(\frac{1}{n}\sum_{i=1}^{n}X_i\geq\mu+t\Big)\leq e^{-nD(\mu+t,\mu)}$$

for any $t\geq 0$ such that $\mu+t\leq 1$.

Proof. Notice that the probability is zero when $\mu+t>1$, since the average can not exceed 1, so $\mu+t\leq 1$ is not really a constraint here. Using the convexity of the exponential function, we can write for $x\in[0,1]$ that

$$e^{\lambda x}=e^{\lambda(x\cdot 1+(1-x)\cdot 0)}\leq xe^{\lambda}+(1-x)e^{\lambda\cdot 0}=1-x+xe^{\lambda},$$

which implies that

$$\mathbb{E}e^{\lambda X}\leq 1-\mathbb{E}X+\mathbb{E}Xe^{\lambda}=1-\mu+\mu e^{\lambda}.$$

Using this, we get the following bound for any $\lambda\geq 0$,

$$\begin{aligned}\mathbb{P}\Big(\sum_{i=1}^{n}X_i\geq n(\mu+t)\Big)&\leq e^{-\lambda n(\mu+t)}\mathbb{E}e^{\lambda\sum X_i}=e^{-\lambda n(\mu+t)}(\mathbb{E}e^{\lambda X})^n\\&\leq e^{-\lambda n(\mu+t)}(1-\mu+\mu e^{\lambda})^n.\end{aligned}$$

The derivative of the right hand side in λ is equal to zero at:

$$-n(\mu+t)e^{-\lambda n(\mu+t)}(1-\mu+\mu e^{\lambda})^n+n(1-\mu+\mu e^{\lambda})^{n-1}\mu e^{\lambda}e^{-\lambda n(\mu+t)}=0,$$

$$-(\mu+t)(1-\mu+\mu e^{\lambda})+\mu e^{\lambda}=0,$$

$$e^{\lambda}=\frac{(1-\mu)(\mu+t)}{\mu(1-\mu-t)}\geq 1,$$

so the critical point $\lambda\geq 0$, as required. Substituting this back into the above bound,

$$\left(\left(\frac{\mu(1-\mu-t)}{(1-\mu)(\mu+t)}\right)^{\mu+t}\left(1-\mu+\frac{(1-\mu)(\mu+t)}{1-\mu-t}\right)\right)^n$$
$$=\left(\left(\frac{\mu}{\mu+t}\right)^{\mu+t}\left(\frac{1-\mu}{1-\mu-t}\right)^{1-\mu-t}\right)^n$$
$$=\exp\left(-n\left((\mu+t)\log\frac{\mu+t}{\mu}+(1-\mu-t)\log\frac{1-\mu-t}{1-\mu}\right)\right),$$

finishes the proof. □

We can also apply this bound to $Z_i = 1-X_i$, with the mean $\mu_Z = 1-\mu$, to get

$$\mathbb{P}\Big(\frac{1}{n}\sum_{i=1}^n X_i \le \mu - t\Big) = \mathbb{P}\Big(\frac{1}{n}\sum_{i=1}^n Z_i \ge \mu_Z + t\Big)$$
$$\le e^{-nD(\mu_z+t,\mu_Z)} = e^{-nD(1-\mu+t,1-\mu)}.$$

These inequalities show that, no matter how small $t>0$ is, the probability that the average $\overline{X}_n$ deviates from the expectation μ by more than t in either direction decreases exponentially fast with n. Of course, the same conclusion applies to any bounded random variables, $|X_i| \le M$, by shifting and rescaling the interval $[-M,M]$ into the interval $[0,1]$.

Even though the Hoeffding-Chernoff inequality applies to all bounded random variables, in real-world applications in engineering, computer science, etc., one would like to improve the control of the probability by incorporating other measures of closeness of the random variable X to the mean μ, for example, the *variance*

$$\sigma^2 = \mathrm{Var}(X) = \mathbb{E}(X-\mu)^2 = \mathbb{E}X^2 - (\mathbb{E}X)^2.$$

The following inequality is classical. Let us denote

$$\phi(x) = (1+x)\log(1+x) - x.$$

We will center the random variables (X_n) and instead work with $Z_n = X_n - \mu$.

Theorem 2.3 (Bennett's inequality). *Consider i.i.d.* $(Z_n)_{n\ge1}$ *such that* $\mathbb{E}Z=0$, $\mathbb{E}Z^2=\sigma^2$ *and* $|Z|<M$. *Then,*

$$\mathbb{P}\Big(\frac{1}{n}\sum_{i=1}^n Z_i \ge t\Big) \le \exp\left(-\frac{n\sigma^2}{M^2}\phi\left(\frac{tM}{\sigma^2}\right)\right),$$

for all $t\ge 0$.

Proof. As above, using that (Z_i) are i.i.d.,

$$\mathbb{P}\Big(\sum_{i=1}^n Z_i \ge nt\Big) \le e^{-\lambda nt}\left(\mathbb{E}e^{\lambda Z}\right)^n.$$

Using Taylor series expansion and the fact that $\mathbb{E}Z = 0$, $\mathbb{E}Z^2 = \sigma^2$ and $|Z| \le M$, we can write

$$\begin{aligned}
\mathbb{E}e^{\lambda Z} &= \mathbb{E}\sum_{k=0}^{\infty}\frac{(\lambda Z)^k}{k!} = \sum_{k=0}^{\infty}\lambda^k\frac{\mathbb{E}Z^k}{k!} \\
&= 1+\sum_{k=2}^{\infty}\frac{\lambda^k}{k!}\mathbb{E}Z^2Z^{k-2} \le 1+\sum_{k=2}^{\infty}\frac{\lambda^k}{k!}M^{k-2}\sigma^2 \\
&= 1+\frac{\sigma^2}{M^2}\sum_{k=2}^{\infty}\frac{\lambda^k M^k}{k!} = 1+\frac{\sigma^2}{M^2}\left(e^{\lambda M}-1-\lambda M\right) \\
&\le \exp\left(\frac{\sigma^2}{M^2}\left(e^{\lambda M}-1-\lambda M\right)\right)
\end{aligned}$$

where in the last inequality we used $1+x \le e^x$. Therefore,

$$\mathbb{P}\left(\sum_{i=1}^{n} Z_i \ge nt\right) \le \exp n\left(-\lambda t+\frac{\sigma^2}{M^2}\left(e^{\lambda M}-1-\lambda M\right)\right).$$

Now, we optimize this bound over $\lambda \ge 0$. We find the critical point:

$$-t+\frac{\sigma^2}{M^2}\left(Me^{\lambda M}-M\right)=0,$$

$$e^{\lambda M}=\frac{tM}{\sigma^2}+1,$$

$$\lambda=\frac{1}{M}\log\left(1+\frac{tM}{\sigma^2}\right).$$

Since this $\lambda \ge 0$, plugging it into the above bound,

$$\begin{aligned}
&\exp n\left(-\frac{t}{M}\log\left(1+\frac{tM}{\sigma^2}\right)+\frac{\sigma^2}{M^2}\left(\frac{tM}{\sigma^2}+1-1-\log\left(1+\frac{tM}{\sigma^2}\right)\right)\right) \\
&= \exp n\left(\frac{\sigma^2}{M^2}\left(\frac{tM}{\sigma^2}-\log\left(1+\frac{tM}{\sigma^2}\right)-\frac{tM}{\sigma^2}\log\left(1+\frac{tM}{\sigma^2}\right)\right)\right) \\
&= \exp n\left(\frac{\sigma^2}{M^2}\left(\frac{tM}{\sigma^2}-\left(1+\frac{tM}{\sigma^2}\right)\log\left(1+\frac{tM}{\sigma^2}\right)\right)\right) \\
&= \exp\left(-\frac{n\sigma^2}{M^2}\phi\left(\frac{tM}{\sigma^2}\right)\right),
\end{aligned}$$

finishes the proof. □

To simplify the bound in Bennett's inequality, one can notice that the function $f(x) := \phi(x) - \frac{x^2}{2(1+x/3)} \ge 0$, because

$$f(0) = f'(0) = 0 \text{ and } f''(x) = \frac{x^2(x+9)}{(x+1)(x+3)^3} \ge 0,$$

which implies that

$$\mathbb{P}\Big(\frac{1}{n}\sum_{i=1}^{n} Z_i \geq t\Big) \leq \exp\Big(-\frac{nt^2}{2(\sigma^2+tM/3)}\Big).$$

Combining with the same inequality for $(-Z_i)$, and recalling that $Z_i = X_i - \mu$, we obtain another classical inequality, *Bernstein's inequality*:

$$\mathbb{P}\Big(\Big|\frac{1}{n}\sum_{i=1}^{n} X_i - \mu\Big| \geq t\Big) \leq 2\exp\Big(-\frac{nt^2}{2(\sigma^2+tM/3)}\Big).$$

For small t, the denominator is of order $\sim 2\sigma^2$, and we get a better control of the probability when the variance is small.

Poisson approximation of the Binomial. As a motivation for one of the exercises below, let us prove a classical approximation of the Binomial distribution by Poisson. Consider independent Bernoulli random variables $X_i \sim B(p_i)$ for $i \leq n$ for some $p_i \in (0,1)$. Recall that, for $\lambda > 0$, Poisson distribution Π_λ is defined by

$$\Pi_\lambda(\{k\}) = \frac{\lambda^k}{k!}e^{-\lambda} \text{ for integer } k \geq 0.$$

The following holds.

Theorem 2.4. *If $S_n = X_1 + \ldots + X_n$ and $\lambda = p_1 + \ldots + p_n$ then, for any subset of integers $B \subseteq \mathbb{Z}$,*

$$|\mathbb{P}(S_n \in B) - \Pi_\lambda(B)| \leq \sum_{i=1}^{n} p_i^2.$$

In particular, if $\lambda > 0$ is fixed and all $p_i = \frac{\lambda}{n}$ then the right hand side is $\frac{\lambda^2}{n}$ is small for large n, which shows that Binomial distribution of the sum S_n is well approximated by Poisson with the same mean.

Proof. The proof is based on a construction on "the same probability space". Let us construct Bernoulli r.v. $X_i \sim B(p_i)$ and Poisson r.v. $X_i^* \sim \Pi_{p_i}$ on the same probability space as follows. Let us consider the standard probability space $([0,1],\mathscr{B},\mathbb{P})$ with the Lebesque measure $\mathbb{P}$. Let us fix i for a moment and define a random variable

$$X_i = X_i(x) = \begin{cases} 0, & 0 \leq x \leq 1-p_i, \\ 1, & 1-p_i < x \leq 1. \end{cases}$$

Clearly, $X_i \sim B(p_i)$. Let us construct X_i^* as follows. For $k \geq 0$, let us denote

$$c_k = \sum_{\ell=0}^{k} \frac{(p_i)^\ell}{\ell!}e^{-p_i}$$

and let

$$X_i^* = X_i^*(x) = \begin{cases} 0, & 0 \le x \le c_0, \\ 1, & c_0 < x \le c_1, \\ 2, & c_1 < x \le c_2, \\ \ldots \end{cases}$$

Clearly, X_i^* has the Poisson distribution Π_{p_i}. When do we have $X_i \neq X_i^*$? Since $1 - p_i \le e^{-p_i} = c_0$, this can only happen for $1 - p_i < x \le c_0$ and $c_1 < x \le 1$, which implies that

$$\mathbb{P}(X_i \neq X_i^*) = e^{-p_i} - (1 - p_i) + (1 - e^{-p_i} - p_i e^{-p_i}) = p_i(1 - e^{-p_i}) \le p_i^2.$$

Then we construct pairs (X_i, X_i^*) for $i \le n$ on separate coordinates of the product space $[0,1]^n$ with the product Lebesgue measure, thus, making them independent for $i \le n$. It is well-known (and easy to check) that a sum of independent Poisson random variables is again Poisson with the parameter given by the sum of individual parameters and, therefore, $\sum_{i \le n} X_i^*$ has the Poisson distribution Π_λ, where $\lambda = p_1 + \ldots + p_n$. Finally, we use the union bound to conclude that

$$\mathbb{P}(S_n \neq S_n^*) \le \sum_{i=1}^n \mathbb{P}(X_i \neq X_i^*) \le \sum_{i=1}^n p_i^2,$$

which finishes the proof. □

Exercise 2.1.1. Prove that, for $0 < \mu \le 1/2$ and $0 \le t < \mu$,

$$D(1 - \mu + t, 1 - \mu) \ge \frac{t^2}{2\mu(1 - \mu)}.$$

Hint: compare second derivatives.

Exercise 2.1.2. In the setting of Theorem 2.2, show that, for $s > 0$,

$$\mathbb{P}\Big(\frac{1}{\sqrt{n}} \sum_{i=1}^n (X_i - \mu) \ge s\Big) \le \exp\Big(-\frac{s^2}{2\mu(1 - \mu)} + \mathscr{O}\big(n^{-1/2}\big)\Big).$$

Notice that, by the previous exercise, if $\mu \ge 1/2$ then the same inequality holds without $\mathscr{O}\big(n^{-1/2}\big)$ term. (Once we learn the Central Limit Theorem, this will show that the Hoeffding-Chernoff inequality is almost sharp for i.i.d. Bernoulli $B(\mu)$ random variables in the regime $t = \frac{s}{\sqrt{n}}$ for $s \gg 1$.)

Exercise 2.1.3. Fix $\lambda > 0$ and $s > 0$. (a) In the setting of Theorem 2.2, if $\mu = \frac{\lambda}{n}$, show that

$$\mathbb{P}\Big(\sum_{i=1}^n X_i \ge \lambda + s\Big) \le \exp\Big(s - (\lambda + s)\log\frac{\lambda + s}{\lambda} + \mathscr{O}\big(n^{-1}\big)\Big).$$

(b) If X has Poisson distribution $\mathbb{P}(X = k) = \frac{\lambda^k}{k!} e^{-\lambda}$, $k = 0, 1, 2, \ldots$, show that

$$\mathbb{P}\Big(X \geq \lambda + s\Big) \leq \exp\Big(s - (\lambda + s)\log\frac{\lambda + s}{\lambda}\Big)$$

and, when $\lambda + s$ is integer,

$$\mathbb{P}\Big(X \geq \lambda + s\Big) \geq \frac{1}{e\sqrt{\lambda + s}}\exp\Big(s - (\lambda + s)\log\frac{\lambda + s}{\lambda}\Big).$$

Hint: For the lower bound, use Stirling's formula in the form $k! \leq ek^{k+1/2}e^{-k}$. (Together with Theorem 2.4 above, this exercise shows that the Hoeffding-Chernoff inequality is almost sharp in the regime $\mu = \frac{\lambda}{n}, t = \frac{s}{n}$ for $s \gg 1$.)

Exercise 2.1.4. Let $X_1, \ldots, X_n$ be independent flips of a fair coin, i.e. $\mathbb{P}(X_i = 0) = \mathbb{P}(X_i = 1) = 1/2$. If $\overline{X}_n$ is their average show that for $t \geq 0$

$$\mathbb{P}\big(|\overline{X}_n - 1/2| > t\big) \leq 2e^{-2nt^2}.$$

Hint: use the first problem above twice.

Exercise 2.1.5. Suppose that the random variables $X_1, \ldots, X_n, X_1', \ldots, X_n'$ are independent and, for all $i \leq n$, X_i and X_i' have the same distribution. Prove that

$$\mathbb{P}\Big(\sum_{i=1}^n (X_i - X_i') > \Big(2t\sum_{i=1}^n (X_i - X_i')^2\Big)^{1/2}\Big) \leq e^{-t}.$$

Hint: think about a way to introduce Rademacher random variables ε_i into the problem and then use Hoeffding's inequality.

Exercise 2.1.6 (Bernstein's inequality). Let $X_1, \ldots, X_n$ be i.i.d. random variables such that $|X_i - \mu| \leq M$, $\mathbb{E}X = \mu$ and $\mathrm{var}(X) = \sigma^2$. If $\overline{X}_n$ is their average, make a change of variables in the Bernstein inequality to show that for $t > 0$,

$$\mathbb{P}\Big(\overline{X}_n \geq \mu + \sqrt{\frac{2\sigma^2 t}{n}} + \frac{2Mt}{3n}\Big) \leq e^{-t}.$$

Exercise 2.1.7. Suppose that the chances of winning the jackpot in a lottery are $1 : 139{,}000{,}000$. Assuming that $100{,}000{,}000$ people played independently of each other, estimate the probability that 3 of them will have to share the jackpot. Give a bound on the quality of your estimate.

2.2 Laws of large numbers

In this section, we will study two types of convergence of the average to the mean, in probability and almost surely. Consider a sequence of random variables $(Y_n)_{n\geq 1}$ on some probability space $(\Omega, \mathscr{A}, \mathbb{P})$. We say that Y_n converges *in probability* to a random variable Y (and write $Y_n \xrightarrow{p} Y$) if, for all $\varepsilon > 0$,

$$\lim_{n\to\infty} \mathbb{P}(|Y_n - Y| \geq \varepsilon) = 0.$$

We say that Y_n converges to Y *almost surely*, or *with probability* 1, if

$$\mathbb{P}\big(\omega : \lim_{n\to\infty} Y_n(\omega) = Y(\omega)\big) = 1.$$

Let us begin with an easier case.

Theorem 2.5 (Weak law of large numbers). *Consider a sequence of random variables $(X_n)_{n\geq 1}$ that are centered, $\mathbb{E}X_n = 0$, have uniformly bounded second moments, $\mathbb{E}X_n^2 \leq K < \infty$, and are uncorrelated, $\mathbb{E}X_iX_j = 0$ for $i \neq j$. Then*

$$\overline{X}_n = \frac{1}{n}\sum_{i\leq n} X_i \to 0$$

in probability.

Proof. By Chebyshev's inequality, also using that $\mathbb{E}X_iX_j = 0$,

$$\begin{aligned}\mathbb{P}\big(|\overline{X}_n - 0| \geq \varepsilon\big) &= \mathbb{P}\big(\overline{X}_n^2 \geq \varepsilon^2\big) \leq \frac{\mathbb{E}\overline{X}_n^2}{\varepsilon^2}\\ &= \frac{1}{n^2\varepsilon^2}\mathbb{E}(X_1 + \cdots + X_n)^2 = \frac{1}{n^2\varepsilon^2}\sum_{i=1}^n \mathbb{E}X_i^2 \leq \frac{nK}{n^2\varepsilon^2} = \frac{K}{n\varepsilon^2} \to 0,\end{aligned}$$

as $n \to \infty$, which finishes the proof. □

Of course, if $(X_n)_{n\geq 1}$ are independent then they are automatically uncorrelated, since $\mathbb{E}X_iX_j = \mathbb{E}X_i\mathbb{E}X_j = 0$. Before we move on to the almost sure convergence, let us give one more application of the above argument to the problem of approximation of continuous functions. Consider an i.i.d. sequence $(X_n)_{n\geq 1}$ with the distribution $\mathbb{P}$ on $\mathbb{R}$ that depends on some parameter $\theta \in \Theta \subseteq \mathbb{R}$, and suppose that

$$\mathbb{E}X_i = \theta, \mathrm{Var}(X_i) = \sigma^2(\theta).$$

Then the following holds.

Theorem 2.6. *If $u : \mathbb{R} \to \mathbb{R}$ is continuous and bounded and $\sigma^2(\theta)$ is bounded on compacts then $\mathbb{E}u(\overline{X}_n) \to u(\theta)$ uniformly on compacts.*

Proof. For any $\varepsilon > 0$, we can write

$$\begin{aligned}|\mathbb{E}u(\overline{X}_n)-u(\theta)| &\le \mathbb{E}|u(\overline{X}_n)-u(\theta)|\\ &= \mathbb{E}|u(\overline{X}_n)-u(\theta)|\big(\mathrm{I}(|\overline{X}_n-\theta|\le\varepsilon)+\mathrm{I}(|\overline{X}_n-\theta|>\varepsilon)\big)\\ &\le \max_{|x-\theta|\le\varepsilon}|u(x)-u(\theta)|+2\|u\|_\infty\mathbb{P}(|\overline{X}_n-\theta|>\varepsilon).\end{aligned}$$

The last probability can be bounded as in the previous theorem,

$$\mathbb{P}(|\overline{X}_n-\theta|>\varepsilon)\le\frac{\sigma^2(\theta)}{n\varepsilon^2},$$

and, therefore,

$$|\mathbb{E}u(\overline{X}_n)-u(\theta)|\le\max_{|x-\theta|\le\varepsilon}|u(x)-u(\theta)|+\frac{2\|u\|_\infty\sigma^2(\theta)}{n\varepsilon^2}.$$

The statement of the theorem should now be obvious. □

Example 2.2.1. Let (X_i) be i.i.d. with the Bernoulli distribution $B(\theta)$ with probability of success $\theta\in[0,1]$,

$$\mathbb{P}(X_i=1)=\theta,\ \mathbb{P}(X_i=0)=1-\theta,$$

and let $u:[0,1]\to\mathbb{R}$ be continuous. Then, by the above theorem, the following linear combination of the *Bernstein polynomials*,

$$B_n(\theta):=\mathbb{E}u(\overline{X}_n)=\sum_{k=0}^n u\Big(\frac{k}{n}\Big)\mathbb{P}\Big(\sum_{i=1}^n X_i=k\Big)=\sum_{k=0}^n u\Big(\frac{k}{n}\Big)\binom{n}{k}\theta^k(1-\theta)^{n-k}$$

approximate $u(\theta)$ uniformly on $[0,1]$. This is an explicit example of polynomials in the Weierstrass theorem that approximate a continuous function on $[0,1]$. □

Example 2.2.2. Suppose (X_i) have the Poisson distribution $\Pi(\theta)$ with the parameter $\theta>0$,

$$\mathbb{P}_\theta(X_i=k)=\frac{\theta^k}{k!}e^{-\theta}\text{ for integer }k\ge0.$$

Then, it is well known (and easy to check) that $\mathbb{E}X_i=\theta,\sigma^2(\theta)=\theta$, and the sum $X_1+\ldots+X_n$ has Poisson distribution $\Pi(n\theta)$. Therefore, if u is bounded and continuous on $[0,+\infty)$ then

$$\mathbb{E}u(\overline{X}_n)=\sum_{k=0}^\infty u\Big(\frac{k}{n}\Big)\mathbb{P}\Big(\sum_{i=1}^n X_i=k\Big)=\sum_{k=0}^\infty u\Big(\frac{k}{n}\Big)\frac{(n\theta)^k}{k!}e^{-n\theta}\to u(\theta)$$

uniformly on compact sets. □

Before we turn to the almost sure convergence results, let us note that convergence in probability is weaker than a.s. convergence. For example, consider a probability space which is a circle of circumference 1 with the uniform measure on it. Consider a sequence of r.v. on this probability space defined by

$$X_k(x) = \mathrm{I}\Big(x \in \Big[1 + \frac{1}{2} + \cdots + \frac{1}{k}, 1 + \cdots + \frac{1}{k+1}\Big) \bmod 1\Big).$$

Then, $X_k \to 0$ in probability, since for $0 < \varepsilon < 1$,

$$\mathbb{P}(|X_k - 0| \geq \varepsilon) = \frac{1}{k+1} \to 0.$$

However, X_k does not have an almost sure limit, because the series $\sum_{k\geq 1} 1/k$ diverges and, as a result, each point x on the sphere will fall into the above intervals infinitely many times, i.e. it will satisfy $X_k(x) = 1$ for infinitely many k. Let us begin with the following lemma.

Lemma 2.3. *Consider a sequence $(p_i)_{i\geq 1}$ such that $p_i \in [0,1)$. Then*

$$\prod_{i\geq 1}(1 - p_i) = 0 \Longleftrightarrow \sum_{i\leq 1} p_i = +\infty.$$

Proof. "$\Longleftarrow$". Using that $1 - p \leq e^{-p}$ we get

$$\prod_{i\leq n}(1 - p_i) \leq \exp\Big(-\sum_{i\leq n} p_i\Big) \to 0 \text{ as } n \to \infty.$$

"$\Longrightarrow$". We can assume that $p_i \leq 1/2$ for $i \geq m$ for large enough m, because, otherwise, the series obviously diverges. Since $1 - p \geq e^{-2p}$ for $p \leq 1/2$ we have

$$\prod_{m\leq i\leq n}(1 - p_i) \geq \exp\Big(-2\sum_{m\leq i\leq n} p_i\Big)$$

and the result follows. □

The following result plays a key role in Probability Theory. Consider a sequence $(A_n)_{n\geq 1}$ of events $A_n \in \mathscr{A}$ on the probability space $(\Omega, \mathscr{A}, \mathbb{P})$. We will denote by

$$A_n \text{ i.o.} := \bigcap_{n\geq 1}\bigcup_{m\geq n} A_m$$

the event that A_n occur *infinitely often*, which consists of all $\omega \in \Omega$ that belong to infinitely many events in the sequence $(A_n)_{n\geq 1}$. Then the following holds.

Lemma 2.4 (Borel-Cantelli lemma).

(1) If $\sum_{n\geq 1} \mathbb{P}(A_n) < \infty$ then $\mathbb{P}(A_n \text{ i.o.}) = 0$.
(2) If A_n are independent and $\sum_{n\geq 1} \mathbb{P}(A_n) = +\infty$ then $\mathbb{P}(A_n \text{ i.o.}) = 1$.

Proof. (1) If $B_n = \bigcup_{m\geq n} A_m$ then $B_n \supseteq B_{n+1}$ and, by the continuity of measure,

$$\mathbb{P}(A_n \text{ i.o.}) = \mathbb{P}\Big(\bigcap_{n\geq 1} B_n\Big) = \lim_{n\to\infty} \mathbb{P}(B_n).$$

On the other hand,

$$\mathbb{P}(B_n) = \mathbb{P}\Big(\bigcup_{m\geq n} A_m\Big) \leq \sum_{m\geq n} \mathbb{P}(A_m),$$

which goes to zero as $n \to +\infty$, because the series $\sum_{m\geq 1}\mathbb{P}(A_m) < \infty$.

(2) We can write

$$\mathbb{P}(\Omega\setminus B_n) = \mathbb{P}\Big(\Omega\setminus \bigcup_{m\geq n} A_m\Big) = \mathbb{P}\Big(\bigcap_{m\geq n}(\Omega\setminus A_m)\Big)$$
$$\{\text{by independence}\} = \prod_{m\geq n}\mathbb{P}(\Omega\setminus A_m) = \prod_{m\geq n}(1-\mathbb{P}(A_m)) = 0,$$

by Lemma 2.3, since $\sum_{m\geq n}\mathbb{P}(A_m) = +\infty$. Therefore, $\mathbb{P}(B_n) = 1$ and $\mathbb{P}(A_n \text{ i.o.}) = \mathbb{P}(\bigcap_{n\geq 1} B_n) = 1$. $\square$

Let us show how this implies the strong law of large number for bounded random variables. Recall that in the case when $\mu = \mathbb{E}X_1, \sigma^2 = \mathrm{Var}(X_1)$ and $|X_1-\mu| \leq M$, the average satisfies Bernstein's inequality proved in the last section,

$$\mathbb{P}\big(|\overline{X}_n - \mu| \geq t\big) \leq 2\exp\Big(-\frac{nt^2}{2(\sigma^2+tM/3)}\Big).$$

If we take

$$t = t_n = \Big(\frac{8\sigma^2\log n}{n}\Big)^{1/2}$$

then, for n large enough such that $t_nM/3 \leq \sigma^2$, we have

$$\mathbb{P}\big(|\overline{X}_n - \mu| \geq t_n\big) \leq 2\exp\Big(-\frac{nt_n^2}{4\sigma^2}\Big) = \frac{2}{n^2}.$$

Since the series $\sum_{n\geq 1} 2n^{-2}$ converges, the Borel-Cantelli lemma implies that

$$\mathbb{P}\big(|\overline{X}_n - \mu| \geq t_n \text{ i.o.}\big) = 0.$$

This means that for $\mathbb{P}$-almost all $\omega \in \Omega$, the difference $|\overline{X}_n(\omega) - \mu|$ will become smaller than t_n for large enough $n \geq n_0(\omega)$. If we recall the definition of almost sure convergence, this means that $\overline{X}_n$ converges to μ almost surely — the so-called *strong law of large number*. Next, we will show that this holds even for unbounded random variables under a minimal assumption that the expectation $\mu = \mathbb{E}X_1$ exists.

Strong law of large numbers. The following simple observation will be useful: if a random variable $X \geq 0$ then

$$\mathbb{E}X = \int_0^\infty \mathbb{P}(X \geq x)\,dx.$$

Indeed, if F is the law of X on $\mathbb{R}$ then

$$\mathbb{E}X = \int_0^\infty x dF(x) = \int_0^\infty \int_0^x 1\, ds dF(x) = \int_0^\infty \int_s^\infty 1\, dF(x) ds = \int_0^\infty \mathbb{P}(X \geq s)\, ds.$$

For $X \geq 0$ such that $\mathbb{E}X < \infty$ this implies

$$\sum_{i\geq 1} \mathbb{P}(X \geq i) \leq \int_0^\infty \mathbb{P}(X \geq s)\, ds = \mathbb{E}X < \infty. \tag{2.2}$$

As before, let $(X_n)_{n\geq 1}$ be i.i.d. random variables on the same probability space.

Theorem 2.7 (Strong law of large numbers). *If* $\mathbb{E}|X_1| < \infty$ *then*

$$\overline{X}_n = \frac{1}{n}\sum_{i=1}^n X_i \to \mu = \mathbb{E}X_1 \textit{ almost surely.}$$

Proof. The proof will proceed in several steps.

Step 1. First, without loss of generality we can assume that $X_i \geq 0$. Indeed, for signed random variables we can decompose $X_i = X_i^+ - X_i^-$, where

$$X_i^+ = X_i \mathrm{I}(X_i \geq 0) \text{ and } X_i^- = |X_i|\mathrm{I}(X_i < 0)$$

and the general result would follow, since

$$\frac{1}{n}\sum_{i=1}^n X_i = \frac{1}{n}\sum_{i=1}^n X_i^+ - \frac{1}{n}\sum_{i=1}^n X_i^- \to \mathbb{E}X_1^+ - \mathbb{E}X_1^- = \mathbb{E}X_1.$$

Thus, from now on we assume that $X_i \geq 0$.

Step 2. (Truncation) Next, we can replace X_i by $Y_i = X_i \mathrm{I}(X_i \leq i)$ using the Borel-Cantelli lemma. Since

$$\sum_{i\geq 1} \mathbb{P}(X_i \neq Y_i) = \sum_{i\geq 1} \mathbb{P}(X_i > i) \leq \mathbb{E}X_1 < \infty,$$

the Borel-Cantelli lemma implies that $\mathbb{P}(\{X_i \neq Y_i\} \text{ i.o.}) = 0$. This means that for some (random) i_0 and for $i \geq i_0$ we have $X_i = Y_i$ and, therefore,

$$\lim_{n\to\infty} \frac{1}{n}\sum_{i=1}^n X_i = \lim_{n\to\infty} \frac{1}{n}\sum_{i=1}^n Y_i \text{ a.s.}$$

It remains to show that if $T_n = \sum_{i=1}^n Y_i$ then $T_n/n \to \mathbb{E}X$ almost surely.

Step 3. (Limits over subsequences) We will first prove almost sure convergence along the subsequences of the type $n(k) = \lfloor \alpha^k \rfloor$ for $\alpha > 1$. For any $\varepsilon > 0$,

$$\sum_{k\geq 1} \mathbb{P}\big(|T_{n(k)} - \mathbb{E}T_{n(k)}| \geq \varepsilon n(k)\big)$$

$$\le \sum_{k\ge 1} \frac{1}{\varepsilon^2 n(k)^2} \mathrm{Var}(T_{n(k)}) = \sum_{k\ge 1} \frac{1}{\varepsilon^2 n(k)^2} \sum_{i\le n(k)} \mathrm{Var}(Y_i)$$
$$\le \sum_{k\ge 1} \frac{1}{\varepsilon^2 n(k)^2} \sum_{i\le n(k)} \mathbb{E}Y_i^2 = \frac{1}{\varepsilon^2} \sum_{i\ge 1} \mathbb{E}Y_i^2 \sum_{k:n(k)\ge i} \frac{1}{n(k)^2}.$$

To bound the last sum, let us note that

$$\frac{\alpha^k}{2} \le n(k) = \lfloor \alpha^k \rfloor \le \alpha^k$$

and if $k_0 = \min\{k : \alpha^k \ge i\}$ then

$$\sum_{n(k)\ge i} \frac{1}{n(k)^2} \le \sum_{\alpha^k \ge i} \frac{4}{\alpha^{2k}} = \frac{4}{\alpha^{2k_0}(1-\frac{1}{\alpha^2})} \le \frac{K}{i^2},$$

where $K = K(\alpha)$ depends on α only. Therefore, we showed that

$$\sum_{k\ge 1} \mathbb{P}\big(|T_{n(k)} - \mathbb{E}T_{n(k)}| \ge \varepsilon n(k)\big) \le K \sum_{i\ge 1} \mathbb{E}Y_i^2 \frac{1}{i^2} = K \sum_{i\ge 1} \frac{1}{i^2} \int_0^i x^2\, dF(x),$$

where F is the law of X. We can bound the last sum as follows

$$\sum_{i\ge 1} \frac{1}{i^2} \int_0^i x^2\, dF(x) = \sum_{i\ge 1} \frac{1}{i^2} \sum_{m<i} \int_m^{m+1} x^2\, dF(x) = \sum_{m\ge 0} \sum_{i>m} \frac{1}{i^2} \int_m^{m+1} x^2\, dF(x)$$
$$\le \sum_{m\ge 0} \frac{2}{m+1} \int_m^{m+1} x^2\, dF(x) \le 2 \sum_{m\ge 0} \int_m^{m+1} x\, dF(x) = 2\mathbb{E}X < \infty.$$

We proved that

$$\sum_{k\ge 1} \mathbb{P}\big(|T_{n(k)} - \mathbb{E}T_{n(k)}| \ge \varepsilon n(k)\big) < \infty$$

and the Borel-Cantelli lemma implies that

$$\mathbb{P}\big(|T_{n(k)} - \mathbb{E}T_{n(k)}| \ge \varepsilon n(k) \text{ i.o.}\big) = 0.$$

If we take a sequence $\varepsilon_m = m^{-1}$, this implies that

$$\mathbb{P}\big(\exists m \ge 1, |T_{n(k)} - \mathbb{E}T_{n(k)}| \ge m^{-1} n(k) \text{ i.o.}\big) = 0,$$

and this proves that

$$\frac{T_{n(k)}}{n(k)} - \frac{\mathbb{E}T_{n(k)}}{n(k)} \to 0$$

with probability one. On the other hand, it is obvious that

$$\frac{\mathbb{E}T_{n(k)}}{n(k)} = \frac{1}{n(k)} \sum_{i\le n(k)} \mathbb{E}X_1 \mathrm{I}(X_1 \le i) \to \mathbb{E}X$$

as $k \to \infty$, so we proved that

$$\frac{T_{n(k)}}{n(k)} \to \mathbb{E}X \text{ almost surely.}$$

Step 4. Finally, for j such that

$$n(k) \le j < n(k+1) = n(k)\frac{n(k+1)}{n(k)} \le n(k)\alpha^2$$

we can write

$$\frac{1}{\alpha^2}\frac{T_{n(k)}}{n(k)} \le \frac{T_j}{j} \le \alpha^2 \frac{T_{n(k+1)}}{n(k+1)}$$

and, therefore, with probability one,

$$\frac{1}{\alpha^2}\mathbb{E}X \le \liminf_{j\to\infty}\frac{T_j}{j} \le \limsup_{j\to\infty}\frac{T_j}{j} \le \alpha^2\mathbb{E}X.$$

Taking $\alpha = 1+m^{-1}$ and letting $m \to \infty$ proves that $\lim_{j\to\infty} T_j/j = \mathbb{E}X$ almost surely. □

Exercise 2.2.1. Suppose that random variables $(X_n)_{n\ge 1}$ are i.i.d. such that $\mathbb{E}|X_1|^p < \infty$ for some $p > 0$. Show that $\max_{i\le n}|X_i|/n^{1/p}$ goes to zero in probability.

Exercise 2.2.2 (Weak LLN for U-statistics). If $(X_n)_{n\ge 1}$ are i.i.d., $\mathbb{E}X_1 = \mu$ and $\sigma^2 = \mathrm{Var}(X_1) < \infty$, show that

$$\binom{n}{2}^{-1} \sum_{1\le i<j\le n} X_i X_j \to \mu^2$$

in probability as $n \to \infty$.

Exercise 2.2.3. If $u : [0,1]^k \to \mathbb{R}$ is continuous then show that

$$\sum_{0\le j_1,\ldots,j_k\le n} u\Big(\frac{j_1}{n},\ldots,\frac{j_k}{n}\Big)\prod_{i\le k}\binom{n}{j_i} x_i^{j_i}(1-x_i)^{n-j_i} \to u(x_1,\ldots,x_k)$$

as $n\to\infty$, uniformly on $[0,1]^k$.

Exercise 2.2.4. If $\mathbb{E}|X| < \infty$ and $\lim_{n\to\infty}\mathbb{P}(A_n) = 0$, show that $\lim_{n\to\infty}\mathbb{E}XI_{A_n} = 0$. (Hint: use the Borel-Cantelli lemma over some subsequence.)

Exercise 2.2.5. Suppose that A_n is a sequence of events such that

$$\lim_{n\to\infty}\mathbb{P}(A_n) = 0 \text{ and } \sum_{n\ge 1}\mathbb{P}(A_n \setminus A_{n+1}) < \infty.$$

Prove that $\mathbb{P}(A_n \text{ i.o.}) = 0$.

Exercise 2.2.6. Let $\{X_n\}_{n\geq 1}$ be i.i.d. and $S_n = X_1 + \ldots + X_n$. If $S_n/n \to 0$ a.s., show that $\mathbb{E}|X_1| < \infty$. (Hint: use the idea in (2.2) and the Borel-Cantelli lemma.)

Exercise 2.2.7. Suppose that $(X_n)_{n\geq 1}$ are independent random variables. Show that $\mathbb{P}(\sup_{n\geq 1} X_n < \infty) = 1$ if and only if $\sum_{n\geq 1} \mathbb{P}(X_n > M) < \infty$ for some $M > 0$.

Exercise 2.2.8. Consider i.i.d. nonnegative random variables X_i for $i \geq 1$ such that $\mathbb{E}X_1 = +\infty$, and let $S_n = X_1 + \ldots + X_n$. Show that $\lim_{n\to\infty} \frac{S_n}{n} = +\infty$ almost surely.

Exercise 2.2.9. If $(X_n)_{n\geq 1}$ are i.i.d. and not constant with probability one, then $\mathbb{P}(X_n \text{ converges}) = 0$.

Exercise 2.2.10. Let $\{X_n\}_{n\geq 1}$ be independent and exponentially distributed, i.e. with c.d.f. $F(x) = 1 - e^{-x}$ for $x \geq 0$. Show that

$$\mathbb{P}\Big(\limsup_{n\to\infty} \frac{X_n}{\log n} = 1\Big) = 1.$$

Exercise 2.2.11. Suppose that $(X_n)_{n\geq 1}$ are i.i.d. with $\mathbb{E}X_1^+ < \infty$ and $\mathbb{E}X_1^- = -\infty$. Show that $S_n/n \to -\infty$ almost surely.

Exercise 2.2.12. Suppose that $(X_n)_{n\geq 1}$ are i.i.d. such that $\mathbb{E}|X_1| < \infty$ and $\mathbb{E}X_1 = 0$. If (c_n) is a bounded sequence of reals numbers, prove that

$$\frac{1}{n}\sum_{i=1}^{n} c_i X_i \to 0 \text{ almost surely.}$$

Hint: either (a) group the close values of c or (b) examine the proof of the strong law of large numbers.

Exercise 2.2.13. Suppose $(X_n)_{n\geq 1}$ are i.i.d. standard normal. Prove that

$$\mathbb{P}\Big(\limsup_{n\to\infty} \frac{|X_n|}{\sqrt{2\log n}} = 1\Big) = 1.$$

2.3 0 – 1 laws, convergence of random series

Consider a sequence $(X_i)_{i\geq 1}$ of random variables on the same probability space $(\Omega, \mathscr{F}, \mathbb{P})$ and let $\sigma\big((X_i)_{i\geq 1}\big)$ be the σ-algebra generated by this sequence. An event $A\in\sigma\big((X_i)_{i\geq 1}\big)$ is called a *tail event* if $A\in\sigma\big((X_i)_{i\geq n}\big)$ for all $n\geq 1$. In other words,

$$A\in\mathscr{T}=\bigcap_{n\geq 1}\sigma\big((X_i)_{i\geq n}\big),$$

where $\mathscr{T}$ is the so-called *tail σ-algebra.* For example, if $A_i\in\sigma(X_i)$ then

$$A_i \text{ i.o.} = \bigcap_{n\geq 1}\bigcup_{i\geq n} A_i$$

is a tail event. It turns out that when $(X_i)_{i\geq 1}$ are independent then all tail events have probability 0 or 1.

Theorem 2.8 (Kolmogorov's 0–1 law). *If $(X_i)_{i\geq 1}$ are independent then $\mathbb{P}(A)=0$ or 1 for all $A\in\mathscr{T}$.*

Proof. For a finite subset $F=\{i_1,\ldots,i_n\}\subset\mathbb{N}$, let us denote by $X_F=(X_{i_1},\ldots,X_{i_n})$. The σ-algebra $\sigma\big((X_i)_{i\geq 1}\big)$ is generated by the algebra of sets

$$\mathscr{A}=\big\{\{X_F\in B\}\;\big|\;\text{finite } F\subset\mathbb{N}, B\in\mathscr{B}(\mathbb{R}^{|F|})\big\}.$$

This is an algebra, because any set operations on finite number of such sets can again be expressed in terms of finitely many random variables X_i. By the Approximation Lemma, we can approximate any set $A\in\sigma\big((X_i)_{i\geq 1}\big)$ by sets in $\mathscr{A}$. Therefore, for any $\varepsilon>0$, there exists a set $A'\in\mathscr{A}$ such that $\mathbb{P}(A\,\triangle\,A')\leq\varepsilon$ and, therefore,

$$|\mathbb{P}(A)-\mathbb{P}(A')|\leq\varepsilon,\;\; |\mathbb{P}(A)-\mathbb{P}(A\cap A')|\leq\varepsilon.$$

By definition, $A'\in\sigma(X_1,\ldots,X_n)$ for large enough n and, since A is a tail event, $A\in\sigma((X_i)_{i\geq n+1})$. The Grouping Lemma implies that A and A' are independent and $\mathbb{P}(A\cap A')=\mathbb{P}(A)\mathbb{P}(A')$. Finally, we get

$$\mathbb{P}(A)\approx\mathbb{P}(A\cap A')=\mathbb{P}(A)\mathbb{P}(A')\approx\mathbb{P}(A)\mathbb{P}(A),$$

and letting $\varepsilon\downarrow 0$ proves that $\mathbb{P}(A)=\mathbb{P}(A)^2$. □

Example 2.3.1. The event $A=\big\{\text{the series } \sum_{i\geq 1}X_i \text{ converges}\big\}$ is a tail event, so it has probability 0 or 1 when X_i's are independent. □

Example 2.3.2. Consider the series $\sum_{i\geq 1}X_i z^i$ on the complex plane, for $z\in\mathbb{C}$. Its radius of convergence is

$$r=\liminf_{i\to\infty}|X_i|^{-\frac{1}{i}}.$$

For any $x \geq 0$, the event $\{r \leq x\}$ is, obviously, a tail event. This implies that $r =$ const with probability 1 when X_i's are independent. □

Next we will prove a stronger result under a more restrictive assumption that the random variables $(X_i)_{i\geq 1}$ are not only independent, but also identically distributed. A set $B \in \mathbb{R}^{\mathbb{N}}$ is called *symmetric* if, for all $n \geq 1$,

$$(x_1, x_2, \ldots, x_n, x_{n+1}, \ldots) \in B \implies (x_n, x_2, \ldots, x_{n-1}, x_1, x_{n+1}, \ldots) \in B.$$

In other words, the set B is symmetric under the permutations of finitely many coordinates. We will call an event $A \in \sigma((X_i)_{i\geq 1})$ *symmetric* if it is of the form $\{(X_i)_{i\geq 1} \in B\}$ for some symmetric set $B \in \mathbb{R}^{\mathbb{N}}$ in the cylindrical σ-algebra $\mathscr{B}^\infty$ on $\mathbb{R}^{\mathbb{N}}$. For example, any event in the tail σ-algebra $\mathscr{T}$ is symmetric.

Theorem 2.9 (Savage-Hewitt 0–1 law). *If $(X_i)_{i\geq 1}$ are i.i.d. and $A \in \sigma((X_i)_{i\geq 1})$ is symmetric then $\mathbb{P}(A) = 0$ or 1.*

Proof. By the Approximation Lemma, for any $\varepsilon > 0$, there exists $n \geq 1$ and $A_n \in \sigma(X_1, \ldots, X_n)$ such that $\mathbb{P}(A_n \triangle A) \leq \varepsilon$. Of course, we can write this set as

$$A_n = \big\{(X_1, \ldots, X_n) \in B_n\big\} \in \sigma(X_1, \ldots, X_n)$$

for some Borel set B_n on $\mathbb{R}^n$. Let us denote

$$A'_n = \big\{(X_{n+1}, \ldots, X_{2n}) \in B_n\big\} \in \sigma(X_{n+1}, \ldots, X_{2n}).$$

This set is independent of A_n, so $\mathbb{P}(A_n \cap A'_n) = \mathbb{P}(A_n)\mathbb{P}(A'_n)$. Given a sequence $x = (x_1, x_2, \ldots) \in \mathbb{R}^{\mathbb{N}}$, let us define an operator

$$\Gamma x = (x_{n+1}, \ldots, x_{2n}, x_1, \ldots, x_n, x_{2n+1}, \ldots)$$

that switches the first n coordinates with the second n coordinates. Denote $X = (X_i)_{i\geq 1}$ and recall that $A = \{X \in B\}$ for some symmetric set $B \in \mathbb{R}^{\mathbb{N}}$ that, by definition, satisfies $\Gamma B = \big\{\Gamma x : x \in B\big\} = B$. Now we can write

$$\begin{aligned}
\mathbb{P}(A'_n \triangle A) &= \mathbb{P}\big(\big\{(X_{n+1}, \ldots, X_{2n}) \in B_n\big\} \triangle \{X \in B\}\big) \\
&= \mathbb{P}\big(\big\{(X_{n+1}, \ldots, X_{2n}) \in B_n\big\} \triangle \{\Gamma X \in \Gamma B\}\big) \\
&= \mathbb{P}\big(\big\{(X_{n+1}, \ldots, X_{2n}) \in B_n\big\} \triangle \{\Gamma X \in B\}\big) \\
\{\text{using that } X_i\text{'s are i.i.d.}\} &= \mathbb{P}\big(\big\{(X_1, \ldots, X_n) \in B_n\big\} \triangle \{X \in B\}\big) \\
&= \mathbb{P}(A_n \triangle A) \leq \varepsilon.
\end{aligned}$$

This implies that $\mathbb{P}\big((A_n \cap A'_n) \triangle A\big) \leq 2\varepsilon$ and we can conclude that

$$\mathbb{P}(A) \approx \mathbb{P}(A_n),\ \mathbb{P}(A) \approx \mathbb{P}(A_n \cap A'_n) = \mathbb{P}(A_n)\mathbb{P}(A'_n) \approx \mathbb{P}(A)^2.$$

Letting $\varepsilon \downarrow 0$ again implies that $\mathbb{P}(A) = \mathbb{P}(A)^2$. □

Example 2.3.3. Let $S_n = X_1 + \ldots + X_n$ and let

$$r = \limsup_{n\to\infty} \frac{S_n - a_n}{b_n}.$$

The event $\{r \le x\}$ is symmetric, since changing the order of a finite set of coordinates does not affect S_n for large enough n. As a result, $\mathbb{P}(r \le x) = 0$ or 1, which implies that $r =$ const with probability 1. □

Random series. We already saw above that, by Kolmogorov's 0–1 law, the series $\sum_{i\ge 1} X_i$ for independent $(X_i)_{i\ge 1}$ converges with probability 0 or 1. This means that either $S_n = X_1 + \ldots + X_n$ converges to its limit S with probability one, or with probability one it does not converge. We know that almost sure convergence is stronger that convergence in probability, so in the case when with probability one S_n does not converge, is it still possible that it converges to some random variable in probability? The answer is no, because we will now prove that for random series convergence in probability implies almost sure convergence. We will need the following.

Theorem 2.10 (Kolmogorov's inequality). *Suppose that $(X_i)_{i\ge 1}$ are independent and $S_n = X_1 + \ldots + X_n$. If for all $j \le n$,*

$$\mathbb{P}(|S_n - S_j| \ge a) \le p < 1, \tag{2.3}$$

then, for $x > a$,

$$\mathbb{P}\Big(\max_{1\le j\le n} |S_j| \ge x\Big) \le \frac{1}{1-p}\mathbb{P}(|S_n| > x - a).$$

Proof. First of all, let us notice that this inequality is obvious without the maximum, because (2.3) is equivalent to $1 - p \le \mathbb{P}(|S_n - S_j| < a)$ and we can write

$$\begin{aligned}(1-p)\mathbb{P}\big(|S_j| \ge x\big) &\le \mathbb{P}\big(|S_n - S_j| < a\big)\mathbb{P}\big(|S_j| \ge x\big) \\ &= \mathbb{P}\big(|S_n - S_j| < a, |S_j| \ge x\big) \le \mathbb{P}\big(|S_n| > x - a\big).\end{aligned}$$

The equality in the middle holds because the events $\{|S_j| \ge x\}$ and $\{|S_n - S_j| < a\}$ are independent, since the first depends only on $X_1, \ldots, X_j$ and the second only on $X_{j+1}, \ldots, X_n$. The last inequality holds, because if $|S_n - S_j| < a$ and $|S_j| \ge x$ then, by triangle inequality, $|S_n| > x - a$.

To deal with the maximum, instead of looking at an arbitrary partial sum S_j, we will look at the first partial sum that crosses the level x. We define this first time by

$$\tau = \min\{j \le n : |S_j| \ge x\}$$

and let $\tau = n + 1$ if all $|S_j| < x$. Notice that the event $\{\tau = j\}$ also depends only on $X_1, \ldots, X_j$, so we can again write

$$
\begin{aligned}
(1-p)\mathbb{P}(\tau=j) &\le \mathbb{P}(|S_n-S_j|<a)\mathbb{P}(\tau=j)\\
&= \mathbb{P}(|S_n-S_j|<a,\tau=j) \le \mathbb{P}(|S_n|>x-a,\tau=j).
\end{aligned}
$$

The last inequality is true, because when $\tau=j$ we have $|S_j|\ge x$ and

$$
\{|S_n-S_j|<a,\tau=j\}\subseteq\{|S_n|>x-a,\tau=j\}.
$$

It remains to add up over $j\le n$ to get

$$
(1-p)\mathbb{P}(\tau\le n)\le\mathbb{P}(|S_n|>x-a,\tau\le n)\le\mathbb{P}(|S_n|>x-a)
$$

and notice that $\{\tau\le n\}=\{\max_{j\le n}|S_j|\ge x\}$. □

We will need one more simple lemma.

Lemma 2.5. *A sequence $Y_n\to Y$ almost surely if and only if $M_n=\max_{i\ge n}|Y_i-Y|\to 0$ in probability.*

Proof. The "only if" direction is obvious, so we only need to prove the "if" part. Since the sequence M_n is decreasing, it converges to some limit $M_n\downarrow M\ge 0$ everywhere. Since for all $\varepsilon>0$,

$$
\mathbb{P}(M\ge\varepsilon)\le\mathbb{P}(M_n\ge\varepsilon)\to 0 \text{ as } n\to\infty,
$$

this means that $\mathbb{P}(M=0)=1$ and $M_n\to 0$ almost surely. Of course, this implies that $Y_n\to Y$ almost surely. □

We are now ready to prove the result mentioned above.

Theorem 2.11. *If the series $\sum_{i\ge1}X_i$ converges in probability then it converges almost surely.*

Proof. Suppose that the partial sums S_n converge to some random variable S in probability, i.e., for any $\varepsilon>0$, for large enough $n\ge n_0(\varepsilon)$ we have $\mathbb{P}(|S_n-S|\ge\varepsilon)\le\varepsilon$. If $k\ge j\ge n\ge n_0(\varepsilon)$ then

$$
\mathbb{P}(|S_k-S_j|\ge 2\varepsilon)\le\mathbb{P}(|S_k-S|\ge\varepsilon)+\mathbb{P}(|S_j-S|\ge\varepsilon)\le 2\varepsilon.
$$

Next, we apply Kolmogorov's inequality with $x=4\varepsilon$, $a=2\varepsilon$ and $p=2\varepsilon$ to the partial sums $X_{n+1}+\ldots+X_j$ to get

$$
\mathbb{P}\Big(\max_{n\le j\le k}|S_j-S_n|\ge 4\varepsilon\Big)\le\frac{1}{1-2\varepsilon}\mathbb{P}(|S_k-S_n|\ge 2\varepsilon)\le\frac{2\varepsilon}{1-2\varepsilon}\le 3\varepsilon,
$$

for small ε. The events $\{\max_{n\le j\le k}|S_j-S_n|\ge 4\varepsilon\}$ are increasing as $k\uparrow\infty$ and, by the continuity of measure,

$$
\mathbb{P}\Big(\max_{n\le j}|S_j-S_n|\ge 4\varepsilon\Big)\le 3\varepsilon.
$$

Finally, since $\mathbb{P}(|S_n - S| \geq \varepsilon) \leq \varepsilon$ we get

$$\mathbb{P}\Big(\max_{n\leq j} |S_j - S| \geq 5\varepsilon\Big) \leq 4\varepsilon.$$

This means that the maximum $\max_{n\leq j} |S_j - S| \to 0$ in probability and, by previous lemma, $S_n \to S$ almost surely. □

Let us give one easy-to-check criterion for convergence of random series. Again, we will need one auxiliary result.

Lemma 2.6. *Random sequence $(Y_n)_{n\geq 1}$ converges in probability to some (proper) limit Y if and only if it is Cauchy in probability, which means that*

$$\lim_{n,m\to\infty} \mathbb{P}(|Y_n - Y_m| \geq \varepsilon) = 0$$

for all $\varepsilon > 0$.

Proof. Again, the "only if" direction is obvious and we only need to prove the "if" part. Given $\varepsilon = \ell^{-2}$, we can find $m(\ell)$ large enough such that, for $n, m \geq m(\ell)$,

$$\mathbb{P}\Big(|Y_n - Y_m| \geq \frac{1}{\ell^2}\Big) \leq \frac{1}{\ell^2}. \tag{2.4}$$

Without loss of generality, we can assume that $m(\ell+1) \geq m(\ell)$ so that

$$\mathbb{P}\Big(|Y_{m(\ell+1)} - Y_{m(\ell)}| \geq \frac{1}{\ell^2}\Big) \leq \frac{1}{\ell^2}.$$

Then,

$$\sum_{\ell\geq 1} \mathbb{P}\Big(|Y_{m(\ell+1)} - Y_{m(\ell)}| \geq \frac{1}{\ell^2}\Big) \leq \sum_{\ell\geq 1} \frac{1}{\ell^2} < \infty$$

and, by the Borel-Cantelli lemma,

$$\mathbb{P}\Big(|Y_{m(\ell+1)} - Y_{m(\ell)}| \geq \frac{1}{\ell^2} \text{ i.o.}\Big) = 0.$$

As a result, for large enough (random) ℓ and for $k > \ell$,

$$|Y_{m(k)} - Y_{m(\ell)}| \leq \sum_{i\geq \ell} \frac{1}{i^2} < \frac{1}{\ell - 1}.$$

This means that, with probability one, $(Y_{m(\ell)})_{\ell\geq 1}$ is a Cauchy sequence and there exists an almost sure limit $Y = \lim_{\ell\to\infty} Y_{m(\ell)}$. Together with (2.4) this implies that $Y_n \to Y$ in probability. □

Theorem 2.12. *If $(X_i)_{i\geq 1}$ is a sequence of independent random variables such that $\mathbb{E}X_i = 0$ and $\sum_{i\geq 1} \mathbb{E}X_i^2 < \infty$ then the series $\sum_{i\geq 1} X_i$ converges almost surely.*

Proof. It is enough to prove convergence in probability. For $m < n$,

$$\mathbb{P}(|S_n - S_m| \geq \varepsilon) \leq \frac{1}{\varepsilon^2}\mathbb{E}(S_n - S_m)^2 = \frac{1}{\varepsilon^2}\sum_{m<i\leq n} \mathbb{E}X_i^2 \to 0$$

as $n, m \to \infty$, since the series $\sum_{i\geq 1} \mathbb{E}X_i^2$ converges. This means that $(S_n)_{n\geq 1}$ is Cauchy in probability and, by previous lemma, S_n converges to some limit S in probability. □

Example 2.3.4. Consider the random series $\sum_{i\geq 1} \varepsilon_i/i^\alpha$ where $\mathbb{P}(\varepsilon_i = \pm 1) = 1/2$. We have

$$\sum_{i\geq 1} \mathbb{E}\left(\frac{\varepsilon_i}{i^\alpha}\right)^2 = \sum_{i\geq 1} \frac{1}{i^{2\alpha}} < \infty \text{ if } \alpha > \frac{1}{2},$$

so the series converges almost surely for $\alpha > 1/2$. □

One famous application of Theorem 2.12 is the following form of the strong law of large numbers.

Theorem 2.13 (Kolmogorov's strong law of large numbers). *Let $(X_i)_{i\geq 1}$ be independent random variables such that $\mathbb{E}X_i = 0$ and $\mathbb{E}X_i^2 < \infty$. Suppose that $b_i \leq b_{i+1}$ and $b_i \uparrow \infty$. If $\sum_{i\geq 1} \mathbb{E}X_i^2/b_i^2 < \infty$ then $\lim_{n\to\infty} b_n^{-1} S_n = 0$ almost surely.*

This follows immediately from Theorem 2.12 and the following well-known calculus lemma.

Lemma 2.7 (Kronecker's lemma). *Suppose that a sequence $(b_i)_{i\geq 1}$ is such that all $b_i > 0$ and $b_i \uparrow \infty$. Given another sequence $(x_i)_{i\geq 1}$, if the series $\sum_{i\geq 1} x_i/b_i$ converges then $\lim_{n\to\infty} b_n^{-1} \sum_{i=1}^n x_i = 0$.*

Proof. Because the series converges, $r_n := \sum_{i\geq n+1} x_i/b_i \to 0$ as $n \to \infty$. Notice that we can write $x_n = b_n(r_{n-1} - r_n)$ and, therefore,

$$\sum_{i=1}^n x_i = \sum_{i=1}^n b_i(r_{i-1} - r_i) = \sum_{i=1}^{n-1} (b_{i+1} - b_i)r_i + b_1 r_0 - b_n r_n.$$

Since $r_i \to 0$, given $\varepsilon > 0$, we can find n_0 such that for $i \geq n_0$ we have $|r_i| \leq \varepsilon$ and $|\sum_{i=n_0+1}^{n-1} (b_{i+1} - b_i)r_i| \leq \varepsilon b_n$. Therefore,

$$\left|b_n^{-1}\sum_{i=1}^n x_i\right| \leq \left|b_n^{-1}\sum_{i=1}^{n_0} (b_{i+1} - b_i)r_i\right| + \varepsilon + b_n^{-1} b_1 |r_0| + |r_n|.$$

Letting $n \to \infty$ and then $\varepsilon \downarrow 0$ finishes the proof. □

A couple of examples of application of Kolmogorov's strong law of large numbers will be given in the exercises below.

Exercise 2.3.1. Let $\{S_n : n \geq 0\}$ be a simple random walk which starts at zero, $S_0 = 0$, and at each step moves to the right with probability p and to the left with probability $1-p$. Show that the event $\{S_n = 0 \text{ i.o.}\}$ has probability 0 or 1. (*Hint:* Hewitt-Savage $0-1$ law.)

Exercise 2.3.2. In the setting of the previous problem, show: (a) if $p \neq 1/2$ then $\mathbb{P}(S_n = 0 \text{ i.o.}) = 0$; (b) if $p = 1/2$ then $\mathbb{P}(S_n = 0 \text{ i.o.}) = 1$. *Hint:* use the fact that the events

$$\Big\{\liminf_{n\to\infty} S_n \leq -1/2\Big\}, \Big\{\limsup_{n\to\infty} S_n \geq 1/2\Big\}$$

are symmetric.

Exercise 2.3.3. Given any sequence of random variables $(Y_i)_{i\geq 1}$, the event $A = \big\{\lim_{i\to\infty} Y_i \text{ exists}\big\}$ and any $\varepsilon > 0$, show that there exist integers $k,n,m \geq 1$ such that an event

$$A_{k,n,m} = \bigcap_{n\leq i<j\leq m} \Big\{\big|Y_j - Y_i\big| \leq \frac{1}{k}\Big\}$$

approximates A in a sense that $\mathbb{P}(A \,\triangle\, A_{k,n,m}) \leq \varepsilon$. *Hint:* use continuity of measure and Cauchy's convergence test.

Exercise 2.3.4. Suppose that $(X_n)_{n\geq 1}$ are i.i.d. with $\mathbb{E}X_1 = 0$ and $\mathbb{E}X_1^2 = 1$. Prove that for $\delta > 0$,

$$\frac{1}{n^{1/2}(\log n)^{1/2+\delta}} \sum_{i=1}^{n} X_i \to 0$$

almost surely. Hint: use Kolmogorov's strong law of large numbers.

Exercise 2.3.5. Let (X_n) be i.i.d. random variables with continuous distribution F. We say that X_n is a *record value* if $X_n > X_i$ for $i < n$. Let I_n be the indicator of the event that X_n is a record value.

(a) Show that the random variables $(I_n)_{n\geq 1}$ are independent and $\mathbb{P}(I_n = 1) = 1/n$. Hint: if $R_n \in \{1,\ldots,n\}$ is the rank of X_n among the first n random variables $(X_i)_{i\leq n}$, prove that (R_n) are independent.

(b) If $S_n = I_1 + \ldots + I_n$ is the number of records up to time n, prove that $S_n/\log n \to 1$ almost surely. Hint: use Kolmogorov's strong law of large numbers.

Exercise 2.3.6. Suppose that two sequences of random variables $X_n : \Omega_1 \to \mathbb{R}$ and $Y_n : \Omega_2 \to \mathbb{R}$ for $n \geq 1$ on two different probability spaces have the same distributions in the sense that all their finite dimensional distributions are the same, $\mathscr{L}((X_i)_{i\leq n}) = \mathscr{L}((Y_i)_{i\leq n})$ for all $n \geq 1$. If X_n converges almost surely to some random variable X on Ω_1 as $n \to \infty$, prove that Y_n also converges almost surely on Ω_2.

2.4 Stopping times, Wald's identity, and strong Markov property

In this section, we will have our first encounter with two concepts, *stopping times* and *Markov property*, in the setting of the sums of independent random variables. Later, stopping times will play an important role in the study of martingales, and Markov property will appear again in the setting of the Brownian motion.

Consider a sequence $(X_i)_{i\geq 1}$ of independent random variables and an integer valued random variable $\tau \in \{1,2,\ldots\}$. We say that τ is *independent of the future* if $\{\tau \leq n\}$ is independent of $\sigma((X_i)_{i\geq n+1})$. Suppose that τ is independent of the future and $\mathbb{E}|X_i| < \infty$ for all $i \geq 1$. We can formally write

$$\begin{aligned}\mathbb{E}S_\tau &= \sum_{k\geq 1} \mathbb{E}S_\tau \mathrm{I}(\tau = k) = \sum_{k\geq 1} \mathbb{E}S_k \mathrm{I}(\tau = k) \\ &= \sum_{k\geq 1}\sum_{n\leq k} \mathbb{E}X_n \mathrm{I}(\tau = k) \overset{(*)}{=} \sum_{n\geq 1}\sum_{k\geq n} \mathbb{E}X_n \mathrm{I}(\tau = k) = \sum_{n\geq 1} \mathbb{E}X_n \mathrm{I}(\tau \geq n).\end{aligned}$$

In (*) we can interchange the order of summation if, for example, the double sequence is absolutely summable, by the Fubini-Tonelli theorem. Since τ is independent of the future, the event $\{\tau \geq n\} = \{\tau \leq n-1\}^c$ is independent of $\sigma(X_n)$ and we get

$$\mathbb{E}S_\tau = \sum_{n\geq 1} \mathbb{E}X_n \mathbb{P}(\tau \geq n). \tag{2.5}$$

This implies the following.

Theorem 2.14 (Wald's identity). *If $(X_i)_{i\geq 1}$ are i.i.d., $\mathbb{E}|X_1| < \infty$ and $\mathbb{E}\tau < \infty$, then $\mathbb{E}S_\tau = \mathbb{E}X_1\mathbb{E}\tau$.*

Proof. By (2.5) we have,

$$\mathbb{E}S_\tau = \sum_{n\geq 1} \mathbb{E}X_n \mathbb{P}(\tau \geq n) = \mathbb{E}X_1 \sum_{n\geq 1} \mathbb{P}(\tau \geq n) = \mathbb{E}X_1 \mathbb{E}\tau.$$

The reason we can interchange the order of summation in (*) is because under our assumptions the double sequence is absolutely summable since

$$\sum_{n\geq 1}\sum_{k\geq n} \mathbb{E}|X_n| \mathrm{I}(\tau = k) = \sum_{n\geq 1} \mathbb{E}|X_n| \mathrm{I}(\tau \geq n) = \mathbb{E}|X_1|\mathbb{E}\tau < \infty,$$

so we can apply the Fubini-Tonelli theorem. □

We say that τ is a *stopping time* if $\{\tau \leq n\} \in \sigma(X_1,\ldots,X_n)$ for all n. Clearly, a stopping time is independent of the future. One example of a stopping time is $\tau = \min\{k \geq 1 : S_k \geq 1\}$, since

$$\{\tau \leq n\} = \bigcup_{k\leq n} \{S_k \geq 1\} \in \sigma(X_1,\ldots,X_n).$$

Given a stopping time τ, we would like to describe all events that depend on τ and the sequence $X_1,\ldots,X_\tau$ up to this stopping time. A formal definition of the σ-algebra σ_τ generated by the sequence up to a stopping time τ is the following:

$$\sigma_\tau = \big\{A : A\cap\{\tau\le n\}\in\sigma(X_1,\ldots,X_n) \text{ for all } n\ge 1\big\}.$$

When τ is a stopping time, one can easily check that this is a σ-algebra, and the meaning of the definition is that, if we know that $\tau\le n$ then the corresponding part of the event A is expressed only in terms of $X_1,\ldots,X_n$.

Theorem 2.15 (Markov property). *Suppose that $(X_i)_{i\ge1}$ are i.i.d. and τ is a stopping time. Then the sequence $T_\tau=(X_{\tau+1},X_{\tau+2},\ldots)$ is independent of the σ-algebra σ_τ and*

$$T_\tau \stackrel{d}{=} (X_1,X_2,\ldots),$$

where $\stackrel{d}{=}$ means the equality in distribution.

In words, this means that the sequence $T_\tau=(X_{\tau+1},X_{\tau+2},\ldots)$ after the stopping time is an independent copy of the entire sequence, also independent of everything that happens before the stopping time.

Proof. Consider an event $A\in\sigma_\tau$ and $B\in\mathscr{B}^\infty$, the cylindrical σ-algebra on $\mathbb{R}^{\mathbb{N}}$. We can write,

$$\begin{aligned}\mathbb{P}\big(A\cap\{T_\tau\in B\}\big) &= \sum_{n\ge1}\mathbb{P}\big(A\cap\{\tau=n\}\cap\{T_\tau\in B\}\big)\\ &= \sum_{n\ge1}\mathbb{P}\big(A\cap\{\tau=n\}\cap\{T_n\in B\}\big),\end{aligned}$$

where $T_n=(X_{n+1},X_{n+2},\ldots)$. By the definition of the σ-algebra σ_τ,

$$A\cap\{\tau=n\}=A\cap\{\tau\le n\}\setminus A\cap\{\tau\le n-1\}\in\sigma(X_1,\ldots,X_n).$$

On the other hand, $\{T_n\in B\}\in\sigma(X_{n+1},\ldots)$ and, therefore, is independent of $A\cap\{\tau=n\}$. Using this and the fact that $(X_i)_{i\ge1}$ are i.i.d.,

$$\begin{aligned}\mathbb{P}\big(A\cap\{T_\tau\in B\}\big) &= \sum_{n\ge1}\mathbb{P}\big(A\cap\{\tau=n\}\big)\mathbb{P}\big(T_n\in B\big)\\ &= \sum_{n\ge1}\mathbb{P}\big(A\cap\{\tau=n\}\big)\mathbb{P}\big(T_1\in B\big)=\mathbb{P}\big(A\big)\mathbb{P}\big(T_1\in B\big),\end{aligned}$$

and this finishes the proof. □

Let us give one interesting application of the Markov property and Wald's identity that will yield another proof of the Strong Law of Large Numbers.

Theorem 2.16. *Suppose that $(X_i)_{i\ge1}$ are i.i.d. such that $\mathbb{E}X_1>0$. If $Z=\inf_{n\ge1}S_n$ then $\mathbb{P}(Z>-\infty)=1$.*

This means that partial sums can not drift down to $-\infty$ if the mean $\mathbb{E}X_1 > 0$. Of course, this is obvious by the strong law of large number, but we want to prove this independently, since this will give another proof of the SLLN.

Proof. Let us define

$$\tau_1 = \min\{k \geq 1 : S_k \geq 1\},\ Z_1 = \min_{k \leq \tau_1} S_k,\ S_k^{(2)} = S_{\tau_1+k} - S_{\tau_1},$$

$$\tau_2 = \min\{k \geq 1 : S_k^{(2)} \geq 1\},\ Z_2 = \min_{k \leq \tau_2} S_k^{(2)},\ S_k^{(3)} = S_{\tau_2+k}^{(2)} - S_{\tau_2}^{(2)},$$

and recursively,

$$\tau_n = \min\{k \geq 1 : S_k^{(n)} \geq 1\},\ Z_n = \min_{k \leq \tau_n} S_k^{(n)},\ S_k^{(n+1)} = S_{\tau_n+k}^{(n)} - S_{\tau_n}^{(n)}.$$

We mentioned above that τ_1 is a stopping time. It is easy to check that Z_1 is σ_{τ_1}-measurable and, by the Markov property, it is independent of the sequence $T_{\tau_1} = (X_{\tau_1+1}, X_{\tau_1+2}, \ldots)$, which has the same distribution as the original sequence. Since τ_2 and Z_2 are defined exactly the same way as τ_1 and Z_1, only in terms of this new sequence T_{τ_1}, Z_2 is an independent copy of Z_1. Now, it should be obvious that $(Z_n)_{n\geq 1}$ are i.i.d. random variables. Clearly,

$$Z = \inf_{k\geq 1} S_k = \inf\{Z_1, S_{\tau_1} + Z_2, S_{\tau_1+\tau_2} + Z_3, \ldots\},$$

and, since, by construction, $S_{\tau_1+\cdots+\tau_{k-1}} \geq k-1$,

$$\{Z < -N\} \subseteq \bigcup_{k\geq 1} \{S_{\tau_1+\ldots+\tau_{k-1}} + Z_k \leq -N\} \subseteq \bigcup_{k\geq 1} \{k - 1 + Z_k \leq -N\}.$$

Therefore,

$$\begin{aligned}
\mathbb{P}(Z < -N) &\leq \sum_{k\geq 1} \mathbb{P}(k - 1 + Z_k \leq -N) = \sum_{k\geq 1} \mathbb{P}(Z_k \leq -N - k + 1) \\
&= \sum_{k\geq 1} \mathbb{P}(Z_1 \leq -N - k + 1) = \sum_{j\geq N} \mathbb{P}(Z_1 \leq -j) \leq \sum_{j\geq N} \mathbb{P}(|Z_1| \geq j) \to 0
\end{aligned}$$

as $N \to \infty$, if we can show that $\mathbb{E}|Z_1| < \infty$, since

$$\sum_{j\geq 1} \mathbb{P}(|Z_1| \geq j) \leq \mathbb{E}|Z_1| < \infty.$$

By Wald's identity,

$$\mathbb{E}|Z_1| \leq \mathbb{E} \sum_{i\leq \tau_1} |X_i| = \mathbb{E}|X_1|\mathbb{E}\tau_1 < \infty$$

if we can show that $\mathbb{E}\tau_1 < \infty$. This is left as an exercise below. We proved that $\mathbb{P}(Z < -N) \to 0$ as $N \to \infty$, which, of course, implies that $\mathbb{P}(Z > -\infty) = 1$. □

This result can be used to give another proof of the Strong Law of Large Numbers.

Theorem 2.17. *If* $(X_i)_{i\geq 1}$ *are i.i.d. and* $\mathbb{E}X_1 = 0$ *then* $S_n/n \to 0$ *almost surely.*

Proof. Given $\varepsilon > 0$ we define $X_i^\varepsilon = X_i + \varepsilon$ so that $\mathbb{E}X_1^\varepsilon = \varepsilon > 0$. By the above result, $\inf_{n\geq 1}(S_n + n\varepsilon) > -\infty$ with probability one. This means that for all $n \geq 1$, $S_n + n\varepsilon \geq -M > -\infty$ for some random variable M. Dividing both sides by n and letting $n \to \infty$ we get

$$\liminf_{n\to\infty} \frac{S_n}{n} \geq -\varepsilon$$

with probability one. We can then let $\varepsilon \downarrow 0$ over some sequence. Similarly, we prove that

$$\limsup_{n\to\infty} \frac{S_n}{n} \leq 0$$

with probability one, which finishes the proof. □

Exercise 2.4.1. Let $(X_i)_{i\geq 1}$ be i.i.d. and $\mathbb{E}X_1 > 0$. Given $a > 0$, show that $\mathbb{E}\tau < \infty$ for $\tau = \inf\{k \geq 1 : S_k > a\}$. (*Hint:* truncate X_i's and τ and use Wald's identity).

Exercise 2.4.2. Let $S_0 = 0, S_n = \sum_{i=1}^n X_i$ be a random walk with i.i.d. (X_i) such that $\mathbb{P}(X_i = 1) = p$, $\mathbb{P}(X_i = -1) = 1 - p$ for $p > 1/2$. Consider integer $b \geq 1$ and let $\tau = \min\{n \geq 1 : S_n = b\}$. Show that for $0 < s \leq 1$,

$$\mathbb{E}s^\tau = \Big(\frac{1-(1-4pqs^2)^{1/2}}{2qs}\Big)^b$$

and compute $\mathbb{E}\tau$.

Exercise 2.4.3. Suppose that we play a game with i.i.d. outcomes $(X_n)_{n\geq 1}$ such that $\mathbb{E}|X_1| < \infty$. If we play n rounds, we gain the largest of the first n outcomes. In addition, to play each round we have to pay amount $c > 0$, so after n rounds our total profit (or loss) is

$$Y_n = \max_{1\leq m\leq n} X_m - cn.$$

In this problem we will find the best strategy to play the game, in some sense.

1. Given $a \in \mathbb{R}$, let $p = \mathbb{P}(X_1 > a) > 0$ and consider the stopping time $T = \inf\{n : X_n > a\}$. Compute $\mathbb{E}Y_T$. (*Hint:* sum over sets $\{T = n\}$.)
2. Consider α such that $\mathbb{E}(X_1 - \alpha)^+ = c$. For $a = \alpha$ show that $\mathbb{E}Y_T = \alpha$.
3. Show that $Y_n \leq \alpha + \sum_{m=1}^n ((X_m - \alpha)^+ - c)$ (for any α, actually).
4. Use Wald's identity to conclude that for any stopping time τ such that $\mathbb{E}\tau < \infty$ we have $\mathbb{E}Y_\tau \leq \alpha$.

This means that stopping at time T results in the best expected profit α among all stopping times τ with $\mathbb{E}\tau < \infty$.

2.5 Azuma inequality

So far we have considered sums of independent random variables. As a digression, we will now give one example of a concentration inequality for general functions $f = f(X_1,\ldots,X_n)$ of n independent random variables. We do not assume that these random variables are identically distributed, but we will assume the following *stability condition* on the function f:

$$\big|f(x_1,\ldots,x_i,\ldots,x_n) - f(x_1,\ldots,x_i',\ldots,x_n)\big| \le a_i \tag{2.6}$$

for all $i \le n$, for some constants $a_1,\ldots,a_n$. This means that changing the ith coordinate of the function f while keeping all the other coordinates fixed can change its value by not more than a_i. In particular, the function f is bounded.

Theorem 2.18 (Azuma's inequality). *If (2.6) holds then*

$$\mathbb{P}\big(f - \mathbb{E}f \ge t\big) \le \exp\Big(-\frac{t^2}{2\sum_{i=1}^n a_i^2}\Big) \tag{2.7}$$

for any $t \ge 0$.

Proof. For $i = 1,\ldots,n$, let $\mathbb{E}_i$ denote the expectation in $X_{i+1},\ldots,X_n$ with the random variables $X_1,\ldots,X_i$ fixed. One can think of $(X_1,\ldots,X_n)$ as defined on a product space with the product measure, and $\mathbb{E}_i$ denotes the integration over the last $n-i$ coordinates. Let us denote $Y_i = \mathbb{E}_i f - \mathbb{E}_{i-1} f$ and note that $\mathbb{E}_n f = f$ and $\mathbb{E}_0 f = \mathbb{E}f$. Then we can write

$$f - \mathbb{E}f = \sum_{i=1}^n Y_i$$

(this is called martingale-difference representation) and as before, for $\lambda \ge 0$,

$$\mathbb{P}\Big(f - \mathbb{E}f \ge t\Big) = \mathbb{P}\Big(\sum_{i=1}^n Y_i \ge t\Big) \le e^{-\lambda t}\mathbb{E}e^{\lambda Y_1 + \ldots + \lambda Y_n}.$$

Notice that $\mathbb{E}_{i-1} Y_i = \mathbb{E}_{i-1} f - \mathbb{E}_{i-1} f = 0$. Also, the stability condition implies that $|Y_i| \le a_i$. Since $Y_1,\ldots,Y_{n-1}$ do not depend on X_n (only Y_n does), if we average in X_n first, we can write

$$\mathbb{E}e^{\lambda Y_1 + \ldots + \lambda Y_n} = \mathbb{E}\big(e^{\lambda Y_1 + \ldots + \lambda Y_{n-1}}\mathbb{E}_{n-1}e^{\lambda Y_n}\big).$$

If we apply Lemma 2.2 to Y_n/a_n viewed as a function of X_n, we get

$$\mathbb{E}_{n-1}e^{\lambda Y_n} = \mathbb{E}_{n-1}e^{\lambda a_n (Y_n/a_n)} \le e^{\lambda^2 a_n^2/2}$$

and, therefore,

$$\mathbb{E}e^{\lambda Y_1 + \ldots + \lambda Y_n} \le e^{(\lambda a_n)^2/2}\mathbb{E}e^{\lambda Y_1 + \ldots + \lambda Y_{n-1}}.$$

By induction on n, we get

$$\mathbb{E}e^{\lambda Y_1+\ldots+\lambda Y_n} \le e^{\lambda^2 \sum_{i=1}^n a_i^2/2}$$

and

$$\mathbb{P}\Big(\sum_{i=1}^n Y_i \ge t\Big) \le \exp\Big(-\lambda t + \frac{\lambda^2}{2}\sum_{i=1}^n a_i^2\Big).$$

Optimizing over $\lambda \ge 0$ finishes the proof. □

Notice that in the above proof, we did not use the fact that X_i's are random variables. They could be random vectors or arbitrary random elements taking values in some measurable spaces. We only used the assumption that they are independent. Keeping this in mind, let us give one example of application of Azuma's inequality.

Example 2.5.1 (Chromatic number of the Erdős–Rényi graph). Consider the Erdős–Rényi random graph $G(n,p)$ on n vertices, where each edge is present with probability p independently of other edges. Let $\chi(G(n,p))$ be the *chromatic number* of this graph, which is the smallest number of colours needed to colour the vertices so that no two adjacent vertices share the same colour. Let us denote the set of vertices by $\{v_1,\ldots,v_n\}$ and let

$$e_{i,j} = \mathrm{I}\big(\text{the edge between } v_i \text{ and } v_j \text{ is present}\big).$$

Then, all $e_{i,j}$ are independent Bernoulli $B(p)$ random variables, by the definition of the Erdős–Rényi random graph. For $i=2,\ldots,n$, let us denote by $X_i = (e_{1,i}, e_{2,i}\ldots, e_{i-1,i})$ the vector of indicators of edges between the vertex v_i and vertices $v_1,\ldots,v_{i-1}$. The vectors $X_2,\ldots,X_n$ are independent because they consist of indicators of disjoint sets of edges, and the chromatic number is a function

$$\chi(G(n,p)) = f(X_2,\ldots,X_n),$$

since these vectors include indicators of all edges in the graph. To apply Azuma's inequality, we need to determine the stability constants $a_2,\ldots,a_n$. Notice that if we replace X_i with another value $X_i' = (e_{1,i}', e_{2,i}'\ldots, e_{i-1,i}')$, this means that we modify some edges between v_i and $v_1,\ldots,v_{i-1}$. The chromatic number can not increase by more than 1, because we can always assign a new colour to the vertex v_i, so

$$f(X_2,\ldots,X_i',\ldots,X_n) - f(X_2,\ldots,X_i,\ldots,X_n) \le 1.$$

By the same logic,

$$f(X_2,\ldots,X_i,\ldots,X_n) - f(X_2,\ldots,X_i',\ldots,X_n) \le 1,$$

and this shows that the stability condition (2.6) holds with $a_i = 1$. Therefore, Azuma's inequality implies (when applied to both f and $-f$)

$$\mathbb{P}\Big(\big|\chi(G(n,p)) - \mathbb{E}\chi(G(n,p))\big| \ge t\Big) \le 2e^{-\frac{t^2}{2(n-1)}}.$$

For example, if we take $t=\sqrt{2n\log n}$, we get

$$\mathbb{P}\Big(\big|\chi(G(n,p))-\mathbb{E}\chi(G(n,p))\big|\le\sqrt{2n\log n}\Big)\ge 1-\frac{2}{n},$$

so, with high probability, the chromatic number will be within $\sqrt{2n\log n}$ from its expected value $\mathbb{E}\chi(G(n,p))$. It is known (but non-trivial) that this expected value

$$\mathbb{E}\chi(G(n,p))\sim\frac{n}{2\log n}\log\frac{1}{1-p},$$

so the deviation $\sqrt{2n\log n}$ is of a smaller order compared to the expectation. This shows that the chromatic number is typically of the same order as its expectation, which gives us an example of the law of large numbers for a very non-trivial functional of independent random variables. □

Example 2.5.2 (Balls in boxes). Suppose that we throw n balls into m boxes at random, so that the probability that a ball lands in any given box is $1/m$, and independently of each other. Let N be the number of non-empty boxes. If $X_i\in\{1,\ldots,m\}$ is the box number in which the ith ball lands then $X_1,\ldots,X_n$ are independent random variables and $N=\mathrm{card}\{X_1,\ldots,X_n\}$ is the number of distinct boxes hit.

It is easy to compute, writing $N=\sum_{i\le m}\mathrm{I}(\text{box }\#i\text{ not empty})$, that

$$\mathbb{E}N=m\Big(1-\Big(1-\frac{1}{m}\Big)^n\Big). \tag{2.8}$$

When n is large and $m=\alpha n$ for some fixed $\alpha>0$ then

$$\mathbb{E}N\sim n\alpha\big(1-e^{-1/\alpha}\big).$$

Changing the value of one box X_i changes N by at most 1, so the stability constants are all equal to $a_i=1$. As in the previous example,

$$\mathbb{P}\big(|N-\mathbb{E}N|\ge t\big)\le 2e^{-\frac{t^2}{2n}}, \tag{2.9}$$

by Azuma's inequality, and

$$\mathbb{P}\big(|N-\mathbb{E}N|\le\sqrt{2n\log n}\big)\ge 1-\frac{2}{n}.$$

So the typical number of non-empty boxes in this case is close to its expectation, relatively speaking. □

Example 2.5.3 (Max-Cut of sparse random graph). This example is quite similar to the previous one, only balls and boxes have a different meaning and, instead of the number of non-empty boxes, we consider a much more complicated function.

Let $V = \{v_1, \ldots, v_n\}$ be the set of vertices of a graph and E be its set of edges. Max-Cut problem is to divide vertices into two groups in a way that maximizes the number of edges between the two groups. This is a very important problem in computer science that has many applications, for example, to layout of electronic circuitry, and as a reformulations of various combinatorial optimization problems. For example, imagine that we have a group of n people and edges represent people that dislike each other. Then our goal is to separate them into two groups in such a way that as many 'enemies' as possible are in the opposite groups.

Here we will consider a specific model of a random graph, and will show that the maximal number of edges between the two groups concentrates around its expectation. We will select edges randomly, but using a different procedure than in the Erdős–Rényi graph. We will take

$$m = dn$$

possible edges for some fixed $d > 0$ and, as usual, we think of n as being large. Then we place each of these m edges at random among the set of $\binom{n}{2}$ possible edges, independently of each other. Each pair of vertices (a possible location to place an edge) represents a box, and edges represent balls. The random variables $X_1, \ldots, X_m$ that describe between which vertices each edge is placed take $\binom{n}{2}$ possible values and are all independent. This model produces what is called a *sparse* graph, because the number of edges $m = nd$ is small relative to $\binom{n}{2}$ and there is a fixed average number of edges per vertex, d. As in the above example, let

$$e_{i,j} = \mathrm{I}\big(\text{the edge between } v_i \text{ and } v_j \text{ is present}\big).$$

It is possible that more than one edge is placed between two vertices, in which case we keep one of them.

The function that we will consider on the above sparse random graph is called Max-Cut. It is defined as follows. If we want to split all vertices into two groups, one way to encode this is to assign each vertex one of the two labels $\{-1, 1\}$. Then all vertices with the same label belong to the same group. In other words, each vector

$$\sigma = \big(\sigma_1, \ldots, \sigma_n\big) \in \{-1, 1\}^n$$

describes a possible *cut* of the graph into two groups. Let $E(\sigma)$ be the number of present edges connecting the vertices in opposite groups,

$$E(\sigma) = \mathrm{card}\Big\{i < j : e_{i,j} = 1, \sigma_i \neq \sigma_j\Big\}. \tag{2.10}$$

Another way to represent $E(\sigma)$ is as follows. Notice that $\sigma_i \neq \sigma_j$ only if $\sigma_i\sigma_j = -1$, otherwise, $\sigma_i\sigma_j = 1$. Therefore, $\mathrm{I}(\sigma_i \neq \sigma_j) = (1 - \sigma_i\sigma_j)/2$ and

$$E(\sigma) = \frac{1}{2}\sum_{i<j} e_{i,j}\big(1 - \sigma_i\sigma_j\big). \tag{2.11}$$

Then the value M of the *Max-Cut* corresponds to a way to cut the graph so that the number of edges between the two groups is as large as possible,

$$M = \max_{\sigma} E(\sigma). \tag{2.12}$$

Recall that M is a random variable that depends on the positions $X_1, \ldots, X_m$ of our possible $m = dn$ edges.

If we change the placement X_i of one edge, this can change the maximum M by at most 1, because, for each possible cut (configuration σ), moving one edge might increase or decrease the number of edges between the two groups by at most 1. This means that the stability condition in Azuma's inequality holds with $a_i = 1$ and, therefore,

$$\mathbb{P}\big(|M - \mathbb{E}M| \geq t\big) \leq 2e^{-\frac{t^2}{2m}} = 2e^{-\frac{t^2}{2dn}}. \tag{2.13}$$

If we take $t = \sqrt{2dn\log n}$, we get

$$\mathbb{P}\Big(|M - \mathbb{E}M| \leq \sqrt{2dn\log n}\Big) \geq 1 - \frac{2}{n}.$$

Is the deviation $\sqrt{2dn\log n}$ of smaller order than $\mathbb{E}M$? It turns out that there exists a cut such that at least half of all edges are between the vertices that belong to opposite groups. To see this, choose labels $\sigma_1, \ldots, \sigma_n$ to be i.i.d. Rademacher, which means that we assign each vertex to one of the two groups $\{-1, +1\}$ at random. Then

$$M = \max_{\sigma} E(\sigma) \geq \mathbb{E}E(\sigma) = \frac{1}{2}\sum_{i<j} e_{i,j}(1 - \mathbb{E}\sigma_i\sigma_j) = \frac{1}{2}\sum_{i<j} e_{i,j},$$

which is exactly one half of all the present edges. Taking the expectations on both sides, we can write

$$\begin{aligned}\mathbb{E}M &\geq \frac{1}{2}\sum_{i<j}\mathbb{P}\big(\text{there is an edge between } v_i \text{ and } v_j\big)\\ &= \frac{1}{2}\binom{n}{2}\left(1 - \left(1 - \frac{1}{\binom{n}{2}}\right)^{dn}\right),\end{aligned}$$

where on the right hand side we have the analogue of (2.8). Using the inequality $1 - x \leq e^{-x}$, we can write

$$\left(1 - \frac{1}{\binom{n}{2}}\right)^{dn} \leq e^{-dn/\binom{n}{2}} \leq e^{-\frac{2d}{n}} = 1 - \frac{2d}{n} + \frac{2d^2}{n^2} + O\Big(\frac{1}{n^3}\Big),$$

where in the second line we used Taylor's theorem for e^x at zero. Plugging this in the above inequality, we get

$$\mathbb{E}M \geq \frac{1}{2}\binom{n}{2}\left(\frac{2d}{n} - \frac{2d^2}{n^2} + O\left(\frac{1}{n^3}\right)\right) = \frac{dn}{2} - \frac{d+d^2}{2} + O\left(\frac{1}{n}\right).$$

This shows that (up to the smaller order terms) $\mathbb{E}M$ is at least $dn/2$. This means that the deviation $\sqrt{2dn\log n}$ in Azuma's inequality above is of a smaller order, so the typical Max-Cut value M is relatively close to its expectation.

By the way, it is a major open problem to compute the limit $\lim_{n\to\infty}\mathbb{E}M/n$ in terms of d. □

Example 2.5.4 (Hamming cube). For any $x,y \in \{0,1\}^n$, the number of coordinates of x and y that differ,

$$\rho(x,y) = \sum_{i=1}^{n} \mathrm{I}(x_i \neq y_i), \tag{2.14}$$

is called the *Hamming distance* between x and y. The function ρ is called the Hamming metric on $\{0,1\}^n$. We will now use Azuma's inequality to show that, given a subset $A \subseteq \{0,1\}^n$ that contains a positive proportion of all points,

$$\mathrm{card}(A) > \varepsilon 2^n \tag{2.15}$$

for some $\varepsilon > 0$, most points in $\{0,1\}^n$ are relatively close to the set A, namely, within the Hamming distance of order $\sqrt{n}$. In other words, for most points in $\{0,1\}^n$, only of order $\sqrt{n}$ coordinates are different from one of the points in A.

Given $t > 0$, let us denote by

$$B(A,t) = \Big\{x \in \{0,1\}^n : \rho(x,y) \leq t \text{ for some } y \in A\Big\} \tag{2.16}$$

the set of all points within Hamming distance t from A. For $\varepsilon,\delta \in (0,1)$, let us denote

$$t_{\varepsilon,\delta} = \sqrt{2\log\frac{1}{\varepsilon}} + \sqrt{2\log\frac{1}{\delta}}.$$

Then the following lemma shows that $t_{\varepsilon,\delta}\sqrt{n}$-neighbourhood of A contains at least $(1-\delta)$ proportion of all points in $\{0,1\}^n$.

Lemma 2.8. *For $\varepsilon,\delta \in (0,1)$, if* $\mathrm{card}(A) > \varepsilon 2^n$ *then*

$$\mathrm{card}\, B\big(A, t_{\varepsilon,\delta}\sqrt{n}\big) \geq (1-\delta)2^n. \tag{2.17}$$

Proof. Let $X = (X_1,\ldots,X_n)$ be a vector consisting of i.i.d. Bernoulli $B(1/2)$ random variables, which means that, for any subset $S \subseteq \{0,1\}^n$,

$$\mathbb{P}\big(X \in S\big) = \frac{\mathrm{card}(S)}{2^n}.$$

Let us consider a random variable $Z = \min_{y\in A}\rho(X,y)$ equal to the Hamming distance from X to the set A. Changing one coordinate X_i to X_i' can change Z by at most 1, and Azuma's inequality (applied to Z and $-Z$) implies that

$$\mathbb{P}\big(Z \le \mathbb{E}Z - t\sqrt{n}\big) \le e^{-t^2/2}, \tag{2.18}$$
$$\mathbb{P}\big(Z \ge \mathbb{E}Z + t\sqrt{n}\big) \le e^{-t^2/2}. \tag{2.19}$$

If in the first inequality we take $t_\varepsilon = \sqrt{2\log\frac{1}{\varepsilon}}$ then

$$\mathbb{P}\big(Z \le \mathbb{E}Z - t_\varepsilon\sqrt{n}\big) \le \varepsilon.$$

This direction of Azuma's inequality allows us to conclude that $\mathbb{E}Z \le t_\varepsilon\sqrt{n}$ because of the following observation. If $\mathbb{E}Z - t_\varepsilon\sqrt{n} > 0$ then the above probability would be bigger than $\mathbb{P}(Z=0)$. However, since $Z=0$ if and only if $X \in A$,

$$\mathbb{P}(Z=0) = \mathbb{P}(X \in A) = \frac{\mathrm{card}(A)}{2^n} > \varepsilon,$$

by our assumption, which is a contradiction. Therefore,

$$\mathbb{E}Z \le t_\varepsilon\sqrt{n}.$$

The second inequality then implies

$$\mathbb{P}\big(Z \ge t_\varepsilon\sqrt{n} + t\sqrt{n}\big) \le \mathbb{P}\big(Z \ge \mathbb{E}Z + t\sqrt{n}\big) \le e^{-t^2/2}.$$

In this inequality, we take $t_\delta = \sqrt{2\log\frac{1}{\delta}}$ then

$$\mathbb{P}\big(Z \ge t_\varepsilon\sqrt{n} + t_\delta\sqrt{n}\big) \le \delta,$$

which implies that

$$\mathbb{P}(Z \le t_\varepsilon\sqrt{n} + t_\delta\sqrt{n}) \ge 1-\delta.$$

Since this event is exactly the set $B(A, t_{\varepsilon,\delta}\sqrt{n})$ of all points within the Hamming distance $t_{\varepsilon,\delta}\sqrt{n}$ from A, this finishes the proof. □

Exercise 2.5.1 (Empirical process). Given a family $\mathscr{F}$ of functions $f\colon \mathbb{R} \to [0,1]$ and independent random variables $X_1, \ldots, X_n$, consider

$$Z := \frac{1}{\sqrt{n}} \sup_{f \in \mathscr{F}} \Big| \sum_{i=1}^{n} \big(f(X_i) - \mathbb{E}f(X_i)\big) \Big|.$$

Prove that $\mathbb{P}(|Z - \mathbb{E}Z| \ge t) \le 2e^{-t^2/2}$ for all $t \ge 0$.

Exercise 2.5.2. Suppose we throw n balls into m boxes at random, and let N be the number of boxes with at least two balls in them. Compute $\mathbb{E}N$. What is the limit $\lim_{n\to\infty} \mathbb{E}N/n$ when $m = \alpha n$ for a fixed $\alpha > 0$? What does Azuma's inequality say about $|N - \mathbb{E}N|$?

Exercise 2.5.3 (3-SAT). Suppose that a debate class has a large number of students, n. The professor creates $m = 2n$ debate teams, each composed of 3 members, and he does it by assigning each student to 6 different teams.

Each member of each team is assigned to defend a position from the platform of one of two major political parties, Party A or Party B. The professor does not know the party affiliations of the students, and we suppose that each student belongs to either Party A or Party B with probability $1/2$ each, independently of each other. Given their turn, each team will select only one of its members to defend his or her assigned position. However, if it turns out that all three members have been assigned positions of the opposing party (not the one to which they belong), the team will pass.

Let N be the number of speeches delivered in class. What is the expectation $\mathbb{E}N$? Use Azuma's inequality to show that, with high probability, the number of speeches N will be relatively close to $\mathbb{E}N$.

Exercise 2.5.4 (Independent set). Let $G(n,p)$ be the Erdős-Rényi random graph on the set of n vertices $V = \{1,\ldots,n\}$. A subset of vertices $V' \subseteq V$ in this graph is called an *independent set* if there are no edges between any two vertices in V'. Let X be the largest cardinality of an independent set in $G(n,p)$. Show that

$$\mathbb{P}\big(|X - \mathbb{E}X| \geq t\big) \leq 2\exp\Big(-\frac{t^2}{2n}\Big).$$

Chapter 3
Central Limit Theorem

3.1 Convergence of laws. Selection theorem

In this section we begin the discussion of weak convergence of distributions on metric spaces. Let (S,d) be a metric space with the metric d. Consider a measurable space $(S,\mathscr{B})$ with the Borel σ-algebra $\mathscr{B}$ generated by open sets and let $(\mathbb{P}_n)_{n\geq 1}$ and $\mathbb{P}$ be some probability distributions on $\mathscr{B}$. We define

$$C_b(S) = \big\{f : S \to \mathbb{R} - \text{continuous and bounded}\big\}.$$

We say that $\mathbb{P}_n \to \mathbb{P}$ *weakly* if

$$\lim_{n\to\infty} \int_S f\, d\mathbb{P}_n = \int_S f\, d\mathbb{P} \ \text{ for all } \ f \in C_b(S). \tag{3.1}$$

This is often denoted by $\mathbb{P}_n \xrightarrow{d} \mathbb{P}$ or $\mathbb{P}_n \Longrightarrow \mathbb{P}$. Of course, the idea here is that the measure $\mathbb{P}$ on the Borel σ-algebra is determined uniquely by the integrals of $f \in C_b(S)$, so we compare closeness of the measures by closeness of all these integrals. To see this, suppose that $\int f\, d\mathbb{P} = \int f\, d\mathbb{Q}$ for all $f \in C_b(S)$. Consider any open set U in S and let $F = U^c$. Using that $d(x,F) = 0$ if and only if $x \in F$ (because F is closed), it is easy to see that

$$f_m(x) = \min\big(1, m\,d(x,F)\big) \uparrow \mathrm{I}(x \in U) \ \text{ as } \ m \uparrow \infty.$$

Since $f_m \in C_b(S)$, by monotone convergence theorem we get that $\mathbb{P}(U) = \mathbb{Q}(U)$ and, by Dynkin's theorem, $\mathbb{P} = \mathbb{Q}$. We will also say that random variables $X_n \to X$ *in distribution* if their laws converge weakly. Notice that in this definition the random variables need not be defined on the same probability space, as long as they take values in the same metric space S. We will come back to the study of convergence on general metric spaces later in the course, and in this section we will prove only one general result, the Selection Theorem. Other results will be proved only on $\mathbb{R}$ or $\mathbb{R}^n$ to prepare us for the most famous example of convergence of laws - the Central Limit Theorem (CLT). First of all, let us notice that on the real line the convergence of probability measures can be expressed in terms of their c.d.f.s, as follows.

Theorem 3.1. *If* $S = \mathbb{R}$ *then* $\mathbb{P}_n \to \mathbb{P}$ *weakly if and only if* $F_n(t) = \mathbb{P}_n\big((-\infty,t]\big) \to F(t) = \mathbb{P}\big((-\infty,t]\big)$ *for any point of continuity* t *of the c.d.f.* F.

Proof. "$\Longrightarrow$" Suppose that (3.1) holds. Let us approximate the indicator $\mathrm{I}(x \le t)$ by continuous functions $\varphi_1, \varphi_2 \in C_b(\mathbb{R})$ so that

$$\mathrm{I}(x \le t-\varepsilon) \le \varphi_1(x) \le \mathrm{I}(x \le t) \le \varphi_2(x) \le \mathrm{I}(x \le t+\varepsilon).$$

Then, using (3.1) for φ_1 and φ_2,

$$\begin{aligned} F(t-\varepsilon) &\le \int \varphi_1\, dF = \lim_{n\to\infty} \int \varphi_1\, dF_n \\ &\le \lim_{n\to\infty} F_n(t) \le \lim_{n\to\infty} \int \varphi_2\, dF_n = \int \varphi_2\, dF \le F(t+\varepsilon). \end{aligned}$$

Therefore, for any $\varepsilon > 0$,

$$F(t-\varepsilon) \le \lim_{n\to\infty} F_n(t) \le F(t+\varepsilon).$$

More carefully, we should write lim inf and lim sup but, since t is a point of continuity of F, letting $\varepsilon \downarrow 0$ proves that the limit $\lim_{n\to\infty} F_n(t)$ exists and is equal to $F(t)$.

"$\Longleftarrow$" Let $PC(F)$ be the set of points of continuity of F. Since F is monotone, the set $PC(F)$ is dense in $\mathbb{R}$. Take M large enough such that both $M, -M \in PC(F)$ and $\mathbb{P}((-M,M]^c) \le \varepsilon$. Clearly, for large enough $n \ge 1$ we have $\mathbb{P}_n((-M,M]^c) \le 2\varepsilon$. For any $k > 1$, consider a sequence of points $-M = x_1^k \le x_2^k \le \ldots \le x_k^k = M$ such that all $x_i \in PC(F)$ and $\max_i |x_{i+1}^k - x_i^k| \to 0$ as $k \to \infty$. Given a function $f \in C_b(\mathbb{R})$, consider an approximating function

$$f_k(x) = \sum_{1<i\le k} f(x_i^k)\, \mathrm{I}\big(x \in (x_{i-1}^k, x_i^k]\big) + 0 \cdot \mathrm{I}\big(x \notin (-M,M]\big).$$

Since f in continuous,

$$\delta_k(M) = \sup_{|x|\le M} \big|f_k(x) - f(x)\big| \to 0,\ k \to \infty.$$

Since all $x_i^k \in PC(F)$, by assumption, we can write

$$\begin{aligned} \int f_k\, dF_n &= \sum_{1<i\le k} f_k(x_i^k)\big(F_n(x_i^k) - F_n(x_{i-1}^k)\big) \\ &\xrightarrow{n\to\infty} \sum_{1<i\le k} f_k(x_i^k)\big(F(x_i^k) - F(x_{i-1}^k)\big) = \int f_k\, dF. \end{aligned}$$

On the other hand,

$$\Big|\int f\,dF - \int f_k\,dF\Big| \le \|f\|_\infty \mathbb{P}\big((-M,M]^c\big) + \delta_k(M) \le \|f\|_\infty \varepsilon + \delta_k(M)$$

and, similarly, for large enough $n \ge 1$,

$$\Big|\int f\,dF_n - \int f_k\,dF_n\Big| \le \|f\|_\infty \mathbb{P}_n\big((-M,M]^c\big) + \delta_k(M) \le \|f\|_\infty 2\varepsilon + \delta_k(M).$$

Letting $n \to \infty$, then $k \to \infty$ and, finally, $\varepsilon \downarrow 0$ (or $M \uparrow \infty$), proves that $\int f\,dF_n \to \int f\,dF$. □

Example 3.1.1. If $\mathbb{P}_n(\{n^{-1}\}) = 1$ and $\mathbb{P}(\{0\}) = 1$ then $\mathbb{P}_n \to \mathbb{P}$ weakly, but their c.d.f.s, obviously, do not converge at the point of discontinuity $x = 0$.

Our typical strategy for proving the convergence of probability measures will be based on the following elementary observation.

Lemma 3.1. *If for any sequence $(n(k))_{k\ge 1}$ there exists a subsequence $(n(k(r)))_{r\ge 1}$ such that $\mathbb{P}_{n(k(r))} \to \mathbb{P}$ weakly then $\mathbb{P}_n \to \mathbb{P}$ weakly.*

Proof. Suppose not. Then for some $f \in C_b(S)$ and for some $\varepsilon > 0$ there exists a subsequence $(n(k))$ such that

$$\Big|\int f\,d\mathbb{P}_{n(k)} - \int f\,d\mathbb{P}\Big| > \varepsilon.$$

But this contradicts the fact that for some subsequence $\mathbb{P}_{n(k(r))} \to \mathbb{P}$ weakly. □

As a result, to prove that $\mathbb{P}_n \to \mathbb{P}$ weakly, it is enough to demonstrate two things:

1. First, that the sequence $(\mathbb{P}_n)$ is such that one can always find converging subsequences for any sequence $(\mathbb{P}_{n(k)})$. This describes a sort of "relative compactness" of the sequence $(\mathbb{P}_n)$ and will be a consequence of *uniform tightness* and the Selection Theorem proved below.
2. Second, we need to show that any subsequential limit is always the same, $\mathbb{P}$, and this will be done by some ad hoc methods, for example, using the method of characteristic functions in the case of the CLT.

We say that a sequence of distributions $(\mathbb{P}_n)_{n\ge 1}$ on a metric space (S,d) is *uniformly tight* if, for any $\varepsilon > 0$, there exists a compact $K \subseteq S$ such that $\mathbb{P}_n(K) \ge 1-\varepsilon$ for all n. Checking this property can be difficult and, of course, one needs to understand how compacts look like for a particular metric space. In the case of the CLT on $\mathbb{R}^n$, this will be a trivial task. The following fact is quite fundamental.

Theorem 3.2 (Selection Theorem). *If $(\mathbb{P}_n)_{n\ge 1}$ is a uniformly tight sequence of laws on the metric space (S,d) then there exists a subsequence $(n(k))$ such that $\mathbb{P}_{n(k)}$ converges weakly to some probability law $\mathbb{P}$.*

Let us recall the following well-known result.

Lemma 3.2 (Cantor's diagonalization). *Let A be a countable set and $f_n\colon A \to \mathbb{R}$ for $n \geq 1$. Then there exists a subsequence $(n(k))$ such that $f_{n(k)}(a)$ converges for all $a \in A$, possibly to $\pm\infty$.*

Proof. Let $A = \{a_1, a_2, \ldots\}$. Take $(n^1(k))$ such that $f_{n^1(k)}(a_1)$ converges. Take $(n^2(k)) \subseteq (n^1(k))$ such that $f_{n^2(k)}(a_2)$ converges. Recursively, take $(n^\ell(k)) \subseteq (n^{\ell-1}(k))$ such that $f_{n^\ell(k)}(a_\ell)$ converges. Now consider the sequence $(n^k(k))$. Clearly, $f_{n^k(k)}(a_\ell)$ converges for any ℓ because for $k \geq \ell, n^k(k) \in \{n^\ell(k)\}$ by construction. □

Proof (of Theorem 3.2). We will prove the Selection Theorem for arbitrary metric spaces, since this result will be useful to us later when we study the convergence of laws on general metric spaces. However, when $S = \mathbb{R}$ one can see this in a much more intuitive way, as follows.

(The case $S = \mathbb{R}$). Let A be a dense set of points in $\mathbb{R}$. Given a sequence of probability measures $\mathbb{P}_n$ on $\mathbb{R}$ and their c.d.f.s F_n, by Cantor's diagonalization, there exists a subsequence $(n(k))$ such that $F_{n(k)}(a) \to F(a)$ for all $a \in A$. We then define $\overline{F}(x) = \inf\{F(a) : x < a, a \in A\}$. Since, by construction, F is non-decreasing on A, the function $\overline{F}$ is also non-decreasing. Since

$$\overline{F}(x) = \inf\{F(a) : x < a, a \in A\} = \inf_{x<y} \inf\{F(a) : y < a, a \in A\} = \inf_{x<y} \overline{F}(y),$$

it is also right-continuous. The fact that $\mathbb{P}_n$ are uniformly tight ensures that $\overline{F}(x) \to 0$ or 1 if $x \to -\infty$ or $+\infty$, so $\overline{F}(x)$ is a cumulative distribution function. Although it is possible that $\overline{F}(a) \neq F(a)$ for some $a \in A$, it is clear from the definition of $\overline{F}$ that if x is a point of continuity of $\overline{F}$ then

$$\lim_{A \ni a \uparrow x} F(a) = \overline{F}(x), \quad \lim_{A \ni b \downarrow x} F(b) = \overline{F}(x). \tag{3.2}$$

In order to prove weak convergence of $\mathbb{P}_{n(k)}$ to the measure $\mathbb{P}$ with the c.d.f. $\overline{F}$, let x be a point of continuity of $\overline{F}(x)$ and let $a, b \in A$ such that $a < x < b$. We have,

$$F(a) = \lim_{k\to\infty} F_{n(k)}(a) \leq \liminf_{k\to\infty} F_{n(k)}(x) \leq \limsup_{k\to\infty} F_{n(k)}(x) \leq \lim_{k\to\infty} F_{n(k)}(b) = F(b).$$

Therefore, (3.2) implies that $F_{n(k)}(x) \to \overline{F}(x)$ for all such x and, by Theorem 3.1, this means that the laws $\mathbb{P}_n$ converge to $\mathbb{P}$. Let us now prove the general case on general metric spaces.

(The general case). If K is a compact then, obviously, $C_b(K) = C(K)$. Later in these lectures, when we deal in more detail with convergence on general metric spaces, we will prove the following fact, which is well-known and is a consequence of the Stone-Weierstrass theorem.

$C(K)$ is separable w.r.t. ℓ_∞ norm $||f||_\infty = \sup_{x\in K}|f(x|$.

Since $\mathbb{P}_n$ are uniformly tight, for any $r \geq 1$ we can find a compact K_r such that $\mathbb{P}_n(K_r) > 1 - 1/r$ for all $n \geq 1$. Let $C_r \subset C(K_r)$ be a countable dense subset of $C(K_r)$. By Cantor's diagonalization, there exists a subsequence $(n(k))$ such that $\mathbb{P}_{n(k)}(f)$ converges for all $f \in C_r$ for all $r \geq 1$. Since C_r is dense in $C(K_r)$, this implies that $\mathbb{P}_{n(k)}(f)$ converges for all $f \in C(K_r)$ for all $r \geq 1$. Next, for any $f \in C_b(S)$,

$$\left|\int f\,d\mathbb{P}_{n(k)} - \int_{K_r} f\,d\mathbb{P}_{n(k)}\right| \leq \int_{K_r^c} |f|\,d\mathbb{P}_{n(k)} \leq ||f||_\infty \mathbb{P}_{n(k)}(K_r^c) \leq \frac{||f||_\infty}{r}.$$

This implies that the limit

$$I(f) := \lim_{k\to\infty} \int f\,d\mathbb{P}_{n(k)} \tag{3.3}$$

exists. The question is why this limit is an integral $I(f) = \int f\,d\mathbb{P}$ for some probability measure $\mathbb{P}$? Basically, this is a consequence of the Riesz representation theorem for positive linear functionals on locally compact Hausdorff spaces. One can define the functional $I_r(f)$ on each compact K_r, find a measure on it using the Riesz representation theorem, check that these measures agree on the intersections, and extend to a probability measure on their union. However, instead, we will use a relative of the Riesz representation theorem — the Stone-Daniell theorem from measure theory – which states the following.

Given a set S, a family of function $\alpha = \{f : S \to \mathbb{R}\}$ is called a *vector lattice* if

$$f, g \in \alpha \Longrightarrow cf + g \in \alpha, \forall c \in \mathbb{R} \text{ and } f \wedge g, f \vee g \in \alpha.$$

Vector lattice α is called a *Stone vector lattice* if $f \wedge 1 \in \alpha$ for any $f \in \alpha$. For example, any vector lattice that contains constants is automatically a Stone vector lattice.

A functional $I : \alpha \to \mathbb{R}$ is called a *pre-integral* if

1. $I(cf + g) = cI(f) + I(g)$,
2. $f \geq 0, I(f) \geq 0$,
3. $f_n \downarrow 0, ||f_n||_\infty < \infty \Longrightarrow I(f_n) \to 0$.

See R.M. Dudley "Reals Analysis and Probability" for a proof of the following:

(Stone-Daniell theorem) If α is a Stone vector lattice and I is a pre-integral on α then $I(f) = \int f\,d\mu$ for some unique measure μ on the minimal σ-algebra on which all functions in α are measurable.

We will use this theorem with $\alpha = C_b(S)$ and I defined in (3.3). The first two properties are obvious. To prove the third one, let us consider a sequence such that

$$f_n \downarrow 0,\ 0 \leq f_n(x) \leq f_1(x) \leq ||f_1||_\infty.$$

On any compact K_r, $f_n \downarrow 0$ uniformly, i.e.

$$\|f_n\|_{\infty,K_r} \le \varepsilon_{n,r} \stackrel{n\to\infty}{\longrightarrow} 0.$$

Since

$$\int f_n \, d\mathbb{P}_{n(k)} = \int_{K_r} f_n \, d\mathbb{P}_{n(k)} + \int_{K_r^c} f_n \, d\mathbb{P}_{n(k)} \le \varepsilon_{n,r} + \frac{1}{r}\|f_1\|_\infty,$$

we get

$$I(f_n) = \lim_{k\to\infty} \int f_n d\mathbb{P}_{n(k)} \le \varepsilon_{n,r} + \frac{1}{r}\|f_1\|_\infty.$$

Letting $n \to \infty$ and $r \to \infty$, we get that $I(f_n) \to 0$. By the Stone-Daniell theorem,

$$I(f) = \int f \, d\mathbb{P}$$

for some unique measure $\mathbb{P}$ on $\sigma(C_b(S))$. The choice of $f = 1$ shows that $I(f) = 1 = \mathbb{P}(S)$, which means that $\mathbb{P}$ is a probability measure. Finally, let us show that $\sigma(C_b(S))$ is the Borel σ-algebra $\mathscr{B}$ generated by open sets. Since any $f \in C_b(S)$ is measurable on $\mathscr{B}$ we get $\sigma(C_b(S)) \subseteq \mathscr{B}$. On the other hand, let $F \subseteq S$ be any closed set and take a function $f(x) = \min(1, d(x,F))$. We have, $|f(x) - f(y)| \le d(x,y)$ so $f \in C_b(S)$ and

$$f^{-1}(\{0\}) \in \sigma(C_b(S)).$$

However, since F is closed, $f^{-1}(\{0\}) = \{x : d(x,F) = 0\} = F$ and this proves that $\mathscr{B} \subseteq \sigma(C_b(S))$. □

Conversely, the following holds (we will prove this result later for any complete separable metric space).

Theorem 3.3. *If $\mathbb{P}_n$ converges weakly to $\mathbb{P}$ on $\mathbb{R}^k$ then $(\mathbb{P}_n)_{n\ge1}$ is uniformly tight.*

Proof. For any $\varepsilon > 0$, there exists large enough $M > 0$, such that $\mathbb{P}(|x| > M) < \varepsilon$. Consider a function

$$\varphi(s) = \begin{cases} 0, & s \le M, \\ 1, & s \ge 2M, \\ (s-M)/M, & M \le s \le 2M, \end{cases}$$

and let $\alpha(x) := \varphi(|x|)$ for $x \in \mathbb{R}^k$. Since $\mathbb{P}_n \to \mathbb{P}$ weakly,

$$\begin{aligned} \limsup_{n\to\infty} \mathbb{P}_n\big(|x| > 2M\big) &\le \limsup_{n\to\infty} \int \alpha(x) d\mathbb{P}_n(x) \\ &= \int \alpha(x) d\mathbb{P}(x) \le \mathbb{P}\big(|x| > M\big) \le \varepsilon. \end{aligned}$$

For n large enough, $n \ge n_0$, we get $\mathbb{P}_n(|x| > 2M) \le 2\varepsilon$. For $n < n_0$ choose M_n so that $\mathbb{P}_n(|x| > M_n) \le 2\varepsilon$. Take $M' = \max\{M_1, \ldots, M_{n_0-1}, 2M\}$. As a result, $\mathbb{P}_n(|x| > M') \le 2\varepsilon$ for all $n \ge 1$. □

Finally, let us relate convergence in distribution to other forms of convergence. Consider random variables X and X_n on some probability space $(\Omega, \mathscr{A}, \mathbb{P})$ with values in a metric space (S,d). Let $\mathbb{P}$ and $\mathbb{P}_n$ be their corresponding laws on Borel sets $\mathscr{B}$ in S. Convergence of X_n to X in probability and almost surely is defined exactly the same way as for $S = \mathbb{R}$ by replacing $|X_n - X|$ with $d(X_n, X)$.

Lemma 3.3. *$X_n \to X$ in probability if and only if for any sequence $(n(k))$ there exists a subsequence $(n(k(r)))$ such that $X_{n(k(r))} \to X$ almost surely.*

Proof. "$\Longleftarrow$". Suppose X_n does not converge to X in probability. Then, for small $\varepsilon > 0$, there exists a subsequence $(n(k))_{k\geq 1}$ such that $\mathbb{P}(d(X, X_{n(k)}) \geq \varepsilon) \geq \varepsilon$. This contradicts the existence of a subsequence $X_{n(k(r))}$ that converges to X almost surely.

"$\Longrightarrow$". Given a subsequence $(n(k))$ let us choose $(k(r))$ so that

$$\mathbb{P}\Big(d(X_{n(k(r))}, X) \geq \frac{1}{r}\Big) \leq \frac{1}{r^2}.$$

By the Borel-Cantelli lemma, these events can occur i.o. with probability 0, which means that with probability one, for large enough r,

$$d(X_{n(k(r))}, X) \leq \frac{1}{r},$$

i.e. $X_{n(k(r))} \to X$ almost surely. □

Lemma 3.4. *$X_n \to X$ in probability then $X_n \to X$ in distribution.*

Proof. By Lemma 3.3, for any subsequence $(n(k))$ there exists a subsequence $(n(k(r)))$ such that $X_{n(k(r))} \to X$ almost surely. Given $f \in C_b(\mathbb{R})$, by the dominated convergence theorem, $\mathbb{E}f(X_{n(k(r))}) \to \mathbb{E}f(X)$, i.e. $X_{n(k(r))} \to X$ weakly. By Lemma 3.1, $X_n \to X$ in distribution. □

Exercise 3.1.1. Let X_n be random variables on the same probability space with values in a metric space S. If for some point $s \in S$, $X_n \to s$ in distribution, show that $X_n \to s$ in probability.

Exercise 3.1.2. For the following sequences of laws $\mathbb{P}_n$ on $\mathbb{R}$ having densities f_n, which are uniformly tight? (a) $f_n = \mathrm{I}(0 \leq x \leq n)/n$, (b)$f_n = ne^{-nx}\mathrm{I}(x \geq 0)$, (c) $f_n = e^{-x/n}/n\mathrm{I}(x \geq 0)$.

Exercise 3.1.3. Suppose that $X_n \to X$ in distribution on $\mathbb{R}$ and $Y_n \to c \in \mathbb{R}$ in probability. Show that $X_n Y_n \to cX$ in distribution, assuming that X_n, Y_n are defined on the same probability space.

Exercise 3.1.4. Suppose that random variables (X_n) are independent and $X_n \to X$ in probability. Show that X is almost surely constant.

Exercise 3.1.5. Suppose that random variables X and Y taking values in $[0,1]$ satisfy $\mathbb{E}X^n \geq \mathbb{E}Y^n$ for all integer $n \geq 1$. Does this imply that X is stochastically greater than Y, i.e. $\mathbb{P}(X \geq t) \geq \mathbb{P}(Y \geq t)$ for all t?

3.2 Characteristic functions

In the next section, we will prove one of the most classical results in Probability Theory – the central limit theorem – and one of the main tools will be the so-called characteristic functions. Let $X = (X_1, \ldots, X_k)$ be a random vector on $\mathbb{R}^k$ with the distribution $\mathbb{P}$ and let $t = (t_1, \ldots, t_k) \in \mathbb{R}^k$. The *characteristic function* of X is defined by

$$f(t) = \mathbb{E}e^{i(t,X)} = \int e^{i(t,x)}\,d\mathbb{P}(x).$$

In this section, we will collect various important and useful facts about the characteristic functions. The standard normal distribution $N(0,1)$ on $\mathbb{R}$ with the density

$$p(x) = \frac{1}{\sqrt{2\pi}} e^{-x^2/2}$$

will play a central role, so let us start by computing its characteristic function. First of all, notice that this is indeed a density, since

$$\Big(\frac{1}{\sqrt{2\pi}} \int_{\mathbb{R}} e^{-x^2/2}\,dx\Big)^2 = \frac{1}{2\pi} \iint_{\mathbb{R}^2} e^{-(x^2+y^2)/2}\,dxdy = \frac{1}{2\pi} \int_0^{2\pi} \int_0^{\infty} e^{-r^2/2} r\,dr d\theta = 1.$$

If X has the standard normal distribution $N(0,1)$ then, obviously, $\mathbb{E}X = 0$ and

$$\mathrm{Var}(X) = \mathbb{E}X^2 = \frac{1}{\sqrt{2\pi}} \int_{\mathbb{R}} x^2 e^{-x^2/2}\,dx = \frac{1}{\sqrt{2\pi}} \int_{\mathbb{R}} x\,d(-e^{-x^2/2}) = 1,$$

by integration by parts. To motivate the computation of the characteristic function, let us first notice that for $\lambda \in \mathbb{R}$,

$$\mathbb{E}e^{\lambda X} = \frac{1}{\sqrt{2\pi}} \int e^{\lambda x - \frac{x^2}{2}}\,dx = e^{\frac{\lambda^2}{2}} \int \frac{1}{\sqrt{2\pi}} e^{-\frac{(x-\lambda)^2}{2}}\,dx = e^{\frac{\lambda^2}{2}} \int \frac{1}{\sqrt{2\pi}} e^{-\frac{x^2}{2}}\,dx = e^{\frac{\lambda^2}{2}}.$$

For complex $\lambda = it$, we begin similarly by completing the square,

$$\mathbb{E}e^{itX} = e^{-\frac{t^2}{2}} \int \frac{1}{\sqrt{2\pi}} e^{-\frac{(x-it)^2}{2}}\,dx = e^{-\frac{t^2}{2}} \int_{-it+\mathbb{R}} \varphi(z)\,dz,$$

where we denoted

$$\varphi(z) = \frac{1}{\sqrt{2\pi}} e^{-\frac{z^2}{2}} \text{ for } z \in \mathbb{C}.$$

Since φ is analytic, by Cauchy's theorem, integral over a closed path is equal to 0. Let us take a closed path $-it + x$ for x from $-M$ to $+M$, $M + iy$ for y from $-t$ to 0, x from M to $-M$ and, finally, $-M + iy$ for y from 0 to $-t$. For large M, the function $\varphi(z)$ is very small on the intervals $\pm M + iy$, so letting $M \to \infty$ we get

$$\int_{-it+\mathbb{R}} \varphi(z)\,dz = \int_{\mathbb{R}} \varphi(z)\,dz = 1,$$

and we proved that

$$f(t) = \mathbb{E}e^{itX} = e^{-\frac{t^2}{2}}. \tag{3.4}$$

It is also easy to check that $Y = \sigma(X+\mu)$ has mean $\mu = \mathbb{E}Y$, variance $\sigma^2 = \mathrm{Var}(Y)$ and the density

$$\frac{1}{\sqrt{2\pi}\sigma}\exp\Big(-\frac{(x-\mu)^2}{2\sigma^2}\Big).$$

This is the so-called *normal distribution* $N(\mu,\sigma^2)$ and its characteristic function is given by

$$\mathbb{E}e^{itY} = \mathbb{E}e^{it(\mu+\sigma X)} = e^{it\mu - t^2\sigma^2/2}. \tag{3.5}$$

Our next very important observation is that integrability of X is related to smoothness of its characteristic function.

Lemma 3.5. *If X is a real-valued random variable such that $\mathbb{E}|X|^r < \infty$ for integer $r \geq 1$ then $f(t) \in C^r(\mathbb{R})$ and $f^{(j)}(t) = \mathbb{E}(iX)^j e^{itX}$ for $j \leq r$.*

For a partial converse, see an exercise below, where the existence of $f''(0)$ is shown to imply that $\mathbb{E}X^2 < \infty$. One can similarly show that, for even n, the existence of $f^{(n)}(0)$ implies that $\mathbb{E}X^n < \infty$.

Proof. If $r = 0$, we use the fact that $|e^{itX}| \leq 1$ to conclude that

$$f(t) = \mathbb{E}e^{itX} \to \mathbb{E}e^{isX} = f(s) \text{ if } t \to s,$$

by the dominated convergence theorem. This means that $f \in C(\mathbb{R})$, i.e. characteristic functions are always continuous. If $r = 1$, $\mathbb{E}|X| < \infty$, we can use

$$\left|\frac{e^{itX} - e^{isX}}{t-s}\right| \leq |X|$$

and, therefore, by the dominated convergence theorem,

$$f'(t) = \lim_{s\to t}\mathbb{E}\frac{e^{itX} - e^{isX}}{t-s} = \mathbb{E}iXe^{itX}.$$

Also, by dominated convergence theorem, $\mathbb{E}iXe^{itX} \in C(\mathbb{R})$, which means that $f \in C^1(\mathbb{R})$. We proceed by induction. Suppose that we proved that $f^{(j)}(t) = \mathbb{E}(iX)^j e^{itX}$ and that $r = j+1$, $\mathbb{E}|X|^{j+1} < \infty$. Then, we can use that

$$\left|\frac{(iX)^j e^{itX} - (iX)^j e^{isX}}{t-s}\right| \leq |X|^{j+1},$$

so that

$$f^{(j+1)}(t) = \mathbb{E}(iX)^{j+1}e^{itX} \in C(\mathbb{R})$$

by the dominated convergence theorem. □

Next, we want to show that the characteristic function uniquely determines the distribution. This is usually proved using convolutions. Let X and Y be two independent random vectors on $\mathbb{R}^k$ with the distributions $\mathbb{P}$ and $\mathbb{Q}$. We denote by $\mathbb{P}*\mathbb{Q}$ the *convolution* of $\mathbb{P}$ and $\mathbb{Q}$, which is the distribution $\mathscr{L}(X+Y)$ of the sum $X+Y$. We have,

$$\begin{aligned}\mathbb{P}*\mathbb{Q}(A) = \mathbb{E}\mathrm{I}(X+Y\in A) &= \iint \mathrm{I}(x+y\in A)\,d\mathbb{P}(x)d\mathbb{Q}(y)\\ &= \iint \mathrm{I}(x\in A-y)\,d\mathbb{P}(x)d\mathbb{Q}(y) = \int \mathbb{P}(A-y)\,d\mathbb{Q}(y).\end{aligned}$$

If $\mathbb{P}$ has density p then

$$\begin{aligned}\mathbb{P}*\mathbb{Q}(A) &= \iint \mathrm{I}(x+y\in A)p(x)\,dxd\mathbb{Q}(y) = \iint \mathrm{I}(z\in A)p(z-y)\,dzd\mathbb{Q}(y)\\ &= \int\Bigl(\int_A p(z-y)\,dz\Bigr)d\mathbb{Q}(y) = \int_A\Bigl(\int p(z-y)\,d\mathbb{Q}(y)\Bigr)\,dz,\end{aligned}$$

which means that $\mathbb{P}*\mathbb{Q}$ has density

$$f(x) = \int p(x-y)\,d\mathbb{Q}(y). \tag{3.6}$$

If, in addition, $\mathbb{Q}$ has density q then

$$f(x) = \int p(x-y)q(y)\,dy. \tag{3.7}$$

Let us denote by $N(0,\sigma^2 I)$ the distribution of the random vector $X=(X_1,\ldots,X_k)$ of i.i.d. random variables with the normal distribution $N(0,\sigma^2)$. The density of X is given by

$$\prod_{i=1}^k \frac{1}{\sqrt{2\pi}\sigma}e^{-\frac{1}{2\sigma^2}x_i^2} = \Bigl(\frac{1}{\sqrt{2\pi}\sigma}\Bigr)^k e^{-\frac{1}{2\sigma^2}|x|^2}.$$

Given a distribution $\mathbb{P}$ on $\mathbb{R}^k$, let us denote

$$\mathbb{P}^\sigma = \mathbb{P}*N(0,\sigma^2 I).$$

It turns out that $\mathbb{P}^\sigma$ is always absolutely continuous and its density can be written in terms of the characteristic function of $\mathbb{P}$.

Lemma 3.6. *The convolution $\mathbb{P}^\sigma = \mathbb{P}*N(0,\sigma^2 I)$ has density*

$$p^\sigma(x) = \Bigl(\frac{1}{2\pi}\Bigr)^k \int f(t)e^{-i(t,x)-\frac{\sigma^2}{2}|t|^2}dt$$

where $f(t)=\int e^{i(t,x)}\,d\mathbb{P}(x)$.

Proof. By (3.6), $\mathbb{P}*N(0,\sigma^2 I)$ has density

$$p^\sigma(x) = \Big(\frac{1}{\sqrt{2\pi}\sigma}\Big)^k \int e^{-\frac{1}{2\sigma^2}|x-y|^2}\, d\mathbb{P}(y).$$

Using (3.4), we can write

$$e^{-\frac{1}{2\sigma^2}(x_i-y_i)^2} = \frac{1}{\sqrt{2\pi}} \int e^{-i\frac{1}{\sigma}(x_i-y_i)z_i} e^{-\frac{1}{2}z_i^2}\, dz_i$$

and taking a product over $i \le k$ we get

$$e^{-\frac{1}{2\sigma^2}|x-y|^2} = \Big(\frac{1}{\sqrt{2\pi}}\Big)^k \int e^{-i\frac{1}{\sigma}(x-y,z)} e^{-\frac{1}{2}|z|^2}\, dz.$$

Then we can continue

$$\begin{aligned} p^\sigma(x) &= \Big(\frac{1}{2\pi\sigma}\Big)^k \iint e^{-i\frac{1}{\sigma}(x-y,z)-\frac{1}{2}|z|^2}\, dz d\mathbb{P}(y) \\ &= \Big(\frac{1}{2\pi\sigma}\Big)^k \iint e^{-i\frac{1}{\sigma}(x-y,z)-\frac{1}{2}|z|^2}\, d\mathbb{P}(y) dz \\ &= \Big(\frac{1}{2\pi\sigma}\Big)^k \int f\Big(\frac{z}{\sigma}\Big) e^{-i\frac{1}{\sigma}(x,z)-\frac{1}{2}|z|^2}\, dz. \end{aligned}$$

Making the change of variables $z = t\sigma$ finishes the proof. □

This immediately implies that the characteristic function uniquely determines the distribution.

Theorem 3.4 (Uniqueness). *If* $\int e^{i(t,x)}\, d\mathbb{P}(x) = \int e^{i(t,x)}\, d\mathbb{Q}(x)$ *then* $\mathbb{P} = \mathbb{Q}$.

Proof. By the above Lemma, $\mathbb{P}^\sigma = \mathbb{Q}^\sigma$. If $X \sim \mathbb{P}$ and $g \sim N(0,I)$ then $X + \sigma g \to X$ almost surely as $\sigma \downarrow 0$ and, therefore, $\mathbb{P}^\sigma \to \mathbb{P}$ weakly. Similarly, $\mathbb{Q}^\sigma \to \mathbb{Q}$. □

This immediately implies the following *stability property* of the normal distribution.

Lemma 3.7. *If* X_j *for* $j = 1,2$ *are independent and have normal distributions* $N(\mu_j, \sigma_j^2)$ *then* $X_1 + X_2$ *has normal distribution* $N(\mu_1 + \mu_2, \sigma_1^2 + \sigma_2^2)$.

Proof. By independence and (3.5), the characteristic function of the sum is equal to

$$\mathbb{E}e^{it(X_1+X_2)} = \mathbb{E}e^{itX_1}\mathbb{E}e^{itX_2} = e^{it\mu_1 - t^2\sigma_1^2/2} e^{it\mu_2 - t^2\sigma_2^2/2} = e^{it(\mu_1+\mu_2) - t^2(\sigma_1^2+\sigma_2^2)/2}.$$

This is the characteristic function of the normal distribution $N(\mu_1 + \mu_2, \sigma_1^2 + \sigma_2^2)$, so the claim follows from the Uniqueness Theorem. One can also prove this by a straightforward computation using the formula (3.7) for the density of the convolution, which is left as an exercise below. □

Lemma 3.6 gives some additional information when the characteristic function is integrable.

Lemma 3.8 (Fourier inversion formula). *If* $\int |f(t)|\,dt < \infty$ *then* $\mathbb{P}$ *has density*

$$p(x) = \Big(\frac{1}{2\pi}\Big)^k \int f(t)e^{-i(t,x)}\,dt.$$

Proof. Since

$$f(t)e^{-i(t,x)-\frac{1}{2}\sigma^2|t|^2} \to f(t)e^{-i(t,x)}$$

pointwise as $\sigma \downarrow 0$ and

$$\Big|f(t)e^{-i(t,x)-\frac{1}{2}\sigma^2|t|^2}\Big| \le |f(t)|,$$

the integrability of $f(t)$ implies, by the dominated convergence theorem, that $p^\sigma(x) \to p(x)$. Since $\mathbb{P}^\sigma \to \mathbb{P}$ weakly, for any $g \in C_b(\mathbb{R}^k)$,

$$\int g(x)p^\sigma(x)\,dx \to \int g(x)\,d\mathbb{P}(x).$$

On the other hand, since

$$|p^\sigma(x)| \le \Big(\frac{1}{2\pi}\Big)^k \int |f(t)|dt < \infty,$$

by the dominated convergence theorem, for any compactly supported $g \in C_c(\mathbb{R}^k)$,

$$\int g(x)p^\sigma(x)\,dx \to \int g(x)p(x)\,dx.$$

Therefore, for any such $g \in C_c(\mathbb{R}^k)$,

$$\int g(x)\,d\mathbb{P}(x) = \int g(x)p(x)\,dx.$$

Of course, this means that $p(x)$ is the density of $\mathbb{P}$. □

Because of the uniqueness theorem, the convergence of characteristic functions implies convergence of distributions under some mild additional assumptions.

Lemma 3.9. *If* $(\mathbb{P}_n)$ *is uniformly tight on* $\mathbb{R}^k$ *and the characteristic functions converge to some function* f,

$$f_n(t) = \int e^{i(t,x)}d\mathbb{P}_n(x) \to f(t),$$

then $f(t) = \int e^{i(t,x)}\,d\mathbb{P}(x)$ *is a characteristic function of some distribution* $\mathbb{P}$ *and* $\mathbb{P}_n \to \mathbb{P}$ *weakly.*

Proof. For any sequence $(n(k))$, by the Selection Theorem, there exists a subsequence $(n(k(r)))$ such that $\mathbb{P}_{n(k(r))}$ converges weakly to some distribution $\mathbb{P}$. Since $e^{i(t,x)}$ is bounded and continuous,

$$\int e^{i(t,x)}\,d\mathbb{P}_{n(k(r))} \to \int e^{i(t,x)}\,d\mathbb{P}(x)$$

as $r \to \infty$ and, therefore, f is the characteristic function of $\mathbb{P}$. By the uniqueness theorem, the distribution $\mathbb{P}$ does not depend on the sequence $(n(k))$. By Lemma 3.1, $\mathbb{P}_n \to \mathbb{P}$ weakly. □

Even though in many cases uniform tightness of $(\mathbb{P}_n)$ is not difficult to check directly, it also follows automatically from the continuity of the limit of characteristic functions at zero. To show this, we will need the following bound.

Lemma 3.10. *If X is a real-valued random variable then*

$$\mathbb{P}\Big(|X| > \frac{1}{u}\Big) \le \frac{7}{u}\int_0^u (1 - \Re\mathfrak{e} f(t))\,dt.$$

Proof. Since $\Re\mathfrak{e} f(t) = \int \cos(tx)\,d\mathbb{P}(x)$, we can write

$$\begin{aligned}
\frac{1}{u}\int_0^u (1-\Re\mathfrak{e} f(t))\,dt &= \frac{1}{u}\int_0^u\int_{\mathbb{R}} (1-\cos tx)\,d\mathbb{P}(x)dt \\
&= \frac{1}{u}\int_{\mathbb{R}}\int_0^u (1-\cos tx)\,dtd\mathbb{P}(x) \\
&= \int_{\mathbb{R}} \Big(1-\frac{\sin xu}{xu}\Big)\,d\mathbb{P}(x) \\
&\ge \int_{|xu|\ge 1} \Big(1-\frac{\sin xu}{xu}\Big)\,d\mathbb{P}(x) \\
&\ge (1-\sin 1)\int_{|xu|\ge 1} 1\,d\mathbb{P}(x) \ge \frac{1}{7}\mathbb{P}\Big(|X| \ge \frac{1}{u}\Big),
\end{aligned}$$

where in the last inequality we used that $\frac{\sin y}{y} < \frac{\sin 1}{1}$ if $y > 1$. □

Theorem 3.5 (Levy's continuity theorem). *Let (X_n) be a sequence of random variables on $\mathbb{R}^k$. Suppose that*

$$f_n(t) = \mathbb{E}e^{i(t,X_n)} \to f(t)$$

and $f(t)$ is continuous at 0 along each axis. Then there exists a probability distribution $\mathbb{P}$ such that

$$f(t) = \int e^{i(t,x)}\,d\mathbb{P}(x)$$

and $\mathbb{P}_n = \mathscr{L}(X_n) \to \mathbb{P}$ weakly.

Proof. By Lemma 3.9, we only need to show that $\{\mathscr{L}(X_n)\}$ is uniformly tight. If we denote $X_n = (X_{n,1},\ldots,X_{n,k})$ then the characteristic functions of the ith coordinate:

$$f_n^i(t_i) := f_n(0,\ldots,t_i,0,\ldots 0) = \mathbb{E}e^{it_iX_{n,i}} \to f(0,\ldots,t_i,\ldots 0) =: f^i(t_i).$$

Since $f_n(0) = 1$, we have $f(0) = 1$. By the continuity of f^i at 0, for any $\varepsilon > 0$ we can find $\delta > 0$ such that for all $i \le k$, $|f^i(t_i) - 1| \le \varepsilon$ if $|t_i| \le \delta$. By the dominated convergence theorem,

$$\begin{aligned}\lim_{n\to\infty} \frac{1}{\delta}\int_0^\delta \bigl(1 - \mathfrak{Re} f_n^i(t_i)\bigr)\,dt_i &= \frac{1}{\delta}\int_0^\delta \bigl(1 - \mathfrak{Re} f^i(t_i)\bigr)\,dt_i \\ &\le \frac{1}{\delta}\int_0^\delta \bigl|1 - f^i(t_i)\bigr|\,dt_i \le \varepsilon.\end{aligned}$$

Together with the previous lemma this implies that

$$\mathbb{P}\Bigl(|X_{n,i}| > \frac{1}{\delta}\Bigr) \le \frac{7}{\delta}\int_0^\delta \bigl(1 - \mathfrak{Re} f_n^i(t_i)\bigr)\,dt_i \le 7\cdot 2\varepsilon,$$

for large enough n. Since $|X_n|^2 = \sum_{i=1}^k |X_{n,i}|^2$, the union bound implies that

$$\mathbb{P}\Bigl(|X_n| > \frac{\sqrt{k}}{\delta}\Bigr) \le 14k\varepsilon,$$

which means that the sequence $\mathbb{P}_n = \mathscr{L}(X_n)$ is uniformly tight. □

Using characteristic functions, we can prove a couple more basic properties of convergence in distributions.

Lemma 3.11 (Continuous Mapping). *Suppose that $\mathbb{P}_n \to \mathbb{P}$ weakly on X and $G : X \to Y$ is a continuous map. Then $\mathbb{P}_n \circ G^{-1} \to \mathbb{P} \circ G^{-1}$ on Y. In other words, if $Z_n \to Z$ in distribution then $G(Z_n) \to G(Z)$ in distribution.*

Proof. This is obvious, because for any function $f \in C_b(Y)$ we have $f \circ G \in C_b(X)$ and, therefore, $\mathbb{E}f(G(Z_n)) \to \mathbb{E}f(G(Z))$, which is the definition of convergence $G(Z_n) \to G(Z)$ in distribution. □

Lemma 3.12. *If $\mathbb{P}_n \to \mathbb{P}$ on $\mathbb{R}^k$ and $\mathbb{Q}_n \to \mathbb{Q}$ on $\mathbb{R}^m$ then $\mathbb{P}_n \times \mathbb{Q}_n \to \mathbb{P} \times \mathbb{Q}$ on $\mathbb{R}^{k+m}$.*

Proof. By Fubini's theorem, the characteristic function

$$\begin{aligned}\int e^{i(t,x)}\,d\mathbb{P}_n \times \mathbb{Q}_n(x) &= \int e^{i(t_1,x_1)}\,d\mathbb{P}_n \int e^{i(t_2,x_2)}d\mathbb{Q}_n \\ \to \int e^{i(t_1,x_1)}\,d\mathbb{P} \int e^{i(t_2,x_2)}\,d\mathbb{Q} &= \int e^{i(t,x)}\,d\mathbb{P}\times\mathbb{Q}.\end{aligned}$$

Using Theorem 3.5 finishes the proof. □

Corollary 3.1. *If both $\mathbb{P}_n \to \mathbb{P}$ and $\mathbb{Q}_n \to \mathbb{Q}$ on $\mathbb{R}^k$ then $\mathbb{P}_n * \mathbb{Q}_n \to \mathbb{P} * \mathbb{Q}$.*

Proof. Since the function $G : \mathbb{R}^{k+k} \to \mathbb{R}^k$ given by $G(x,y) = x+y$ is continuous, by the continuous mapping lemma,

$$\mathbb{P}_n * \mathbb{Q}_n = (\mathbb{P}_n \times \mathbb{Q}_n) \circ G^{-1} \to (\mathbb{P} \times \mathbb{Q}) \circ G^{-1} = \mathbb{P} * \mathbb{Q},$$

which finishes the proof. □

Exercise 3.2.1. Prove Lemma 3.7 using the formula (3.7) for the density of the convolution.

Exercise 3.2.2. If X is a random variable such that $\mathbb{E}e^{itX} = e^{-t^4}$ for all $t \in \mathbb{R}$, show that $\mathbb{E}X^2 < \infty$. *Hint:* use Lemma 3.10.

Exercise 3.2.3. Does there exist a random variable X such that $\mathbb{E}e^{itX} = e^{-t^4}$?

Exercise 3.2.4. Suppose that $f(t) = \mathbb{E}e^{itX}$ is differentiable near 0 and $f'(t)$ is differentiable at zero, i.e. $f''(0)$ is well defined. Prove that $\mathbb{E}X^2 < \infty$. *Hint:* consider $\varphi(t) = |f(t)|^2 = f(t)f(-t)$, which is the characteristic function of $X_1 - X_2$ for two independent copies X_1, X_2 of X, apply l'Hospital's rule to $(1-\varphi(t))/t^2$ and then use Fatou's lemma to show that $\mathbb{E}(X_1 - X_2)^2 < \infty$; conclude using Fubini's theorem.

Exercise 3.2.5. Prove that a probability distribution on $\mathbb{R}^k$ is uniquely determined by the probabilities of half-spaces.

Exercise 3.2.6. Prove that if $\mathbb{P} * \mathbb{Q} = \mathbb{P}$ on $\mathbb{R}^k$ then $\mathbb{Q}(\{0\}) = 1$. (Hint: use that a characteristic function is continuous, in particular, in the neighborhood of zero.)

Exercise 3.2.7. Find the characteristic function of the law on $\mathbb{R}$ with density $\max(1-|x|,0)$. Then use the Fourier inversion formula to conclude that $\max(1-|t|,0)$ is a characteristic function of some distribution.

Exercise 3.2.8. Let f be a continuous function on $\mathbb{R}$ with $f(0) = 1$, $f(t) = f(-t)$ for all t, and such that f restricted to $[0,\infty)$ is convex, with $f(t) \to 0$ as $t \to \infty$. Show that f is a characteristic function. (*Hint*: Approximate the general case by piecewise linear. Then use that $\max(1-|t|,0)$ is a characteristic function and that characteristic functions are closed under convex combinations and scaling $f(t) \to f(at)$ for $a \in \mathbb{R}$.)

Exercise 3.2.9. Does there exist a random variable X such that $\mathbb{E}e^{itX} = e^{-|t|}$?

Exercise 3.2.10. Show that there are distributions $\mathbb{P}_1 \neq \mathbb{P}_2$ whose characteristic functions coincide on some interval (a,b). One the other hand, show that if $\mathbb{P}$ is compactly supported then its characteristic function on any open interval (a,b) uniquely determines $\mathbb{P}$. (*Hint*: Show that in this case $f(z) = \int e^{zx}\, d\mathbb{P}(x)$ is analytic on $\mathbb{C}$.)

Exercise 3.2.11. If f and f_n are characteristic functions of probability measures on $\mathbb{R}^k$ and $f_n \to f$ pointwise, show that $f_n \to f$ uniformly on compacts.

3.3 Central limit theorem for i.i.d. random variables

We have seen in the Hoeffding-Chernoff inequality in Section 2.1 that if $X_1,\ldots,X_n$ are independent flips of a fair coin, $\mathbb{P}(X_i=0)=\mathbb{P}(X_i=1)=1/2$, and $\bar{X}_n$ is their average then

$$\mathbb{P}\big(|\bar{X}_n-1/2|>t\big)\le 2e^{-2nt^2}.$$

If we denote

$$Z_n=2\sqrt{n}(\bar{X}_n-1/2)=\frac{S_n-n\mu}{\sqrt{n\sigma^2}},$$

where $\mu=\mathbb{E}X_1=1/2$ and $\sigma^2=\mathrm{Var}(X_1)=1/4$, then the Hoeffding-Chernoff inequality can be rewritten as

$$\mathbb{P}(|Z_n|\ge t)\le 2e^{-t^2/2}.$$

We will see that this inequality is rather accurate for large sample size n, since we will show that

$$\lim_{n\to\infty}\mathbb{P}(|Z_n|\ge t)=\frac{2}{\sqrt{2\pi}}\int_t^{\infty}e^{-x^2/2}\,dx\overset{t\to\infty}{\sim}\frac{2}{t\sqrt{2\pi}}e^{-t^2/2}.$$

The asymptotic equivalence as $t\to\infty$ is a simple exercise, and the large n limit is a consequence of a more general result,

$$\lim_{n\to\infty}\mathbb{P}(Z_n\le z)=\Phi(z)=\frac{1}{\sqrt{2\pi}}\int_{-\infty}^{z}e^{-x^2/2}\,dx, \tag{3.8}$$

for all $z\in\mathbb{R}$. In other words, Z_n converges in distribution to a standard normal distribution with the density $e^{-x^2/2}/\sqrt{2\pi}$ and the c.d.f. $\Phi(z)$. Moreover, this holds for any i.i.d. sequence (X_n) with $\mu=\mathbb{E}X_1$ and $\sigma^2=\mathrm{Var}(X_1)<\infty$, and

$$Z_n=\frac{S_n-n\mu}{\sqrt{n\sigma^2}}=\frac{1}{\sqrt{n}}\sum_{i=1}^{n}\frac{X_i-\mu}{\sigma}. \tag{3.9}$$

Before we use characteristic functions to prove this result as well as its multivariate version, let us describe a more intuitive approach via the so-called *Lindeberg's method*, which emphasizes a little bit more explicitly the main reason behind the central limit theorem, namely, the stability property of the normal distribution proved in Lemma 3.7.

Theorem 3.6. *Consider an i.i.d. sequence $(X_i)_{i\ge 1}$ such that $\mathbb{E}X_1=\mu$, $\mathrm{Var}(X)=\sigma^2$ and $\mathbb{E}|X_1|^3<\infty$. Then the distribution of Z_n defined in (3.9) converges weakly to standard normal distribution $N(0,1)$.*

Remark 3.1. One can easily modify the proof we give below to get rid of the unnecessary assumption $\mathbb{E}|X_1|^3<\infty$. This will appear as an exercise in the next section, where we will state and prove a more general version of the Lindeberg's CLT for non i.i.d. random variables.

Proof. First of all, notice that the random variables $(X_i - \mu)/\sigma$ have mean 0 and variance 1. Therefore, it is enough to prove the result for

$$Z_n = \frac{1}{\sqrt{n}} \sum_{i=1}^{n} X_i$$

under the assumption that $\mathbb{E}X_1 = 0, \mathbb{E}X_1^2 = 1$. Let $(g_i)_{i\geq 1}$ be independent standard normal random variables. Then, by the stability property,

$$Z = \frac{1}{\sqrt{n}} \sum_{i=1}^{n} g_i$$

also has standard normal distribution $N(0,1)$. If, for $1 \leq m \leq n+1$, we define

$$T_m = \frac{1}{\sqrt{n}}(g_1 + \ldots + g_{m-1} + X_m + \ldots + X_n)$$

then, for any bounded function $f : \mathbb{R} \to \mathbb{R}$, we can write

$$\big|\mathbb{E}f(Z_n) - \mathbb{E}f(Z)\big| = \Big|\sum_{m=1}^{n} \big(\mathbb{E}f(T_m) - \mathbb{E}f(T_{m+1})\big)\Big| \leq \sum_{m=1}^{n} \big|\mathbb{E}f(T_m) - \mathbb{E}f(T_{m+1})\big|.$$

If we denote

$$S_m = \frac{1}{\sqrt{n}}(g_1 + \ldots + g_{m-1} + X_{m+1} + \ldots + X_n)$$

then $T_m = S_m + X_m/\sqrt{n}$ and $T_{m+1} = S_m + g_m/\sqrt{n}$. Suppose now that f has uniformly bounded third derivative. Then, by Taylor's formula,

$$\left|f(T_m) - f(S_m) - \frac{f'(S_m)X_m}{\sqrt{n}} - \frac{f''(S_m)X_m^2}{2n}\right| \leq \frac{\|f'''\|_\infty |X_m|^3}{6n^{3/2}}$$

and

$$\left|f(T_{m+1}) - f(S_m) - \frac{f'(S_m)g_m}{\sqrt{n}} - \frac{f''(S_m)g_m^2}{2n}\right| \leq \frac{\|f'''\|_\infty |g_m|^3}{6n^{3/2}}.$$

Notice that S_m is independent of X_m and g_m and, therefore,

$$\mathbb{E}f'(S_m)X_m = \mathbb{E}f'(S_m)\mathbb{E}X_m = 0 = \mathbb{E}f'(S_m)\mathbb{E}g_m = \mathbb{E}f'(S_m)g_m$$

and

$$\mathbb{E}f''(S_m)X_m^2 = \mathbb{E}f''(S_m)\mathbb{E}X_m^2 = \mathbb{E}f''(S_m) = \mathbb{E}f''(S_m)\mathbb{E}g_m^2 = \mathbb{E}f''(S_m)g_m^2.$$

As a result, taking expectations and subtracting the above inequalities, we get

$$\left|\mathbb{E}f(T_m) - \mathbb{E}f(T_{m+1})\right| \le \frac{\|f'''\|_\infty(\mathbb{E}|X_1|^3 + \mathbb{E}|g_1|^3)}{6n^{3/2}}.$$

Adding up over $1 \le m \le n$, we proved that

$$\left|\mathbb{E}f(Z_n) - \mathbb{E}f(Z)\right| \le \frac{\|f'''\|_\infty(\mathbb{E}|X_1|^3 + \mathbb{E}|g_1|^3)}{6\sqrt{n}}$$

and $\lim_{n\to\infty}\mathbb{E}f(Z_n) = \mathbb{E}f(Z)$. We proved in Section 3.1 that this convergence for $f \in C_b(\mathbb{R})$ implies convergence of the c.d.f.s in (3.8), but the same proof also works using functions with bounded third derivative. We will leave it as an exercise at the end of this section. □

Now, let us prove the CLT using the method of characteristic functions.

Theorem 3.7 (Central Limit Theorem). *Consider an i.i.d. sequence $(X_i)_{i\ge1}$ such that $\mathbb{E}X_1 = 0, \mathbb{E}X_1^2 = 1$. Then the distribution of $Z_n = S_n/\sqrt{n}$ converges weakly to standard normal distribution $N(0,1)$.*

Proof. We begin by showing that the characteristic function of $S_n/\sqrt{n}$ converges to the characteristic function of the standard normal distribution $N(0,1)$,

$$\lim_{n\to\infty}\mathbb{E}e^{it\frac{S_n}{\sqrt{n}}} = e^{-\frac{1}{2}t^2},$$

for all $t \in \mathbb{R}$. By independence,

$$\mathbb{E}e^{it\frac{S_n}{\sqrt{n}}} = \prod_{i=1}^n \mathbb{E}e^{\frac{itX_i}{\sqrt{n}}} = \left(\mathbb{E}e^{\frac{itX_1}{\sqrt{n}}}\right)^n.$$

Since $\mathbb{E}X_1^2 < \infty$, Lemma 3.5 implies that the characteristic function $f(t) = \mathbb{E}e^{itX_1} \in C^2(\mathbb{R})$ and, therefore,

$$f(t) = f(0) + f'(0)t + \frac{1}{2}f''(0)t^2 + o(t^2) \text{ as } t \to 0.$$

Since

$$f(0) = 1,\ f'(0) = \mathbb{E}iXe^{i\cdot 0\cdot X} = i\mathbb{E}X = 0,\ f''(0) = \mathbb{E}(iX)^2 = -\mathbb{E}X^2 = -1,$$

we get

$$f(t) = 1 - \frac{t^2}{2} + o(t^2).$$

Finally,

$$\mathbb{E}e^{\frac{itS_n}{\sqrt{n}}} = \left(f\left(\frac{t}{\sqrt{n}}\right)\right)^n = \left(1 - \frac{t^2}{2n} + o\left(\frac{t^2}{n}\right)\right)^n \to e^{-\frac{1}{2}t^2}$$

as $n \to \infty$. The result then follows from Levy's continuity theorem in the previous section. Alternatively, we could also use Lemma 3.9 since it is easy to check that the sequence of laws $(\mathscr{L}(S_n/\sqrt{n}))_{n\geq 1}$ is uniformly tight. Indeed, by Chebyshev's inequality,

$$\mathbb{P}\Big(\Big|\frac{S_n}{\sqrt{n}}\Big| > M\Big) \leq \frac{1}{M^2}\mathrm{Var}\Big(\frac{S_n}{\sqrt{n}}\Big) = \frac{1}{M^2} < \varepsilon$$

for large enough M. □

Next, we will prove a multivariate version of the central limit theorem. For a random vector $X = (X_1,\ldots,X_k) \in \mathbb{R}^k$, let $\mathbb{E}X = (\mathbb{E}X_1,\ldots,\mathbb{E}X_k)$ denote its expectation and

$$\mathrm{Cov}(X) = \big(\mathbb{E}X_iX_j\big)_{1\leq i,j\leq k}$$

its *covariance matrix*. Again, the main step is to prove convergence of characteristic functions.

Theorem 3.8. *Let $(X_i)_{i\geq 1}$ be a sequence of i.i.d. random vectors on $\mathbb{R}^k$ such that $\mathbb{E}X_1 = 0$ and $\mathbb{E}|X_1|^2 < \infty$. Then the distribution of $S_n/\sqrt{n}$ converges weakly to the distribution $\mathbb{P}$ with the characteristic function*

$$f_P(t) = e^{-\frac{1}{2}(Ct,t)}, \tag{3.10}$$

where $C = \mathrm{Cov}(X_1)$.

Proof. Consider any $t \in \mathbb{R}^k$. Then $Z_i = (t,X_i)$ are i.i.d. real-valued random variables and, by the central limit theorem on the real line,

$$\mathbb{E}\exp i\Big(t,\frac{S_n}{\sqrt{n}}\Big) = \mathbb{E}\exp i\frac{1}{\sqrt{n}}\sum_{i=1}^{n} Z_i \to \exp\Big(-\frac{\mathrm{Var}(Z_1)}{2}\Big) = \exp\Big(-\frac{1}{2}(Ct,t)\Big)$$

as $n \to \infty$, since

$$\mathrm{Var}\big(t_1X_{1,1} + \cdots + t_kX_{1,k}\big) = \sum_{i,j} t_it_j\mathbb{E}X_{1,i}X_{1,j} = (Ct,t).$$

The sequence $\Big\{\mathscr{L}\Big(\frac{S_n}{\sqrt{n}}\Big)\Big\}$ is uniformly tight on $\mathbb{R}^k$ since

$$\begin{aligned}\mathbb{P}\Big(\Big|\frac{S_n}{\sqrt{n}}\Big| \geq M\Big) &\leq \frac{1}{M^2}\mathbb{E}\Big|\frac{S_n}{\sqrt{n}}\Big|^2 = \frac{1}{nM^2}\mathbb{E}\big|(S_{n,1},\ldots,S_{n,k})\big|^2 \\ &= \frac{1}{nM^2}\sum_{i\leq k}\mathbb{E}S_{n,i}^2 = \frac{1}{M^2}\mathbb{E}|X_1|^2 \overset{M\to\infty}{\longrightarrow} 0,\end{aligned}$$

so we can use Lemma 3.9 to finish the proof. Alternatively, we can just apply Levy's continuity theorem. □

Multivariate Gaussian distributions. The covariance matrix $C = \mathrm{Cov}(X)$ is symmetric and non-negative definite, since

$$(Ct,t) = \mathbb{E}(t,X)^2 \geq 0.$$

The unique distribution with covariance function in (3.10) is called a *multivariate normal distribution with the covariance* C and is denoted by $N(0,C)$. It can also be defined more constructively as follows. Consider an i.i.d. sequence $g_1,\ldots,g_n$ of standard normal $N(0,1)$ random variables and let $g=(g_1,\ldots,g_n)^T$. Given a $k\times n$ matrix A, the covariance matrix of $Ag\in\mathbb{R}^k$ is

$$C=\mathrm{Cov}(Ag)=\mathbb{E}Ag(Ag)^T=A\mathbb{E}gg^TA^T=AA^T.$$

Since we know the characteristic function of the standard normal random variables g_ℓ, the characteristic function of Ag is

$$\begin{aligned}\mathbb{E}\exp i(t,Ag)&=\mathbb{E}\exp i(A^Tt,g)=\mathbb{E}\prod_{\ell\le n}\exp i(A^Tt)_\ell g_\ell=\prod_{\ell\le n}\exp(-((A^Tt)_\ell)^2/2)\\&=\exp\Bigl(-\frac{1}{2}|A^Tt|^2\Bigr)=\exp\Bigl(-\frac{1}{2}t^TAA^Tt\Bigr)=\exp\Bigl(-\frac{1}{2}(Ct,t)\Bigr),\end{aligned}$$

which means that Ag has the distribution $N(0,C)$. On the other hand, given a symmetric non-negative definite matrix C one can always find A such that $C=AA^T$. For example, let $C=QDQ^T$ be its eigenvalue decomposition, for orthogonal matrix Q and diagonal matrix D. Since C is nonnegative definite, the elements of D are nonnegative. Then, one can take $n=k$ and $A=C^{1/2}:=QD^{1/2}Q^T$ or $A=QD^{1/2}$. This means that any normal distribution can be generated as a linear transformation of the vector of i.i.d. standard normal random variables.

Density in the invertible case. Suppose $\det(C)\neq 0$. Take any invertible A such that $C=AA^T$, so that $Ag\sim N(0,C)$. Since the density of g is

$$\prod_{\ell\le k}\frac{1}{\sqrt{2\pi}}\exp\Bigl(-\frac{1}{2}x_\ell^2\Bigr)=\Bigl(\frac{1}{\sqrt{2\pi}}\Bigr)^k\exp\Bigl(-\frac{1}{2}|x|^2\Bigr),$$

for any Borel set $\Omega\subseteq\mathbb{R}^k$ we can write

$$\mathbb{P}(Ag\in\Omega)=\mathbb{P}(g\in A^{-1}\Omega)=\int_{A^{-1}\Omega}\Bigl(\frac{1}{\sqrt{2\pi}}\Bigr)^k\exp\Bigl(-\frac{1}{2}|x|^2\Bigr)\,dx.$$

Let us now make the change of variables $y=Ax$ or $x=A^{-1}y$. Then

$$\mathbb{P}(Ag\in\Omega)=\int_{\Omega}\Bigl(\frac{1}{\sqrt{2\pi}}\Bigr)^k\exp\Bigl(-\frac{1}{2}|A^{-1}y|^2\Bigr)\frac{1}{|\det(A)|}dy.$$

But since

$$\det(C)=\det(AA^T)=\det(A)\det(A^T)=\det(A)^2$$

we have $|\det(A)|=\sqrt{\det(C)}$. Also

$$|A^{-1}y|^2=(A^{-1}y)^T(A^{-1}y)=y^T(A^T)^{-1}A^{-1}y=y^T(AA^T)^{-1}y=y^TC^{-1}y.$$

Therefore, we get

$$\mathbb{P}(Ag \in \Omega) = \int_{\Omega} \Big(\frac{1}{\sqrt{2\pi}}\Big)^k \frac{1}{\sqrt{\det(C)}} \exp\Big(-\frac{1}{2} y^T C^{-1} y\Big) dy.$$

This means that

$$\Big(\frac{1}{\sqrt{2\pi}}\Big)^k \frac{1}{\sqrt{\det(C)}} \exp\Big(-\frac{1}{2} y^T C^{-1} y\Big)$$

is the density of the distribution $N(0,C)$ when C is invertible. □

General case. If $C = QDQ^T$ is the eigenvalue decomposition of C, let us take, for example, $X = QD^{1/2}g$ for i.i.d. standard normal vector g, so that $X \sim N(0,C)$. If $q_1,\ldots,q_k$ are the column vectors of Q then

$$X = QD^{1/2}g = (\lambda_1^{1/2} g_1)q_1 + \ldots + (\lambda_k^{1/2} g_k)q_k.$$

Therefore, in the orthonormal coordinate basis $q_1,\ldots,q_k$, the random vector X has coordinates $\lambda_1^{1/2} g_1,\ldots,\lambda_k^{1/2} g_k$. These coordinates are independent and have normal distributions with variances $\lambda_1,\ldots,\lambda_k$ correspondingly. When $\det(C) = 0$, i.e. C is not invertible, some of its eigenvalues will be zero, say, $\lambda_{n+1} = \ldots = \lambda_k = 0$. Then the distribution will be concentrated on the subspace spanned by vectors $q_1,\ldots,q_n$, and it will not have density on the entire space $\mathbb{R}^k$. It will have the density

$$f(x_1,\ldots,x_n) = \prod_{\ell \le n} \frac{1}{\sqrt{2\pi\lambda_\ell}} \exp\Big(-\frac{x_\ell^2}{2\lambda_\ell}\Big)$$

on the subspace spanned by vectors $q_1,\ldots,q_n$. □

Let us look at some properties of normal distributions.

Lemma 3.13. *If $X \sim N(0,C)$ on $\mathbb{R}^k$ and A is a $m \times k$ matrix then $AX \sim N(0,ACA^T)$ on $\mathbb{R}^m$.*

Proof. The characteristic function of AX is

$$\mathbb{E}\exp i(t,AX) = \mathbb{E}\exp i(A^T t,X) = \exp\Big(-\frac{1}{2}(CA^T t,A^T t)\Big) = \exp\Big(-\frac{1}{2}(ACA^T t,t)\Big),$$

and the statement follows by definition. □

Lemma 3.14. *X is normal on $\mathbb{R}^k$ if and only if (t,X) is normal on $\mathbb{R}$ for all $t \in \mathbb{R}^k$.*

Proof. "$\Longrightarrow$". The c.f. of real-valued random variable (t,X) is

$$\begin{aligned} f(\lambda) &= \mathbb{E}\exp i\lambda(t,X) = \mathbb{E}\exp i(\lambda t,X) \\ &= \exp\Big(-\frac{1}{2}(C\lambda t,\lambda t)\Big) = \exp\Big(-\frac{1}{2}\lambda^2(Ct,t)\Big), \end{aligned}$$

which implies that (t,X) has normal distribution $N(0,(Ct,t))$ with variance (Ct,t).

"$\Longleftarrow$". If (t,X) is normal then

$$\mathbb{E}\exp i(t,X) = \exp\Big(-\frac{1}{2}(Ct,t)\Big),$$

because the variance of (t,X) is (Ct,t). This means that X is normal $N(0,C)$. □

Lemma 3.15. *Let $Z = (X,Y)$, where $X = (X_1,\ldots,X_i)$ and $Y = (Y_1,\ldots,Y_j)$, and suppose that Z is normal on $\mathbb{R}^{i+j}$. Then X and Y are independent if and only* $\mathrm{Cov}(X_m,Y_n) = 0$ *for all m,n.*

Proof. One way is obvious. The other way around, suppose that

$$C = \mathrm{Cov}(Z) = \begin{bmatrix} D & 0 \\ 0 & F \end{bmatrix}.$$

If $t = (t_1,t_2)$ then the characteristic function of Z is

$$\begin{aligned}\mathbb{E}\exp i(t,Z) &= \exp\Big(-\frac{1}{2}(Ct,t)\Big)\\ &= \exp\Big(-\frac{1}{2}(Dt_1,t_1)-\frac{1}{2}(Ft_2,t_2)\Big) = \mathbb{E}\exp i(t_1,X)\,\mathbb{E}\exp i(t_2,Y),\end{aligned}$$

which is precisely the characteristic function of independent X and Y. By the uniqueness theorem, X and Y are independent. □

Exercise 3.3.1. Finish the proof of Theorem 3.6. Namely, show that (3.8) holds if $\lim_{n\to\infty}\mathbb{E}f(Z_n) = \mathbb{E}f(Z)$ for functions f with bounded third derivative.

Exercise 3.3.2. 24% of the residents in a community are members of a minority group but among the 96 people called for jury duty only 13 are. Does this data indicate that minorities are less likely to be called for jury duty?

Exercise 3.3.3. Given $0 < t_1 < \ldots < t_k < \infty$, show that the Gaussian distribution $N(0,(t_i\wedge t_j)_{1\le i,j\le k})$ has density on $\mathbb{R}^k$. (Hint: either check that the covariance is non-degenerate, or look up the definition of the Brownian motion.)

Exercise 3.3.4. Given a vector of i.i.d. standard normal random variables $g = (g_1,\ldots,g_k)^T$ and orthogonal $k\times k$ matrix V, show that $Y = Vg$ also has i.i.d. standard normal coordinates.

Exercise 3.3.5. If g is standard normal, prove that $\mathbb{E}gF(g) = \mathbb{E}F'(g)$ for nice enough functions $F:\mathbb{R}\to\mathbb{R}$.

Exercise 3.3.6. If $g = (g_1,\ldots,g_k)$ has arbitrary normal distribution $N(0,C)$ on $\mathbb{R}^k$, prove that

$$\mathbb{E}g_1F(g) = \sum_{i=1}^{n}(\mathbb{E}g_1g_i)\mathbb{E}\frac{\partial F}{\partial x_i}(g),$$

where $C_{1i} = \mathbb{E}g_1g_i$, for nice enough functions $F:\mathbb{R}^k\to\mathbb{R}$. Hint: what can you say about g_1 and $g_i - (C_{1i}/C_{11})g_1$ for $1\le i\le k$?

3.4 Central limit theorem for triangular arrays

Instead of considering i.i.d. sequences, for each $n \geq 1$ we will now consider a vector $(X_{n1},\ldots,X_{nn})$ of independent random variables, not necessarily identically distributed. This setting is called the *triangular arrays*. Let us denote

$$\mu_{ni} = \mathbb{E}X_{ni},\ \sigma_{ni}^2 = \mathrm{Var}(X_{ni}) < \infty,\ D_n^2 = \sum_{i=1}^{n} \sigma_{ni}^2. \tag{3.11}$$

Consider

$$Z_n = \frac{S_n - \mathbb{E}S_n}{\sqrt{\mathrm{Var}(S_n)}} = \frac{1}{D_n}\sum_{i=1}^{n}(X_{ni} - \mu_{ni}). \tag{3.12}$$

We have $\mathbb{E}Z_n = 0$ and $\mathrm{Var}(Z_n) = 1$, but the variance of each summand σ_{ni}^2/D_n^2 is not necessarily $1/n$ as in the case of i.i.d. random variables. One condition that ensures that Z_n satisfies the CLT is

$$\lim_{n\to\infty} \frac{1}{D_n^3}\sum_{i=1}^{n} \mathbb{E}\big|X_{ni} - \mu_{ni}\big|^3 = 0, \tag{3.13}$$

which is called the *Lyapunov condition.*

Theorem 3.9. *Under the Lyapunov condition (3.13), the distribution of Z_n in (3.12) converges weakly to standard Gaussian $N(0,1)$.*

The proof is identical to Theorem 3.6 in the previous section. One small change is to consider $a_{ni} = \sigma_{ni}/D_n$ and represent $g = \sum_{i=1}^n a_{ni}g_i$. At some point, when controlling one of the error terms in the Taylor expansion, one also has to use that

$$\sigma_{ni}^3 = \left(\mathbb{E}\big|X_{ni} - \mu_{ni}\big|^2\right)^{3/2} \leq \mathbb{E}\big|X_{ni} - \mu_{ni}\big|^3.$$

Theorem 3.9 also follows directly from the next more general result.

Theorem 3.10 (Lindeberg's CLT). *For each $n \geq 1$, consider a vector $(X_{ni})_{1\leq i\leq n}$ of independent random variables such that*

$$\mathbb{E}X_{ni} = 0,\ \mathrm{Var}(S_n) = \sum_{i=1}^{n} \mathbb{E}(X_{ni})^2 = 1.$$

Suppose that the following Lindeberg's condition *is satisfied:*

$$\sum_{i=1}^{n} \mathbb{E}(X_{ni})^2 \mathrm{I}(|X_{ni}| > \varepsilon) \to 0 \text{ as } n\to\infty \text{ for all } \varepsilon > 0. \tag{3.14}$$

Then the distribution $\mathscr{L}\left(\sum_{i\leq n} X_{ni}\right) \to N(0,1)$ weakly.

Remark 3.2. We will prove this result again using characteristic functions, but one can also use the same argument as in Theorem 3.6 in the previous section. We will leave it as an exercise below. Also, notice that for i.i.d. random variables this implies the CLT without the assumption $\mathbb{E}|X_1|^3 < \infty$, since in that case $X_{ni} = X_i/\sqrt{n}$.

Proof. First of all, the sequence $\mathscr{L}\big(\sum_{i\le n} X_{ni}\big)$ is uniformly tight, because by Chebyshev's inequality

$$\mathbb{P}\Big(\Big|\sum_{i\le n} X_{ni}\Big| > M\Big) \le \frac{1}{M^2} \le \varepsilon$$

for large enough M. It remains to show that the characteristic function of S_n coverges to $e^{-\frac{\lambda^2}{2}}$. For simplicity of notation, let us omit the upper index n and write X_i instead of X_{ni}. Since $\mathbb{E}e^{i\lambda S_n} = \prod_{i=1}^n \mathbb{E}e^{i\lambda X_i}$ it is enough to show that

$$\log \mathbb{E}e^{i\lambda S_n} = \sum_{i=1}^n \log\big(1 + (\mathbb{E}e^{i\lambda X_i} - 1)\big) \to -\frac{\lambda^2}{2}. \tag{3.15}$$

It is not difficult to check, by induction on m, that for any $a \in \mathbb{R}$,

$$\Big|e^{ia} - \sum_{k=0}^m \frac{(ia)^k}{k!}\Big| \le \frac{|a|^{m+1}}{(m+1)!}. \tag{3.16}$$

We will leave it as an exercise at the end of the section. Using this for $m = 1$,

$$\begin{aligned}\big|\mathbb{E}e^{i\lambda X_i} - 1\big| &= \big|\mathbb{E}e^{i\lambda X_i} - 1 - i\lambda \mathbb{E}X_i\big| \le \frac{\lambda^2}{2}\mathbb{E}X_i^2 \\ &\le \frac{\lambda^2}{2}\varepsilon^2 + \frac{\lambda^2}{2}\mathbb{E}X_i^2 \mathrm{I}(|X_i| > \varepsilon) \le \lambda^2\varepsilon^2 \le \frac{1}{2}\end{aligned} \tag{3.17}$$

for large n by (3.14) and for small enough ε. Using Taylor's expansion of $\log(1+z)$ it is easy to check that

$$\big|\log(1+z) - z\big| \le |z|^2 \text{ for } |z| \le \frac{1}{2}$$

and, therefore, using this and (3.17),

$$\begin{aligned}&\sum_{i=1}^n \Big|\log\big(1 + (\mathbb{E}e^{i\lambda X_i} - 1)\big) - \big(\mathbb{E}e^{i\lambda X_i} - 1\big)\Big| \le \sum_{i=1}^n \big|\mathbb{E}e^{i\lambda X_i} - 1\big|^2 \\ &\le \sum_{i=1}^n \frac{\lambda^4}{4}\big(\mathbb{E}X_i^2\big)^2 \le \frac{\lambda^4}{4}\Big(\max_{1\le i\le n} \mathbb{E}X_i^2\Big)\sum_{i=1}^n \mathbb{E}X_i^2 = \frac{\lambda^4}{4}\max_{1\le i\le n} \mathbb{E}X_i^2.\end{aligned}$$

This goes to zero, because

$$\max_{1\le i\le n} \mathbb{E}X_i^2 \le \varepsilon^2 + \max_{1\le i\le n} \mathbb{E}X_i^2 \mathrm{I}(|X_i| > \varepsilon) \to \varepsilon^2$$

as $n \to \infty$, by (3.14), and then we let $\varepsilon \downarrow 0$. As a result, to prove (3.15) it remains to show that

$$\sum_{i=1}^{n} \left(\mathbb{E}e^{i\lambda X_i} - 1\right) \to -\frac{\lambda^2}{2}.$$

Using (3.16) for $m = 1$, on the event $\{|X_i| > \varepsilon\}$,

$$\left|e^{i\lambda X_i} - 1 - i\lambda X_i\right| \mathrm{I}(|X_i| > \varepsilon) \le \frac{\lambda^2}{2} X_i^2 \mathrm{I}(|X_i| > \varepsilon)$$

and, therefore,

$$\left|e^{i\lambda X_i} - 1 - i\lambda X_i + \frac{\lambda^2}{2} X_i^2\right| \mathrm{I}(|X_i| > \varepsilon) \le \lambda^2 X_i^2 \mathrm{I}(|X_i| > \varepsilon). \tag{3.18}$$

Using (3.16) for $m = 2$, on the event $\{|X_i| \le \varepsilon\}$,

$$\left|e^{i\lambda X_i} - 1 - i\lambda X_i + \frac{\lambda^2}{2} X_i^2\right| \mathrm{I}(|X_i| \le \varepsilon) \le \frac{\lambda^3}{6} |X_i|^3 \mathrm{I}(|X_i| \le \varepsilon) \le \frac{\lambda^3 \varepsilon}{6} X_i^2. \tag{3.19}$$

Combining the last two equations and using that $\mathbb{E}X_i = 0$,

$$\left|\mathbb{E}e^{i\lambda X_i} - 1 + \frac{\lambda^2}{2} \mathbb{E}X_i^2\right| \le \lambda^2 \mathbb{E}X_i^2 \mathrm{I}(|X_i| > \varepsilon) + \frac{\lambda^3 \varepsilon}{6} \mathbb{E}X_i^2.$$

Finally, this implies

$$\begin{aligned}\left|\sum_{i=1}^{n} \left(\mathbb{E}e^{i\lambda X_i} - 1\right) + \frac{\lambda^2}{2}\right| &= \left|\sum_{i=1}^{n} \left(\mathbb{E}e^{i\lambda X_i} - 1\right) + \frac{\lambda^2}{2} \sum_{i=1}^{n} \mathbb{E}X_i^2\right| \\ &\le \frac{\lambda^3 \varepsilon}{6} + \lambda^2 \sum_{i=1}^{n} \mathbb{E}X_i^2 \mathrm{I}(|X_i| > \varepsilon) \to \frac{\lambda^3 \varepsilon}{6}\end{aligned}$$

as $n \to \infty$, using Lindeberg's condition (3.14). It remains to let $\varepsilon \downarrow 0$. □

Theorem 3.11 (Multivariate Lindeberg's CLT). *For each $n \ge 1$, let $(X_{ni})_{1 \le i \le n}$ be independent random vectors in $\mathbb{R}^k$ such that*

$$\mathbb{E}X_{ni} = 0,\ \mathrm{Cov}(S_n) = \sum_{i=1}^{n} \mathrm{Cov}(X_{ni}) \to C \text{ as } n \to \infty.$$

Suppose that the following Lindeberg's condition *is satisfied:*

$$\sum_{i=1}^{n} \mathbb{E}|X_{ni}|^2 \mathrm{I}(|X_{ni}| > \varepsilon) \to 0 \text{ as } n \to \infty \text{ for all } \varepsilon > 0. \tag{3.20}$$

Then the distribution $\mathscr{L}\left(\sum_{i \le n} X_{ni}\right) \to N(0, C)$ weakly.

Just like in the i.i.d. case, using characteristic functions, the proof is reduced to the one-dimensional case. One only needs to observe that, for any $t \in \mathbb{R}^k$, random variables (t, X_{ni}) satisfy the condition (3.14), because $|(t, X_{ni})| \le |t||X_{ni}|$.

Three series theorem. We will now use Lindeberg's CLT above to prove the "three series theorem" for random series. Let us begin with the following observation.

Lemma 3.16. *If $\mathbb{P}, \mathbb{Q}$ are distributions on $\mathbb{R}$ such that $\mathbb{P} * \mathbb{Q} = \mathbb{P}$ then $\mathbb{Q}(\{0\}) = 1$.*

Proof. Let us consider the characteristic functions

$$f_P(t) = \int e^{itx}\, d\mathbb{P}(x),\ f_Q(t) = \int e^{itx}\, d\mathbb{Q}(x).$$

The condition $\mathbb{P} * \mathbb{Q} = \mathbb{P}$ implies that $f_P(t) f_Q(t) = f_P(t)$. Since $f_P(0) = 1$ and $f_P(t)$ is continuous, for small enough $|t| \le \varepsilon$ we have $|f_P(t)| > 0$ and, as a result, $f_Q(t) = 1$. Since

$$f_Q(t) = \int \cos(tx)\, d\mathbb{Q}(x) + i \int \sin(tx)\, d\mathbb{Q}(x),$$

for $|t| \le \varepsilon$ this implies that $\int \cos(tx)\, d\mathbb{Q}(x) = 1$ and, since $\cos(s) \le 1$, this can happen only if

$$\mathbb{Q}\big(\{x : xt = 0 \mod 2\pi\}\big) = 1 \text{ for all } |t| \le \varepsilon.$$

Take s, t such that $|s|, |t| \le \varepsilon$ and s/t is irrational. For x to be in the support of $\mathbb{Q}$ we must have $xs = 2\pi k$ and $xt = 2\pi m$ for some integer k, m. This can happen only if $x = 0$. □

We proved in Section 2.3 that convergence of random series in probability implies almost sure convergence. It turns out that this result can be strengthened, and convergence in distribution also implies convergence in probability and, therefore, almost sure convergence.

Theorem 3.12 (Levy's equivalence theorem). *If (X_i) is a sequence of independent random variables then the series $\sum_{i \ge 1} X_i$ converges almost surely, in probability, or in distribution, at the same time.*

Proof. Of course, we only need to show that convergence in distribution implies convergence in probability. Suppose that $\mathscr{L}(S_n) \to \mathbb{P}$. Convergence in law implies that $\{\mathscr{L}(S_n)\}$ is uniformly tight, which implies that $\{\mathscr{L}(S_n - S_k)\}_{n,k \ge 1}$ is uniformly tight. We will now show that this implies that, for any $\varepsilon > 0$,

$$\mathbb{P}(|S_n - S_k| > \varepsilon) < \varepsilon \tag{3.21}$$

for $n \ge k \ge N$ for large enough N. Suppose not. Then there exists $\varepsilon > 0$ and sequences $(n(\ell))$ and $(n'(\ell))$ such that $n(\ell) \le n'(\ell)$ and

$$\mathbb{P}(|S_{n'(\ell)} - S_{n(\ell)}| > \varepsilon) \geq \varepsilon.$$

Let us denote $Y_\ell = S_{n'(\ell)} - S_{n(\ell)}$. Since $\{\mathscr{L}(Y_\ell)\}$ is uniformly tight, by the selection theorem, there exists a subsequence $(\ell(r))$ such that $\mathscr{L}(Y_{\ell(r)}) \to \mathbb{Q}$. Since

$$S_{n'(\ell(r))} = S_{n(\ell(r))} + Y_{\ell(r)} \text{ and } \mathscr{L}(S_{n'(\ell(r))}) = \mathscr{L}(S_{n(\ell(r))}) * \mathscr{L}(Y_{\ell(r)}),$$

letting $r \to \infty$ we get that $\mathbb{P} = \mathbb{P} * \mathbb{Q}$. By the above Lemma, $\mathbb{Q}(\{0\}) = 1$, which implies that $\mathbb{P}(|Y_{\ell(r)}| > \varepsilon) \leq \varepsilon$ for large r – a contradiction. By Lemma 2.6, (3.21) implies that S_n converges in probability. □

Theorem 3.13 (Three series theorem). *Let $(X_i)_{i\geq 1}$ be a sequence of independent random variables and let $Z_i = X_i \mathrm{I}(|X_i| \leq 1)$. Then the series $\sum_{i\geq 1} X_i$ converges almost surely if and only if the following three conditions hold:*

(1) $\sum_{i\geq 1} \mathbb{P}(|X_i| > 1) < \infty$,
(2) $\sum_{i\geq 1} \mathbb{E}Z_i$ *converges,*
(3) $\sum_{i\geq 1} \mathrm{Var}(Z_i) < \infty$.

Proof. "$\Longleftarrow$". Suppose (1) – (3) hold. Since

$$\sum_{i\geq 1} \mathbb{P}(|X_i| > 1) = \sum_{i\geq 1} \mathbb{P}(X_i \neq Z_i) < \infty,$$

by the Borel-Cantelli lemma, $\mathbb{P}(X_i \neq Z_i \text{ i.o.}) = 0$, which means that $\sum_{i\geq 1} X_i$ converges if and only if $\sum_{i\geq 1} Z_i$ converges. By (2), it is enough to show that $\sum_{i\geq 1}(Z_i - \mathbb{E}Z_i)$ converges, but this follows from Theorem 2.12 by (3).

"$\Longrightarrow$". If $\sum_{i\geq 1} X_i$ converges almost surely, $\mathbb{P}(|X_i| > 1 \text{ i.o}) = 0$, and, since (X_i) are independent, by the Borel-Cantelli lemma, $\sum_{i\geq 1} \mathbb{P}(|X_i| > 1) < \infty$. This proves (1).

We will prove (3) by contradiction. If $\sum_{i\geq 1} X_i$ converges then, obviously, $\sum_{i\geq 1} Z_i$ converges. Let us denote $S_{mn} = \sum_{m\leq k\leq n} Z_k$. Since $S_{mn} \to 0$ as $m,n \to \infty$, $\mathbb{P}(|S_{mn}| > \delta) \leq \delta$ for any $\delta > 0$ for m,n large enough. Suppose that $\sum_{i\geq 1} \mathrm{Var}(Z_i) = \infty$. Then

$$\sigma_{mn}^2 = \mathrm{Var}(S_{mn}) = \sum_{m\leq k\leq n} \mathrm{Var}(Z_k) \to \infty$$

as $n \to \infty$ for any fixed m. Intuitively, this should not happen, since $S_{mn} \to 0$ in probability but their variance goes to infinity. In principle, one can construct such sequence of random variables, but in our case it will be ruled out by Lindeberg's CLT, as follows. Let us show that, by Lindeberg's theorem,

$$T_{mn} = \frac{S_{mn} - \mathbb{E}S_{mn}}{\sigma_{mn}} = \sum_{m\leq k\leq n} \frac{Z_k - \mathbb{E}Z_k}{\sigma_{mn}}$$

converges in distribution to $N(0,1)$ if $m,n \to \infty$ in such a way that $\sigma_{mn}^2 \to \infty$. We only need to check that

$$\sum_{m\le k\le n} \mathbb{E}\Big(\frac{Z_k-\mathbb{E}Z_k}{\sigma_{mn}}\Big)^2 \mathrm{I}\Big(\Big|\frac{Z_k-\mathbb{E}Z_k}{\sigma_{mn}}\Big|>\varepsilon\Big)\to 0$$

if $m,n\to\infty$ in such a way that $\sigma_{mn}^2\to\infty$. This is obvious, because $|Z_k-\mathbb{E}Z_k|<2$ and $\sigma_{mn}\to\infty$, so the event in the indicator does not occur and the sum is, in fact, equal to 0 for large m,n. Next, since $\mathbb{P}(|S_{mn}|>\delta)\le\delta$, we get

$$\mathbb{P}\big(|S_{mn}|\le\delta\big)=\mathbb{P}\Big(\Big|T_{mn}+\frac{\mathbb{E}S_{mn}}{\sigma_{mn}}\Big|\le\frac{\delta}{\sigma_{mn}}\Big)\ge 1-\delta.$$

But this is impossible, since T_{mn} is approximately standard normal, which can not concentrate near any constant. Therefore, we proved that $\sum_{i\ge1}\mathrm{Var}(Z_i)<\infty$. By Kolmogorov's SLLN this implies that $\sum_{i\ge1}(Z_i-\mathbb{E}Z_i)$ converges almost surely and, since $\sum_{i\ge1}Z_i$ converges almost surely, $\sum_{i\ge1}\mathbb{E}Z_i$ also converges. □

Exercise 3.4.1. Let (X_n) be i.i.d. random variables with continuous distribution F. We say that X_n is a *record value* if $X_n>X_i$ for $i<n$. If S_n is the number of records up to time n, show that Z_n in (3.12) satisfies the CLT.

Exercise 3.4.2. Consider a triangular array of i.i.d. Bernoulli $X_{n1},\ldots,X_{nn}\sim B(p_n)$, where p_n varies with n. If $p_n\to0$ but $np_n\to+\infty$, show that Z_n in (3.12) satisfies the CLT.

Exercise 3.4.3. Let $X_1,X_2,\ldots$ be independent random variables such that $X_i\sim B(p_i)$ for some $p_i\in[0,1]$. If the series $\sum_{i=1}^{\infty}p_i(1-p_i)$ diverges, show that Z_n in (3.12) satisfies the CLT.

Exercise 3.4.4. Adapt the argument of Theorem 3.6 in the previous section to prove Theorem 3.10. *Hint:* when using the Taylor expansion in that proof, mimic equations (3.18) and (3.19).

Exercise 3.4.5. Prove (3.16). (Hint: integrate the inequality to make the induction step.)

Exercise 3.4.6. Consider the random series $\sum_{n\ge1}\varepsilon_n n^{-\alpha}$ where (ε_n) are i.i.d. random signs $\mathbb{P}(\varepsilon_n=\pm1)=1/2$. Give the necessary and sufficient condition on α for this series to converge.

Exercise 3.4.7. Let $(X_n)_{n\ge1}$ be i.i.d. random variables such that $\mathbb{E}X_1=0$ and $0<\mathbb{E}X_1^2<\infty$. Show that $\sum_{n\ge1}X_n/a_n$ converges almost surely if and only if $\sum_{n\ge1}1/a_n^2<\infty$.

Exercise 3.4.8. Gamma distribution with parameters $k,\theta>0$ has density

$$\frac{x^{k-1}e^{-x/\theta}}{\theta^k\Gamma(k)}\mathrm{I}(x>0).$$

Suppose X_n has Gamma(k_n, θ_n) distribution with $k_n = 1 + \sin^2 n$ and $\theta_n = 1/k_n$, and X_n are independent for $n \geq 1$. Either show that

$$\frac{X_1 + \ldots + X_n - n}{\sqrt{n}}$$

converges in distribution to a Gaussian, or reduce this to a question about the existence of a certain limit. *Hint*: to compute the limit, look up strong ergodic theorem.

Exercise 3.4.9. Let $(X_n)_{n\geq 1}$ be i.i.d. random variables with the Poisson distribution with mean $\lambda = 1$. Can you find a_n and b_n such that $a_n \sum_{\ell=1}^{n} X_\ell \log \ell - b_n$ converges in distribution to standard Gaussian?

Exercise 3.4.10. Suppose that X_k for $k \geq 1$ are independent random variables and

$$\mathbb{P}(X_k = \pm k) = \frac{\log k}{2k}, \ \mathbb{P}(X_k = 0) = 1 - \frac{\log k}{k}.$$

If $S_n = \sum_{k=1}^{n} X_k$ and $D_n^2 = \sum_{k=1}^{n} \mathrm{Var}(X_k)$, does the distribution of S_n/D_n converge to $N(0,1)$?

Chapter 4
Metrics on Probability Measures

4.1 Bounded Lipschitz functions

We will now study the class of the so called *bounded Lipschitz functions* on a metric space (S,d), which will play an important role in the next section, when we deal with convergence of laws on metric spaces. For a function $f : S \to \mathbb{R}$, let us define a Lipschitz semi-norm by

$$\|f\|_{\mathrm{L}} = \sup_{x \neq y} \frac{|f(x) - f(y)|}{d(x,y)}.$$

(When the cardinality of S is 1, we set $\|f\|_{\mathrm{L}} = 0$.) Clearly, $\|f\|_{\mathrm{L}} = 0$ if and only if f is constant so $\|f\|_{\mathrm{L}}$ is not a norm, even though it satisfies the triangle inequality. Let us define a *bounded Lipschitz norm* by

$$\|f\|_{\mathrm{BL}} = \|f\|_{\mathrm{L}} + \|f\|_{\infty},$$

where $\|f\|_{\infty} = \sup_{s \in S} |f(s)|$. Let

$$BL(S,d) = \big\{ f : S \to \mathbb{R} \, : \, \|f\|_{\mathrm{BL}} < \infty \big\}$$

be the set of all bounded Lipschitz functions on (S,d). We will now prove several fact about these functions.

Lemma 4.1. *If $f, g \in BL(S,d)$ then $fg \in BL(S,d)$ and $\|fg\|_{\mathrm{BL}} \leq \|f\|_{\mathrm{BL}} \|g\|_{\mathrm{BL}}$.*

Proof. First of all, $\|fg\|_{\infty} \leq \|f\|_{\infty} \|g\|_{\infty}$. We can write,

$$\begin{aligned} |f(x)g(x) - f(y)g(y)| &\leq |f(x)(g(x) - g(y))| + |g(y)(f(x) - f(y))| \\ &\leq \|f\|_{\infty} \|g\|_{\mathrm{L}} d(x,y) + \|g\|_{\infty} \|f\|_{\mathrm{L}} d(x,y) \end{aligned}$$

and, therefore,

$$\|fg\|_{\mathrm{BL}} \leq \|f\|_{\infty} \|g\|_{\infty} + \|f\|_{\infty} \|g\|_{\mathrm{L}} + \|g\|_{\infty} \|f\|_{\mathrm{L}} \leq \|f\|_{\mathrm{BL}} \|g\|_{\mathrm{BL}}.$$

This finishes the proof. □

Let us recall the notations $a \vee b = \max(a,b)$ and $a \wedge b = \min(a,b)$ and let $* = \wedge$ or $\vee$.

Lemma 4.2. *The following inequalities hold:*

(1) $\|f_1 * \cdots * f_k\|_{\mathrm{L}} \le \max_{1\le i\le k} \|f_i\|_{\mathrm{L}}$,
(2) $\|f_1 * \cdots * f_k\|_{\mathrm{BL}} \le 2\max_{1\le i\le k} \|f_i\|_{\mathrm{BL}}$.

Proof. (1) It is enough to consider $k=2$. For specificity, take $* = \vee$. Given $x,y \in S$, suppose that

$$f_1 \vee f_2(x) \ge f_1 \vee f_2(y) = f_1(y).$$

Then

$$\begin{aligned} |f_1 \vee f_2(y) - f_1 \vee f_2(x)| &= f_1 \vee f_2(x) - f_1 \vee f_2(y) \\ &\le \begin{cases} f_1(x) - f_1(y), \text{ if } f_1(x) \ge f_2(x) \\ f_2(x) - f_2(y), \text{ otherwise} \end{cases} \\ &\le \|f_1\|_{\mathrm{L}} \vee \|f_2\|_{\mathrm{L}} d(x,y). \end{aligned}$$

The case of $* = \wedge$ can be considered similarly.

(2) First of all, obviously, $\|f_1 * \cdots * f_k\|_\infty \le \max_{1\le i\le k} \|f_i\|_\infty$. Therefore, using (1),

$$\|f_1 * \cdots * f_k\|_{\mathrm{BL}} \le \max_i \|f_i\|_\infty + \max_i \|f_i\|_{\mathrm{L}} \le 2\max_i \|f_i\|_{\mathrm{BL}}$$

and this finishes the proof. □

Another important fact is the following.

Theorem 4.1 (Extension theorem). *Given a set $A \subseteq S$ and a bounded Lipschitz function $f \in BL(A,d)$ on A, there exists an extension $h \in BL(S,d)$ such that $f = h$ on A and $\|h\|_{\mathrm{BL}} = \|f\|_{\mathrm{BL}}$.*

Proof. Let us first find an extension such that $\|h\|_{\mathrm{L}} = \|f\|_{\mathrm{L}}$. We will check that the following choice works,

$$h(x) := \sup_{s\in A}(f(s) - \|f\|_{\mathrm{L}} d(x,s)).$$

First of all, if $x \in A$ then, by the Lipschitz property on A,

$$f(s) - \|f\|_{\mathrm{L}} d(x,s) \le f(x) \text{ for all } s \in A,$$

which means that $h(x) = f(x)$ for $x \in A$. Next, by the triangle inequality,

$$f(s) - \|f\|_{\mathrm{L}} d(x,s) \le f(s) - \|f\|_{\mathrm{L}} d(y,s) + \|f\|_{\mathrm{L}} d(x,y)$$

for any $x, y \in S$, and taking the supremum over $s \in A$ shows that

$$h(x) \le h(y) + \|f\|_{\mathrm{L}} d(x,y),$$

which means that $\|h\|_{\mathrm{L}} = \|f\|_{\mathrm{L}}$. To extend preserving the bounded Lipschitz norm, take

$$h' = (h \wedge \|f\|_\infty) \vee (-\|f\|_\infty).$$

By part (1) of previous lemma, we see that $\|h'\|_{\mathrm{BL}} = \|f\|_{\mathrm{BL}}$ and, clearly, h' coincides with f on A. □

Let $(C(S), d_\infty)$ denote the space of continuous real-valued functions on S with

$$d_\infty(f,g) = \sup_{x \in S} |f(x) - g(x)|.$$

To prove the next property of bounded Lipschitz functions, let us first recall the following famous generalization of the Weierstrass theorem. We will give the proof for convenience.

Theorem 4.2 (Stone-Weierstrass theorem). *Let (S,d) be a compact metric space and $\mathscr{F} \subseteq C(S)$ is such that*

1. *$\mathscr{F}$ is an algebra, i.e. for all $f, g \in \mathscr{F}, c \in \mathbb{R}$, we have $cf + g \in \mathscr{F}, fg \in \mathscr{F}$.*
2. *$\mathscr{F}$ separates points, i.e. if $x \neq y \in S$ then there exists $f \in \mathscr{F}$ such that $f(x) \neq f(y)$.*
3. *$\mathscr{F}$ contains constants.*

Then $\mathscr{F}$ is dense in $(C(S), d_\infty)$.

Proof. Consider bounded $f \in \mathscr{F}$, i.e. $|f(x)| \le M$. A function $x \to |x|$ defined on the interval $[-M, M]$ can be uniformly approximated by polynomials of x by the Weierstrass theorem on the real line or, for example, using Bernstein's polynomials. Therefore, $|f(x)|$ can be uniformly approximated by polynomials of $f(x)$, and by properties 1 and 3, by functions in $\mathscr{F}$. Therefore, if $\overline{\mathscr{F}}$ is the closure of $\mathscr{F}$ in d_∞ norm then for any $f \in \overline{\mathscr{F}}$ its absolute value $|f| \in \overline{\mathscr{F}}$. Therefore, for any $f, g \in \overline{\mathscr{F}}$ we have

$$\begin{aligned} \min(f,g) &= \frac{1}{2}(f+g) - \frac{1}{2}|f-g| \in \overline{\mathscr{F}}, \\ \max(f,g) &= \frac{1}{2}(f+g) + \frac{1}{2}|f-g| \in \overline{\mathscr{F}}. \end{aligned} \tag{4.1}$$

Given any points $x \neq y$ and $c, d \in \mathbb{R}$ one can always find $f \in \mathscr{F}$ such that $f(x) = c$ and $f(y) = d$. Indeed, by property 2 we can find $g \in \mathscr{F}$ such that $g(x) \neq g(y)$ and, as a result, a system of equations

$$ag(x) + b = c, \ ag(y) + b = d$$

has a solution a,b. Then the function $f = ag+b$ satisfies the above and it is in $\mathscr{F}$ by 1. Take $h \in C(S)$ and fix x. For any y let $f_y \in \mathscr{F}$ be such that

$$f_y(x) = h(x), \ f_y(y) = h(y).$$

By continuity of f_y, for any $y \in S$ there exists an open neighborhood U_y of y such that

$$f_y(s) \geq h(s) - \varepsilon \text{ for } s \in U_y.$$

Since (U_y) is an open cover of the compact S, there exists a finite subcover $U_{y_1}, \ldots, U_{y_N}$. Let us define a function

$$f^x(s) = \max(f_{y_1}(s), \ldots, f_{y_N}(s)) \in \overline{\mathscr{F}} \text{ by (4.1).}$$

By construction, it has the following properties:

$$f^x(x) = h(x), \ f^x(s) \geq h(s) - \varepsilon \ \text{ for all } \ s \in S.$$

Again, by continuity of $f^x(s)$ there exists an open neighborhood U_x of x such that

$$f^x(s) \leq h(s) + \varepsilon \ \text{ for } \ s \in U_x.$$

Take a finite subcover $U_{x_1}, \ldots, U_{x_M}$ and define

$$h'(s) = \min\left(f^{x_1}(s), \ldots, f^{x_M}(s)\right) \in \overline{\mathscr{F}} \text{ by (4.1).}$$

By construction, $h'(s) \leq h(s) + \varepsilon$ and $h'(s) \geq h(s) - \varepsilon$ for all $s \in S$ which means that $d_\infty(h', h) \leq \varepsilon$. Since $h' \in \overline{\mathscr{F}}$, this proves that $\overline{\mathscr{F}}$ is dense in $(C(S), d_\infty)$. □

The above results can be now combined to prove the following property of bounded Lipschitz functions.

Corollary 4.1. *If (S,d) is a compact space then the set of bounded Lipschitz functions $BL(S,d)$ is dense in $(C(S), d_\infty)$.*

Proof. We apply the Stone-Weierstrass theorem with $\mathscr{F} = BL(S,d)$. Property 3 is obvious, property 1 follows from Lemma 4.1 and property 2 follows from the extension Theorem 4.1, since a function defined on two points $x \neq y$ such that $f(x) \neq f(y)$ can be extended to a bounded Lipschitz function on the entire S. □

We will also need another well-known result from analysis. A set $A \subseteq S$ is *totally bounded* if for any $\varepsilon > 0$ there exists a finite ε-cover of A, i.e. a set of points $a_1, \ldots, a_N$ such that $A \subseteq \cup_{i \leq N} B(a_i, \varepsilon)$, where $B(a, \varepsilon) = \{y \in S : d(a,y) \leq \varepsilon\}$ is a ball of radius ε centered at a.

Theorem 4.3 (Arzela-Ascoli). *If (S,d) is a compact metric space then a subset $\mathscr{F} \subseteq C(S)$ is totally bounded in d_∞ metric if and only if $\mathscr{F}$ is equicontinuous and uniformly bounded.*

Equicontinuous means that for any $\varepsilon > 0$ there exists $\delta > 0$ such that if $d(x,y) \le \delta$ then, for all $f \in \mathscr{F}$, $|f(x) - f(y)| \le \varepsilon$. The following fact was used in the proof of the Selection Theorem, which was proved for general metric spaces.

Corollary 4.2. *If (S,d) is a compact space then $C(S)$ is separable in d_∞.*

Proof. By the above theorem, $BL(S,d)$ is dense in $C(S)$. For any integer $n \ge 1$, the set $\{f : \|f\|_{\mathrm{BL}} \le n\}$ is, obviously, uniformly bounded and equicontinuous. By the Arzela-Ascoli theorem, it is totally bounded and, therefore, separable, which can be seen by taking the union of finite $1/m$-covers for all $m \ge 1$. The union

$$\bigcup_{n \ge 1} \{f : \|f\|_{\mathrm{BL}} \le n\} = BL(S,d)$$

is also separable and, since it is dense in $C(S)$, $C(S)$ is separable. □

Exercise 4.1.1. If F is a finite set $\{x_1, \ldots, x_n\}$ and $\mathbb{P}$ is a law with $\mathbb{P}(F) = 1$, show that for any law $\mathbb{Q}$ the supremum $\sup\{\left|\int f \, d(\mathbb{P} - \mathbb{Q})\right| : \|f\|_{\mathrm{L}} \le 1\}$ can be restricted to functions of the form $f(x) = \min_{1 \le i \le n}(c_i + d(x, x_i))$.

4.2 Convergence of laws on metric spaces

Let (S,d) be a metric space and $\mathscr{B}$ - a Borel σ-algebra generated by open sets. Let us recall that $\mathbb{P}_n \to \mathbb{P}$ weakly if

$$\int f\, d\mathbb{P}_n \to \int f\, d\mathbb{P}$$

for all $f \in C_b(S)$ - real-valued bounded continuous functions on S. For a set $A \subseteq S$, we denote by $\bar{A}$ the closure of A, $\mathrm{int}A$ - the interior of A and $\partial A = \bar{A} \setminus \mathrm{int}A$ - the boundary of A. A measurable set A is called a *continuity set* of $\mathbb{P}$ if $\mathbb{P}(\partial A) = 0$.

Theorem 4.4 (Portmanteau theorem). *The following are equivalent.*

(1) $\mathbb{P}_n \to \mathbb{P}$ *weakly.*
(2) $\int f\, d\mathbb{P}_n \to \int f\, d\mathbb{P}$ *for all bounded Lipschitz functions* f.
(3) For any open set $U \subseteq S$, $\liminf_{n\to\infty} \mathbb{P}_n(U) \geq \mathbb{P}(U)$.
(4) For any closed set $F \subseteq S$, $\limsup_{n\to\infty} \mathbb{P}_n(F) \leq \mathbb{P}(F)$.
(5) For any continuity set A *of* $\mathbb{P}$, $\lim_{n\to\infty} \mathbb{P}_n(A) = \mathbb{P}(A)$.

Proof. (1) $\Longrightarrow$ (2). Obvious.

(2) $\Longrightarrow$ (3). Let U be an open set and $F = U^c$. Consider a sequence of bounded Lipschitz functions

$$f_m(s) = \min(1, md(s,F)).$$

Since F is closed, $d(s,F) = 0$ only if $s \in F$ and, therefore, $f_m(s) \uparrow \mathrm{I}(s \in U)$ as $m \to \infty$. Since, for each $m \geq 1$,

$$\mathbb{P}_n(U) \geq \int f_m\, d\mathbb{P}_n$$

taking the limit $n \to \infty$ and using (2), we get that

$$\liminf_{n\to\infty} \mathbb{P}_n(U) \geq \int f_m\, d\mathbb{P}.$$

Now, letting $m \to \infty$ and using the monotone convergence theorem implies that

$$\liminf_{n\to\infty} \mathbb{P}_n(U) \geq \int \mathrm{I}(s \in U)\, d\mathbb{P}(s) = \mathbb{P}(U),$$

which proves (2).

(3) $\Longleftrightarrow$ (4). Obvious by taking complements.

(3)&(4) $\Longrightarrow$ (5). Since $\mathrm{int}A$ is open and $\bar{A}$ is closed, and $\mathrm{int}A \subseteq \bar{A}$, by (2) and (3),

$$\mathbb{P}(\mathrm{int}A) \leq \liminf_{n\to\infty} \mathbb{P}_n(\mathrm{int}A) \leq \limsup_{n\to\infty} \mathbb{P}_n(\bar{A}) \leq \mathbb{P}(\bar{A}).$$

If $\mathbb{P}(\partial A) = 0$ then $\mathbb{P}(\bar{A}) = \mathbb{P}(\mathrm{int} A) = \mathbb{P}(A)$ and, therefore, $\lim_{n\to\infty} \mathbb{P}_n(A) = \mathbb{P}(A)$.

(5) $\Longrightarrow$ (1). Consider $f \in C_b(S)$ and let $F_y = \{s \in S : f(s) = y\}$ be a level set of f. There exist at most countably many y such that $\mathbb{P}(F_y) > 0$. Therefore, for any $\varepsilon > 0$ we can find a sequence $a_1 \le \ldots \le a_N$ such that

$$\max_k (a_{k+1} - a_k) \le \varepsilon,\ \mathbb{P}(F_{a_k}) = 0 \ \text{ for all } k$$

and the range of f is inside the interval (a_1, a_N). Let

$$B_k = \big\{s \in S : a_k \le f(s) < a_{k+1}\big\} \ \text{ and } \ f_\varepsilon(s) = \sum_k a_k \mathrm{I}(s \in B_k).$$

Since f is continuous, $\partial B_k \subseteq F_{a_k} \cup F_{a_{k+1}}$ and $\mathbb{P}(\partial B_k) = 0$. By (4),

$$\int f_\varepsilon \, d\mathbb{P}_n = \sum_k a_k \mathbb{P}_n(B_k) \stackrel{n\to\infty}{\longrightarrow} \sum_k a_k \mathbb{P}(B_k) = \int f_\varepsilon \, d\mathbb{P}.$$

Since, by construction, $|f_\varepsilon(s) - f(s)| \le \varepsilon$, letting $\varepsilon \to 0$ proves that $\int f \, d\mathbb{P}_n \to \int f \, d\mathbb{P}$, so $\mathbb{P}_n \to \mathbb{P}$ weakly. □

Notice that the weak convergence depends on the metric only through the topology, which means that if metrics d and e are equivalent then $C_b(S,d) = C_b(S,e)$ and, thus, a specific choice of metric does not affect weak convergence. On the other hand, spaces of bounded Lipschitz functions $BL(S,d)$ and $BL(S,e)$ could be different and we can sometimes use it to our advantage by choosing the metric with better properties, as we will see in the next result.

Convergence of empirical measures. Let $(\Omega, \mathscr{A}, \mathbb{P})$ be a probability space and $X_1, X_2, \ldots : \Omega \to S$ be an i.i.d. sequence of random variables with values in a metric space (S,d). Let μ be the law of X_i on S. Let us define the random *empirical measures* μ_n on the Borel σ-algebra $\mathscr{B}$ on S by

$$\mu_n(A)(\omega) = \frac{1}{n} \sum_{i=1}^n \mathrm{I}(X_i(\omega) \in A),\ A \in \mathscr{B}.$$

By the strong law of large numbers, for any $f \in C_b(S)$,

$$\int f \, d\mu_n = \frac{1}{n} \sum_{i=1}^n f(X_i) \to \mathbb{E} f(X_1) = \int f \, d\mu \ \text{a.s.}$$

However, the set of measure zero where this convergence is violated depends on f and it is not right away obvious that the convergence holds for all $f \in C_b(S)$ with probability one. We will need the following lemma.

Lemma 4.3. *If (S,d) is separable then there exists a metric e on S such that (S,e) is totally bounded, and e and d define the same topology, i.e. $e(s_n,s) \to 0$ if and only if $d(s_n,s) \to 0$.*

Proof. First of all, it is easy to check that $c = d/(1+d)$ is a metric (that defines the same topology) taking values in $[0,1]$. If $\{s_n\}$ is a dense subset of S, consider the map

$$T(s) = \big(c(s,s_n)\big)_{n\geq 1} : S \to [0,1]^{\mathbb{N}}.$$

The key point here is that the sequence $t_m \to t$ in S, or $\lim_{m\to\infty} c(t_m,t) = 0$, if and only if $\lim_{m\to\infty} c(t_m,s_n) = c(t,s_n)$ for each $n \geq 1$. In other words, $t_m \to t$ if and only if $T(t_m) \to T(t)$ in the Cartesian product $[0,1]^{\mathbb{N}}$ equipped with the product topology. This topology is compact and metrizable by the metric

$$r\big((u_n)_{n\geq 1},(v_n)_{n\geq 1}\big) = \sum_{n\geq 1} 2^{-n}|u_n - v_n|,$$

so we can now define the metric e on S by $e(s,s') = r(T(s),T(s'))$. Because of what we said above, this metric defines the same topology as d. Moreover, it is totally bounded, because the image $T(S)$ of our metric space S by the map T is totally bounded in $([0,1]^{\mathbb{N}},r)$, since this space is compact. □

Theorem 4.5 (Varadarajan). *Let (S,d) be a separable metric space. Then μ_n converges to μ weakly almost surely,*

$$\mathbb{P}\big(\omega : \mu_n(\cdot)(\omega) \to \mu \text{ weakly}\big) = 1.$$

Proof. Let e be the metric in the preceding lemma. Clearly, $C_b(S,d) = C_b(S,e)$ and, as we mentioned above, weak convergence of measures is not affected by this change of metric. If (T,e) is the completion of (S,e) then (T,e) is compact. By the Arzela-Ascoli theorem, $BL(T,e)$ is separable with respect to the d_∞ norm and, therefore, $BL(S,e)$ is also separable. Let (f_m) be a dense subset of $BL(S,e)$. Then, by the strong law of large number,

$$\int f_m\, d\mu_n = \frac{1}{n}\sum_{i=1}^{n} f_m(X_i) \to \mathbb{E} f_m(X_1) = \int f_m\, d\mu \text{ a.s.}$$

Therefore, on the set of probability one, $\int f_m\, d\mu_n \to \int f_m\, d\mu$ for all $m \geq 1$ and, since (f_m) is dense in $BL(S,e)$, on the same set of probability one, this convergence holds for all $f \in BL(S,e)$. By the portmanteau theorem, $\mu_n \to \mu$ weakly on this set of probability one. □

Next, we will introduce two metrics on the set of all probability measures on (S,d) with the Borel σ-algebra $\mathscr{B}$ and, under some mild conditions, prove that they metrize the weak convergence. For a set $A \subseteq S$, let us denote by

$$A^\varepsilon = \big\{y \in S : d(x,y) < \varepsilon \text{ for some } x \in A\big\}$$

its open ε-neighborhood. If $\mathbb{P}$ and $\mathbb{Q}$ are probability distributions on S then

$$\rho(\mathbb{P},\mathbb{Q}) = \inf\big\{\varepsilon > 0 : \mathbb{P}(A) \leq \mathbb{Q}(A^\varepsilon) + \varepsilon \text{ for all } A \in \mathscr{B}\big\}$$

is called the *Levy-Prohorov distance* between $\mathbb{P}$ and $\mathbb{Q}$.

Lemma 4.4. *ρ is a metric on the set of probability laws on $\mathscr{B}$.*

Proof. (1) First, let us show that $\rho(\mathbb{Q},\mathbb{P}) = \rho(\mathbb{P},\mathbb{Q})$. Suppose that $\rho(\mathbb{P},\mathbb{Q}) > \varepsilon$. Then there exists a set A such that $\mathbb{P}(A) > \mathbb{Q}(A^\varepsilon)+\varepsilon$. Taking complements gives

$$\mathbb{Q}(A^{\varepsilon c}) > \mathbb{P}(A^c)+\varepsilon \geq \mathbb{P}(A^{\varepsilon c\varepsilon})+\varepsilon,$$

where the last inequality follows from the fact that $A^c \supseteq A^{\varepsilon c\varepsilon}$:

$$\begin{aligned} a\in A^{\varepsilon c\varepsilon} \Longrightarrow d(a,A^{\varepsilon c}) < \varepsilon \Longrightarrow\ & d(a,b) < \varepsilon \text{ for some } b\in A^{\varepsilon c} \\ & \{\text{since } b\notin A^\varepsilon, d(b,A)\geq \varepsilon\} \\ \Longrightarrow\ & d(a,A) > 0 \Longrightarrow a\notin A \Longrightarrow a\in A^c. \end{aligned}$$

Therefore, for the set $B = A^{\varepsilon c}$, $\mathbb{Q}(B) > \mathbb{P}(B^\varepsilon)+\varepsilon$. This means that $\rho(\mathbb{Q},\mathbb{P}) > \varepsilon$ and, therefore, $\rho(\mathbb{Q},\mathbb{P}) \geq \rho(\mathbb{P},\mathbb{Q})$. By symmetry, $\rho(\mathbb{Q},\mathbb{P}) \leq \rho(\mathbb{P},\mathbb{Q})$ and, therefore, $\rho(\mathbb{Q},\mathbb{P}) = \rho(\mathbb{P},\mathbb{Q})$.

(2) Next, let us show that if $\rho(\mathbb{P},\mathbb{Q}) = 0$ then $\mathbb{P}=\mathbb{Q}$. For any set F and any $n\geq 1$,

$$\mathbb{P}(F) \leq \mathbb{Q}(F^{\frac{1}{n}})+\frac{1}{n}.$$

If F is closed then $F^{\frac{1}{n}} \downarrow F$ as $n\to\infty$ and, by the continuity of measure,

$$\mathbb{P}(F) \leq \lim_{n\to\infty}\mathbb{Q}\big(F^{\frac{1}{n}}\big) = \mathbb{Q}(F).$$

Similarly, $\mathbb{P}(F)\geq\mathbb{Q}(F)$ and, therefore, $\mathbb{P}(F)=\mathbb{Q}(F)$ for all closed sets and, therefore, for all Borel sets.

(3) Finally, let us prove the triangle inequality

$$\rho(\mathbb{P},\mathbb{T}) \leq \rho(\mathbb{P},\mathbb{Q})+\rho(\mathbb{Q},\mathbb{T}).$$

If $\rho(\mathbb{P},\mathbb{Q}) < x$ and $\rho(\mathbb{Q},\mathbb{T}) < y$ then for any set A,

$$\mathbb{P}(A) \leq \mathbb{Q}(A^x)+x \leq \mathbb{T}\big((A^x)^y\big)+y+x \leq \mathbb{T}\big(A^{x+y}\big)+x+y,$$

which means that $\rho(\mathbb{P},\mathbb{T}) \leq x+y$, and minimizing over x,y proves the triangle inequality. □

Given probability distributions $\mathbb{P},\mathbb{Q}$ on the metric space (S,d), we define the *bounded Lipschitz distance* between them by

$$\beta(\mathbb{P},\mathbb{Q}) = \sup\Big\{\Big|\int f\,d\mathbb{P}-\int f\,d\mathbb{Q}\Big| : \|f\|_{\mathrm{BL}}\leq 1\Big\}.$$

In this case, it is much easier to check that this is a metric.

Lemma 4.5. *β is a metric on the set of probability laws on $\mathscr{B}$.*

Proof. It is obvious that $\beta(\mathbb{P},\mathbb{Q}) = \beta(\mathbb{Q},\mathbb{P})$ and the triangle inequality holds. It remains to prove that if $\beta(\mathbb{P},\mathbb{Q}) = 0$ then $\mathbb{P} = \mathbb{Q}$. Given a closed set F and $U = F^c$, it is easy to see that

$$f_m(x) = md(x,F) \wedge 1 \uparrow \mathrm{I}_U$$

as $m \to \infty$. Obviously, $\|f_m\|_{\mathrm{BL}} \le m+1$ and, therefore, $\beta(\mathbb{P},\mathbb{Q}) = 0$ implies that $\int f_m d\mathbb{P} = \int f_m d\mathbb{Q}$. By the monotone converngence theorem, letting $m \to \infty$ proves that $\mathbb{P}(U) = \mathbb{Q}(U)$, so $\mathbb{P} = \mathbb{Q}$. □

Let us now show that on separable metric space, the metrics ρ and β metrize weak convergence. Before we prove this, let us recall the statement of Ulam's theorem proved in Theorem 1.5 in Section 1.1. Namely, every probability law $\mathbb{P}$ on a complete separable metric space (S,d) is tight, which means that for any $\varepsilon > 0$ there exists a compact $K \subseteq S$ such that $\mathbb{P}(S \setminus K) \le \varepsilon$.

Theorem 4.6. *If (S,d) is separable or $\mathbb{P}$ is tight then the following are equivalent:*

(1) $\mathbb{P}_n \to \mathbb{P}$.
(2) For all $f \in BL(S,d)$, $\int f d\mathbb{P}_n \to \int f d\mathbb{P}$.
(3) $\beta(\mathbb{P}_n,\mathbb{P}) \to 0$.
(4) $\rho(\mathbb{P}_n,\mathbb{P}) \to 0$.

Remark 4.1. We will prove the implications (1) $\Longrightarrow$ (2) $\Longrightarrow$ (3) $\Longrightarrow$ (4) $\Longrightarrow$ (1), and the assumption that (S,d) is separable or $\mathbb{P}$ is tight will be used only in one step, to prove (2) $\Longrightarrow$ (3).

Proof. (1) $\Longrightarrow$ (2). This follow by definition, since $BL(S,d) \subseteq C_b(S)$.

(3) $\Longrightarrow$ (4). In fact, we will prove that

$$\rho(\mathbb{P}_n,\mathbb{P}) \le 2\sqrt{\beta(\mathbb{P}_n,\mathbb{P})}. \tag{4.2}$$

Given a Borel set $A \subseteq S$, consider the function

$$f(x) = 0 \vee \Big(1 - \frac{1}{\varepsilon}d(x,A)\Big), \text{ so that } \mathrm{I}_A \le f \le \mathrm{I}_{A^\varepsilon}.$$

Obviously, $\|f\|_{\mathrm{BL}} \le 1 + \varepsilon^{-1}$ and we can write

$$\begin{aligned}
\mathbb{P}_n(A) &\le \int f d\mathbb{P}_n = \int f d\mathbb{P} + \Big(\int f d\mathbb{P}_n - \int f d\mathbb{P}\Big) \\
&\le \mathbb{P}(A^\varepsilon) + (1+\varepsilon^{-1})\sup\Big\{\Big|\int f d\mathbb{P}_n - \int f d\mathbb{P}\Big| : \|f\|_{\mathrm{BL}} \le 1\Big\} \\
&= \mathbb{P}(A^\varepsilon) + (1+\varepsilon^{-1})\beta(\mathbb{P}_n,\mathbb{P}) \le \mathbb{P}(A^\delta) + \delta,
\end{aligned}$$

where $\delta = \max(\varepsilon, (1+\varepsilon^{-1})\beta(\mathbb{P}_n, \mathbb{P}))$. This implies that $\rho(\mathbb{P}_n, \mathbb{P}) \le \delta$. Since ε is arbitrary we can minimize $\delta = \delta(\varepsilon)$ over ε. If we take $\varepsilon = \sqrt{\beta}$ then $\delta = \max(\sqrt{\beta}, \beta + \sqrt{\beta}) = \beta + \sqrt{\beta}$ and

$$\beta \le 1 \Longrightarrow \rho \le 2\sqrt{\beta};\ \ \beta \ge 1 \Longrightarrow \rho \le 1 \le 2\sqrt{\beta}.$$

(4) $\Longrightarrow$ (1). Suppose that $\rho(\mathbb{P}_n, \mathbb{P}) \to 0$, which means that there exists a sequence $\varepsilon_n \downarrow 0$ such that

$$\mathbb{P}_n(A) \le \mathbb{P}(A^{\varepsilon_n}) + \varepsilon_n \ \text{ for all measurable } A \subseteq S.$$

If A is closed, then $\bigcap_{n\ge 1} A^{\varepsilon_n} = A$ and, by the continuity of measure,

$$\limsup_{n\to\infty} \mathbb{P}_n(A) \le \limsup_{n\to\infty} \Big(\mathbb{P}(A^{\varepsilon_n}) + \varepsilon_n\Big) = \mathbb{P}(A).$$

By the Portmanteau Theorem, $\mathbb{P}_n \to \mathbb{P}$.

(2) $\Longrightarrow$ (3). First, let us consider the case when (S,d) is complete. Then, by Ulam's theorem, again, we can find a compact K such that $\mathbb{P}(S \setminus K) \le \varepsilon$. If (S,d) is not separable then we assume that $\mathbb{P}$ is tight and, by definition, we can again find a compact K such that $\mathbb{P}(S \setminus K) \le \varepsilon$. If we consider the function

$$f(x) = 0 \vee \Big(1 - \frac{1}{\varepsilon} d(x,K)\Big), \ \text{ so that } \|f\|_{\mathrm{BL}} \le 1 + \frac{1}{\varepsilon},$$

then

$$\mathbb{P}_n(K^{\varepsilon}) \ge \int f\, d\mathbb{P}_n \stackrel{n\to\infty}{\longrightarrow} \int f\, d\mathbb{P} \ge \mathbb{P}(K) \ge 1 - \varepsilon,$$

which implies that for n large enough, $\mathbb{P}_n(K^{\varepsilon}) \ge 1 - 2\varepsilon$. Let

$$B = \big\{f : \|f\|_{\mathrm{BL}(S,d)} \le 1\big\},\ B_K = \big\{f\big|_K : f \in B\big\} \subseteq C(K),$$

where $f\big|_K$ denotes the restriction of f to K. Since K is compact, by the Arzela-Ascoli theorem, B_K is totally bounded with respect to d_∞ and, given $\varepsilon > 0$, we can find $k \ge 1$and $f_1, \ldots, f_k \in B$ such that, for all $f \in B$,

$$\sup_{x\in K} |f(x) - f_j(x)| \le \varepsilon \ \text{ for some } j \le k.$$

This uniform approximation can also be extended to K^{ε}. Namely, for any $x \in K^{\varepsilon}$ take $y \in K$ such that $d(x,y) \le \varepsilon$. Then

$$\begin{aligned} |f(x) - f_j(x)| &\le |f(x) - f(y)| + |f(y) - f_j(y)| + |f_j(y) - f_j(x)| \\ &\le \|f\|_{\mathrm{L}} d(x,y) + \varepsilon + \|f_j\|_{\mathrm{L}} d(x,y) \le 3\varepsilon. \end{aligned}$$

Therefore, for large enough n, for any $f \in B$,

$$
\begin{aligned}
\left|\int f\,d\mathbb{P}_n - \int f\,d\mathbb{P}\right| &\le \left|\int_{K^\varepsilon} f\,d\mathbb{P}_n - \int_{K^\varepsilon} f\,d\mathbb{P}\right| + \|f\|_\infty\big(\mathbb{P}_n(K^{\varepsilon c}) + \mathbb{P}(K^{\varepsilon c})\big)\\
&\le \left|\int_{K^\varepsilon} f\,d\mathbb{P}_n - \int_{K^\varepsilon} f\,d\mathbb{P}\right| + 2\varepsilon + \varepsilon\\
&\le \left|\int_{K^\varepsilon} f_j\,d\mathbb{P}_n - \int_{K^\varepsilon} f_j\,d\mathbb{P}\right| + 3\varepsilon + 3\varepsilon + 2\varepsilon + \varepsilon\\
&\le \left|\int f_j\,d\mathbb{P}_n - \int f_j\,d\mathbb{P}\right| + 3\varepsilon + 3\varepsilon + 2(2\varepsilon + \varepsilon)\\
&\le \max_{1\le j\le k}\left|\int f_j\,d\mathbb{P}_n - \int f_j\,d\mathbb{P}\right| + 12\varepsilon.
\end{aligned}
$$

Finally,

$$
\beta(\mathbb{P}_n,\mathbb{P}) = \sup_{f\in B}\left|\int f d\mathbb{P}_n - \int f d\mathbb{P}\right| \le \max_{1\le j\le k}\left|\int f_j d\mathbb{P}_n - \int f_j d\mathbb{P}\right| + 12\varepsilon
$$

and, using the assumption (2), $\limsup_{n\to\infty}\beta(\mathbb{P}_n,\mathbb{P}) \le 12\varepsilon$. Letting $\varepsilon \to 0$ finishes the proof.

Now, suppose that (S,d) is separable but not complete. Let (T,d) be the completion of (S,d). We showed in the previous section that every function $f \in BL(S,d)$ can be extended to $\hat{f}$ on (T,d) preserving the norm $\|f\|_{\mathrm{BL}}$. This extension is obviously unique, since S is dense in T, so there is one-to-one correspondence $f \leftrightarrow \hat{f}$ between the unit balls of $BL(S,d)$ and $BL(T,d)$. Next, we can also view the measure $\mathbb{P}$ as a probability measure on (T,d) as follows. If for any Borel set A in (T,d) we let $\hat{\mathbb{P}}(A) = \mathbb{P}(A\cap S)$ (we define $\hat{\mathbb{P}}_n$ similarly) then $\hat{\mathbb{P}}$ is a probability measure on the Borel σ-algebra on (T,d). Note that S might not be a measurable set in its completion (because of this, there is no natural correspondence between measures on S and measures on its completion, T) but one can easily check that the sets $A\cap S$ are Borel in (S,d) (this is because open sets in S are intersections of open sets in T with S). One can also easily see that, with this definition,

$$
\int_S f\,d\mathbb{P} = \int_T \hat{f}\,d\hat{\mathbb{P}},
$$

which means that both statement in (2) and (3) hold simultaneously on (S,d) and (T,d). This proves that (2) $\Rightarrow$ (3) on all separable metric spaces. □

Convergence and uniform tightness. Next, we will make a connection between the above metrics and uniform tightness. Recall that a sequence of laws $(\mathbb{P}_n)$ is *uniformly tight* if, for each $\varepsilon > 0$, there exists a compact set $K \subseteq S$ such that $\mathbb{P}_n(S\setminus K) \le \varepsilon$ for all $n \ge 1$. First, we will show that, in some sense, uniform tightness is necessary for convergence of laws.

Theorem 4.7. *If $\mathbb{P}_n \to \mathbb{P}_0$ weakly and each $\mathbb{P}_n$ is tight for $n \ge 0$, then $(\mathbb{P}_n)_{n\ge 0}$ is uniformly tight.*

In particular, by Ulam's theorem, any convergent sequence of laws on a complete separable metric space is uniformly tight.

Proof. Since $\mathbb{P}_n \to \mathbb{P}_0$ and $\mathbb{P}_0$ is tight, by Theorem 4.6, the Levy-Prohorov metric $\rho(\mathbb{P}_n, \mathbb{P}_0) \to 0$. Given $\varepsilon > 0$, let us take a compact K such that $\mathbb{P}_0(K_0) > 1 - \varepsilon/2$. By the definition of ρ,

$$1 - \frac{\varepsilon}{2} < \mathbb{P}_0(K_0) \leq \mathbb{P}_n\big(K_0^{\rho(\mathbb{P}_n, \mathbb{P}_0) + \frac{1}{n}}\big) + \rho(\mathbb{P}_n, \mathbb{P}_0) + \frac{1}{n}$$

and, therefore, if we denote $a(n) := \rho(\mathbb{P}_n, \mathbb{P}_0) + n^{-1}$ then $\mathbb{P}_n(K_0^{a(n)}) > 1 - \varepsilon$ for large enough $n \geq N$.

A probability measure is always closed regular, so any measurable set A can be approximated by a closed subset F. Since $\mathbb{P}_n$ is tight, we can choose a compact of measure close to one, and intersecting it with the closed subset F, we can approximate any set A by a compact subset. Therefore, for $n \geq N$, there exists a compact $K_n \subseteq K_0^{a(n)}$ such that $\mathbb{P}_n(K_n) > 1 - \varepsilon$. For $1 \leq n < N$, since $\mathbb{P}_n$ is tight, we can also find a compact K_n such that $\mathbb{P}_n(K_n) > 1 - \varepsilon$. If we now take $L = \cup_{n \geq 0} K_n$ then

$$\mathbb{P}_n(L) \geq \mathbb{P}_n(K_n) > 1 - \varepsilon \text{ for all } n \geq 0.$$

It remains to show that L is compact. Consider a sequence (x_n) in L. There are two possibilities. First, if there exists an infinite subsequence $(x_{n(k)})$ that belongs to one of the compacts K_j then it has a converging subsubsequence in K_j and as a result in L. If not, then there exists a subsequence $(x_{n(k)})$ such that $x_{n(k)} \in K_{m(k)}$ and $m(k) \to \infty$ as $k \to \infty$. Since $K_{m(k)} \subseteq K_0^{a(m(k))}$ there exists $y_k \in K_0$ such that $d(x_{n(k)}, y_k) \leq a(m(k))$. Since K_0 is compact, the sequence $y_k \in K_0$ has a converging subsequence $y_{k(r)} \to y \in K_0$ which implies that $d(x_{n(k(r))}, y) \to 0$, i.e. $x_{n(k(r))} \to y \in L$. Therefore, L is compact. □

Next, on complete separable metric spaces, we will complement the Selection Theorem by showing how uniform tightness can be expressed in the above metrics.

Theorem 4.8. *Let (S,d) be a complete separable metric space and $\mathscr{P}$ be a subset of probability laws on S. Then the following are equivalent.*

(1) $\mathscr{P}$ is uniformly tight.

(2) For any sequence $\mathbb{P}_n \in \mathscr{P}$ there exists a converging subsequence $\mathbb{P}_{n(k)} \to \mathbb{P}$ where $\mathbb{P}$ is a law on S.

(3) $\mathscr{P}$ has the compact closure on the space of probability laws equipped with the Levy-Prohorov or bounded Lipschitz metrics ρ or β.

(4) $\mathscr{P}$ is totally bounded with respect to ρ or β.

Remark 4.2. In other words, we are showing that, on complete separable metric spaces, total boundedness on the space of probability measures is equivalent to uniform tightness. The rest is just basic properties of metric space. Also, implications (1) $\Longrightarrow$ (2) $\Longrightarrow$ (3) $\Longrightarrow$ (4) hold without the completeness assumption and the only implication where completeness will be used is (4) $\Longrightarrow$ (1).

Proof. (1) $\Longrightarrow$ (2). Any sequence $\mathbb{P}_n \in \mathscr{P}$ is uniformly tight and, by the Selection Theorem, there exists a converging subsequence.

(2) $\Longrightarrow$ (3). Since (S,d) is separable, by Theorem 4.6, $\mathbb{P}_n \to \mathbb{P}$ if and only if $\rho(\mathbb{P}_n,\mathbb{P})$ or $\beta(\mathbb{P}_n,\mathbb{P}) \to 0$. Every sequence in the closure $\overline{\mathscr{P}}$ can be approximated by a sequence in $\mathscr{P}$. That sequence has a converging subsequence that, obviously, converges to an element in $\overline{\mathscr{P}}$, which means that the closure of $\mathscr{P}$ is compact.

(3) $\Longrightarrow$ (4). Compact sets are totally bounded and, therefore, if the closure $\overline{\mathscr{P}}$ is compact, the set $\mathscr{P}$ is totally bounded.

(4) $\Longrightarrow$ (1). Since $\rho \le 2\sqrt{\beta}$, we will only deal with ρ. For any $\varepsilon > 0$, there exists a finite subset $\mathscr{P}_0 \subseteq \mathscr{P}$ such that $\mathscr{P} \subseteq \mathscr{P}_0^{\varepsilon}$. Since (S,d) is complete and separable, by Ulam's theorem, for each $\mathbb{Q} \in \mathscr{P}_0$ there exists a compact $K_{\mathbb{Q}}$ such that $\mathbb{Q}(K_{\mathbb{Q}}) > 1-\varepsilon$. Therefore,

$$K = \bigcup_{\mathbb{Q}\in\mathscr{P}_0} K_{\mathbb{Q}} \text{ is a compact and } \mathbb{Q}(K) > 1-\varepsilon \text{ for all } \mathbb{Q} \in \mathscr{P}_0.$$

Let F be a finite set such that $K \subseteq F^{\varepsilon}$ (here we will denote by F^{ε} the closed ε-neighborhood of F). Since $\mathscr{P} \subseteq \mathscr{P}_0^{\varepsilon}$, for any $\mathbb{P} \in \mathscr{P}$ there exists $\mathbb{Q} \in \mathscr{P}_0$ such that $\rho(\mathbb{P},\mathbb{Q}) < \varepsilon$ and, therefore,

$$1-\varepsilon \le \mathbb{Q}(K) \le \mathbb{Q}(F^{\varepsilon}) \le \mathbb{P}(F^{2\varepsilon}) + \varepsilon.$$

Thus, $1-2\varepsilon \le \mathbb{P}(F^{2\varepsilon})$ for all $\mathbb{P} \in \mathscr{P}$. Given $\delta > 0$, take $\varepsilon_m = \delta/2^{m+1}$ and find F_m as above, i.e.

$$1 - \frac{\delta}{2^m} \le \mathbb{P}\big(F_m^{\delta/2^m}\big).$$

Then

$$\mathbb{P}\Big(\bigcap_{m\ge1} F_m^{\delta/2^m}\Big) \ge 1 - \sum_{m\ge1}\frac{\delta}{2^m} = 1-\delta.$$

Finally, $L = \bigcap_{m\geq 1} F_m^{\delta/2^m}$ is compact because it is closed and totally bounded by construction, and S is complete. □

Theorem 4.9 (Prokhorov's theorem). *The set of probability laws on a complete separable metric space is complete with respect to the metrics ρ and β.*

Proof. If a sequence of laws is Cauchy with respect to ρ or β then it is totally bounded and, by the previous theorem, it has a converging subsequence. Obviously, a Cauchy sequence will converge to the same limit. □

Exercise 4.2.1. Prove that the space of probability laws $\mathrm{Pr}(S)$ on a compact metric space (S,d) is compact with respect to the metrics ρ and β.

Exercise 4.2.2. If (S,d) is separable, prove that $\mathrm{Pr}(S)$ is also separable with respect to the metrics ρ and β. Hint: think about probability measures concentrated on finite subsets of a dense countable set in S, and use metric β.

Exercise 4.2.3. Let (S,d) be a metric space and $\mathscr{B}$ its Borel σ-algebra. Let $\mathrm{Pr}(S)$ be the set of all probability measures on $(S,\mathscr{B})$ equipped with the topology of weak convergence. Let $\mathscr{F}$ be the Borel σ-algebra on $\mathrm{Pr}(S)$ generated by open sets. Consider the map $T : \mathrm{Pr}(S)\times\mathscr{B}\to[0,1]$ defined by $T(\mathbb{P},A)=\mathbb{P}(A)$. Prove that for any fixed $A\in\mathscr{B}$, the function $T(\cdot,A) : \mathrm{Pr}(S)\to[0,1]$ is $\mathscr{F}$-measurable. *Hint:* First, prove this for open sets $A\subseteq S$.

Exercise 4.2.4. Let $\mathbb{P},\mathbb{Q}$ be two laws on $\mathbb{R}$ with distribution functions F,G respectively. Let

$$\lambda(\mathbb{P},\mathbb{Q}) := \inf\{\varepsilon>0 : F(x-\varepsilon)-\varepsilon\leq G(x)\leq F(x+\varepsilon)+\varepsilon \text{ for all } x\}$$

– Levy's metric. (a) Show that λ is a metric metrizing convergence of laws on $\mathbb{R}$. (b) Show that $\lambda\leq\rho$, but that there exist laws $\mathbb{P}_n,\mathbb{Q}_n$ with $\lambda(\mathbb{P}_n,\mathbb{Q}_n)\to 0$ while $\rho(\mathbb{P}_n,\mathbb{Q}_n)\not\to 0$.

Exercise 4.2.5. Define a specific finite set of laws F on $[0,1]$ such that for every law $\mathbb{P}$ on $[0,1]$ there exists $\mathbb{Q}\in F$ with $\rho(\mathbb{P},\mathbb{Q})<0.1$, where ρ is Prokhorov's metric. *Hint:* reduce the problem to metric β.

Exercise 4.2.6. Let X_j be i.i.d. $N(0,1)$ random variables. Let H be a Hilbert space with orthonormal basis $\{e_j\}_{j\geq 1}$. Let $X=\sum_{j\geq 1} X_j e_j/j$. For any $\varepsilon>0$, find a compact K such that $\mathbb{P}(X\in K)>1-\varepsilon$.

4.3 Strassen's theorem. Relationships between metrics

Metric for convergence in probability. Let $(\Omega, \mathscr{B}, \mathbb{P})$ be a probability space, (S,d) - a separable metric space and $X,Y:\Omega\to S$ - random variables with values in S. The quantity

$$\alpha(X,Y)=\inf\{\varepsilon\geq 0:\mathbb{P}(d(X,Y)>\varepsilon)\leq\varepsilon\}$$

is called the *Ky Fan metric* on the set $\mathscr{L}^0(\Omega,S)$ of classes of equivalences of such random variables, where two random variables are equivalent if they are equal almost surely. If we take a sequence

$$\varepsilon_k\downarrow\alpha=\alpha(X,Y)$$

then $\mathbb{P}(d(X,Y)>\varepsilon_k)\leq\varepsilon_k$ and, since $\mathrm{I}(d(X,Y)>\varepsilon_k)\uparrow\mathrm{I}(d(X,Y)>\alpha)$, by the monotone convergence theorem, $\mathbb{P}(d(X,Y)>\alpha)\leq\alpha$. Thus, the infimum in the definition of $\alpha(X,Y)$ is attained.

Lemma 4.6. *The Ky Fan metric α on $\mathscr{L}^0(\Omega,S)$ metrizes convergence in probability.*

Proof. First of all, clearly, $\alpha(X,Y)=0$ if and only $X=Y$ almost surely. To prove the triangle inequality,

$$\begin{aligned}&\mathbb{P}\big(d(X,Z)>\alpha(X,Y)+\alpha(Y,Z)\big)\\&\leq\mathbb{P}\big(d(X,Y)>\alpha(X,Y)\big)+\mathbb{P}\big(d(Y,Z)>\alpha(Y,Z)\big)\\&\leq\alpha(X,Y)+\alpha(Y,Z),\end{aligned}$$

so that $\alpha(X,Z)\leq\alpha(X,Y)+\alpha(Y,Z)$. This proves that α is a metric. Next, if $\alpha_n=\alpha(X_n,X)\to 0$ then, for any $\varepsilon>0$ and large enough n such that $\alpha_n<\varepsilon$,

$$\mathbb{P}(d(X_n,X)>\varepsilon)\leq\mathbb{P}(d(X_n,X)>\alpha_n)\leq\alpha_n\to 0.$$

Conversely, if $X_n\to X$ in probability then, for any $m\geq 1$ and large enough $n\geq n(m)$,

$$\mathbb{P}\Big(d(X_n,X)>\frac{1}{m}\Big)\leq\frac{1}{m},$$

which means that $\alpha_n\leq 1/m$ and $\alpha_n\to 0$. □

Lemma 4.7. *For $X,Y\in\mathscr{L}^0(\Omega,S)$, the Levy-Prohorov metric ρ satisfies*

$$\rho(\mathscr{L}(X),\mathscr{L}(Y))\leq\alpha(X,Y).$$

Proof. Take $\varepsilon>\alpha(X,Y)$ so that $\mathbb{P}(d(X,Y)\geq\varepsilon)\leq\varepsilon$. For any measurable $A\subseteq S$,

$$\mathbb{P}(X\in A)=\mathbb{P}(X\in A,d(X,Y)<\varepsilon)+\mathbb{P}(X\in A,d(X,Y)\geq\varepsilon)\leq\mathbb{P}(Y\in A^\varepsilon)+\varepsilon$$

which means that $\rho(\mathscr{L}(X),\mathscr{L}(Y)) \le \varepsilon$. Letting $\varepsilon \downarrow \alpha(X,Y)$ proves the result. □

We will now prove that, in some sense, the opposite is also true. Let (S,d) be a metric space and $\mathbb{P},\mathbb{Q}$ be probability laws on S. Suppose that these laws are close in the Levy-Prohorov metric ρ. Can we construct random variables s_1 and s_2, with laws $\mathbb{P}$ and $\mathbb{Q}$, that are defined on the same probability space and are close to each other in the Ky Fan metric α? We will construct a distribution on the product space $S\times S$ such that the coordinates s_1 and s_2 have marginal distributions $\mathbb{P}$ and $\mathbb{Q}$ and the distribution is concentrated on a neighbourhood of the diagonal $s_1 = s_2$, and the size of the neighbourhood is controlled by $\rho(\mathbb{P},\mathbb{Q})$.

The following result will be a key tool in the proof of the main result of this section. Consider two sets X and Y. Given a subset $K\subseteq X\times Y$ and $A\subseteq X$ we define a K-image of A by

$$A^K = \big\{y\in Y : \exists x\in A, (x,y)\in K\big\}.$$

A *K-matching* f of X into Y is a one-to-one function (injection) $f: X\to Y$ such that $(x,f(x))\in K$. We will need the following well-known matching theorem.

Theorem 4.10 (Hall's marriage theorem). *If X,Y are finite and, for all $A\subseteq X$,*

$$\operatorname{card}(A^K) \ge \operatorname{card}(A) \tag{4.3}$$

then there exists a K-matching f of X into Y.

Proof. We will prove the result by induction on $m = \operatorname{card}(X)$. The case of $m=1$ is obvious. For each $x\in X$ there exists $y\in Y$ such that $(x,y)\in K$. If there is a matching f of $X\setminus\{x\}$ into $Y\setminus\{y\}$ then defining $f(x)=y$ extends f to X. If not, then since $\operatorname{card}(X\setminus\{x\}) < m$, by induction assumption, condition (4.3) is violated, i.e. there exists a set $A\subseteq X\setminus\{x\}$ such that $\operatorname{card}(A^K\setminus\{y\}) < \operatorname{card}(A)$. But because we also know that $\operatorname{card}(A^K)\ge\operatorname{card}(A)$ this implies that $\operatorname{card}(A^K)=\operatorname{card}(A)$. Since $\operatorname{card}(A) < m$, by induction there exists a matching of A onto A^K. If there is a matching of $X\setminus A$ into $Y\setminus A^K$ we can combine it with a matching of A and A^K. If not then again, by the induction assumption, there exists $D\subseteq X\setminus A$ such that $\operatorname{card}(D^K\setminus A^K) < \operatorname{card}(D)$. But then

$$\begin{aligned}\operatorname{card}\big((A\cup D)^K\big) &= \operatorname{card}(D^K\setminus A^K) + \operatorname{card}(A^K)\\ &< \operatorname{card}(D) + \operatorname{card}(A) = \operatorname{card}(D\cup A),\end{aligned}$$

which contradicts the assumption (4.3). □

Theorem 4.11 (Strassen). *Suppose that (S,d) is a separable metric space and $\alpha,\beta>0$. Suppose the laws $\mathbb{P}$ and $\mathbb{Q}$ are such that, for all measurable sets $F\subseteq S$,*

$$\mathbb{P}(F) \le \mathbb{Q}(F^\alpha) + \beta \tag{4.4}$$

Then for any $\varepsilon>0$ there exist two non-negative measures η,γ on $S\times S$ such that

1. *$\mu = \eta + \gamma$ is a law on $S \times S$ with marginals $\mathbb{P}$ and $\mathbb{Q}$.*
2. $\eta(d(x,y) > \alpha + \varepsilon) = 0$.
3. $\gamma(S \times S) \leq \beta + \varepsilon$.
4. *μ is a finite sum of product measures.*

Remark 4.3. In the above statement, it is enough to assume that (4.4) holds only for closed sets or only for open sets F; moreover, one can replace an open α-neighbourhood $F^\alpha = \{s \in S : d(s,F) < \alpha\}$ by a closed α-neighbourhood $F^{\alpha]} = \{s \in S : d(s,F) \leq \alpha\}$. This is because the set F^ε is open and $F^{\varepsilon]}$ is closed, and, for example, the condition (4.4) for closed sets implies $\mathbb{P}(F) \leq \mathbb{P}(F^{\varepsilon]}) \leq \mathbb{Q}((F^{\varepsilon]})^\alpha) + \beta \leq \mathbb{Q}(F^{\alpha+2\varepsilon}) + \beta$ for all measurable sets F, which simply replaces α by $\alpha + 2\varepsilon$ in (4.4).

Remark 4.4. Condition (4.4) is a relaxation of the definition of the Levy-Prohorov metric, since one can take different $\alpha, \beta > \rho(\mathbb{P},\mathbb{Q})$. Conditions 1 – 3 mean that we can construct a measure μ on $S \times S$ such that coordinates x,y have marginal distributions $\mathbb{P},\mathbb{Q}$, concentrated within distance $\alpha + \varepsilon$ of each other (condition 2) except for the set of measure at most $\beta + \varepsilon$ (condition 3).

Proof. Case A. The proof will proceed in several steps. We will start with the simplest case which, however, contains the main idea. Given small $\varepsilon > 0$, take $n \geq 1$ such that $n\varepsilon > 1$. Suppose that the laws $\mathbb{P},\mathbb{Q}$ are uniform on finite subsets $M,N \subseteq S$ of equal cardinality,

$$\mathrm{card}(M) = \mathrm{card}(N) = n,\ \mathbb{P}(x) = \mathbb{Q}(y) = \frac{1}{n} < \varepsilon,\ x \in M, y \in N.$$

Using the condition (4.4), we would like to match as many points from M and N as possible, but only points that are within distance α from each other. To use the matching theorem, we will introduce some auxiliary sets U and V that are not too big, with size controlled by the parameter β, and the union of these sets with M and N satisfies a certain matching condition.

Take integer k such that $\beta n \leq k < (\beta + \varepsilon)n$. Let us take sets U and V such that $k = \mathrm{card}(U) = \mathrm{card}(V)$ and U,V are disjoint from M,N. Define

$$X = M \cup U,\ Y = N \cup V.$$

Let us define a subset $K \subseteq X \times Y$ such that $(x,y) \in K$ if and only if one of the following holds:

(i) $x \in U$,
(ii) $y \in V$,
(iii) $d(x,y) \leq \alpha + \varepsilon$ if $x \in M, y \in N$.

This means that small auxiliary sets can be matched with any points, but only close points, $d(x,y) \leq \alpha + \varepsilon$, can be matched in the main sets M and N. Consider a set

$A \subseteq X$ with cardinality $\mathrm{card}(A) = r$. If $A \not\subseteq M$ then by (i), $A^K = Y$ and $\mathrm{card}(A^K) \geq r$. Suppose now that $A \subseteq M$ and, again, we would like to show that $\mathrm{card}(A^K) \geq r$. Using (4.4) and the fact that, by (iii), $A^\alpha \cap N \subseteq A^K \cap N$, we can write

$$\frac{r}{n} = \mathbb{P}(A) \leq \mathbb{Q}(A^\alpha) + \beta = \frac{1}{n}\mathrm{card}(A^\alpha \cap N) + \beta \leq \frac{1}{n}\mathrm{card}(A^K \cap N) + \beta.$$

Therefore,

$$r = \mathrm{card}(A) \leq n\beta + \mathrm{card}(A^K \cap N) \leq k + \mathrm{card}(A^K \cap N) = \mathrm{card}(A^K),$$

since $k = \mathrm{card}(V)$ and $A^K = V \cup (A^K \cap N)$. By the matching theorem, there exists a K-matching f of X into Y. Let

$$T = \{x \in M : f(x) \in N\},$$

i.e. points in M that are matched with points in N at distance $d(x,y) \leq \alpha + \varepsilon$. Clearly, $\mathrm{card}(T) \geq n - k$, since at most k points can be matched with a point in V, and for $x \in T$, by (iii), $d(x, f(x)) \leq \alpha + \varepsilon$. For $x \in M \setminus T$, redefine $f(x)$ to match each x with a different points in N that are not matched with points in T. This defines a matching of M onto N. We define measures η and γ by

$$\eta = \frac{1}{n}\sum_{x \in T} \delta_{(x,f(x))},\quad \gamma = \frac{1}{n}\sum_{x \in M\setminus T} \delta_{(x,f(x))},$$

and let $\mu = \eta + \gamma$. First of all, obviously, μ has marginals $\mathbb{P}$ and $\mathbb{Q}$ because each point in M or N appears in the sum $\eta + \gamma$ only once with the weight $1/n$. Also,

$$\eta\big(d(x, f(x)) > \alpha + \varepsilon\big) = 0,\ \ \gamma(S \times S) \leq \frac{\mathrm{card}(M \setminus T)}{n} \leq \frac{k}{n} < \beta + \varepsilon. \tag{4.5}$$

Finally, both η and γ are finite sums of point masses, which are product measures of point masses.

Case B. Suppose now that $\mathbb{P}$ and $\mathbb{Q}$ are concentrated on finitely many points with rational probabilities. Then we can artificially split all points into "smaller" points of equal probabilities as follows. Let n be such that $n\varepsilon > 1$ and

$$n\mathbb{P}(x), n\mathbb{Q}(x) \in J = \{1, 2, \ldots, n\}.$$

Define a discrete metric on J by $f(i,j) = \frac{\varepsilon}{2}\mathrm{I}(i \neq j)$ and define a metric on $S \times J$ by

$$e\big((x,i),(y,j)\big) = d(x,y) + f(i,j).$$

Define a measure $\mathbb{P}'$ on $S \times J$ as follows. If $\mathbb{P}(x) = \frac{j}{n}$ then

$$\mathbb{P}'\big((x,i)\big) = \frac{1}{n} \ \text{ for } \ i = 1, \ldots, j.$$

Define $\mathbb{Q}'$ similarly. Let us check that the laws $\mathbb{P}',\mathbb{Q}'$ satisfy the assumptions of Case A with $\alpha+\varepsilon$ instead of α. Given a set $F\subseteq S\times J$, define

$$F_1=\big\{x\in S\,\big|\,(x,j)\in F \text{ for some } j\big\}.$$

Using (4.4),

$$\mathbb{P}'(F)\le\mathbb{P}(F_1)\le\mathbb{Q}(F_1^{\alpha})+\beta\le\mathbb{Q}'(F^{\alpha+\varepsilon})+\beta,$$

because $f(i,j)<\varepsilon$. By Case A in (4.5), we can construct $\mu'=\eta'+\gamma'$ with marginals $\mathbb{P}'$ and $\mathbb{Q}'$ such that

$$\eta'\big(e((x,i),(y,j))>\alpha+2\varepsilon\big)=0,\ \gamma'\big((S\times J)\times(S\times J)\big)<\beta+\varepsilon.$$

Let μ,η, and γ be the projections of μ',η', and γ' back onto $S\times S$ by the map $((x,i),(y,j))\to(x,y)$. Then, clearly, $\mu=\eta+\gamma$, μ has marginals $\mathbb{P}$ and $\mathbb{Q}$ and $\gamma(S\times S)<\beta+\varepsilon$. Since

$$e\big((x,i),(y,j)\big)=d(x,y)+f(i,j)\ge d(x,y),$$

we get

$$\eta\big(d(x,y)>\alpha+2\varepsilon\big)\le\eta'\big(e((x,i),(y,j))>\alpha+2\varepsilon\big)=0.$$

Finally, μ is obviously a finite sum of product measures. Of course, we can replace ε by $\varepsilon/2$.

Case C. (General case) Let $\mathbb{P},\mathbb{Q}$ be laws on a separable metric space (S,d). Let A be a maximal set such that for all $x,y\in A$ the distance $d(x,y)\ge\varepsilon$. Such set A is called an *ε-packing* and it is countable, because S is separable, so $A=\{x_i\}_{i\ge1}$. Since A is maximal, for each $x\in S$ there exists $y\in A$ such that $d(x,y)<\varepsilon$, otherwise, we could add x to the set A. Let us create a partition of S using ε-balls around the points x_i :

$$B_1=\{x\in S:d(x,x_1)<\varepsilon\},\ B_2=\{d(x,x_2)<\varepsilon\}\setminus B_1$$

and, iteratively for $k\ge2$,

$$B_k=\{d(x,x_k)<\varepsilon\}\setminus(B_1\cup\cdots\cup B_{k-1}).$$

$\{B_k\}_{k\ge1}$ is a partition of S. Let us discretize measures $\mathbb{P}$ and $\mathbb{Q}$ by projecting them onto $\{x_i\}_{i\ge1}$:

$$\mathbb{P}'(x_k)=\mathbb{P}(B_k),\ \mathbb{Q}'(x_k)=\mathbb{Q}(B_k).$$

Consider any set $F\subseteq S$. For any point $x\in F$, if $x\in B_k$ then $d(x,x_k)<\varepsilon$, i.e. $x_k\in F^{\varepsilon}$ and, therefore,

$$\mathbb{P}(F)\le\mathbb{P}'(F^{\varepsilon}).$$

Also, if $x_k\in F$ then $B_k\subseteq F^{\varepsilon}$ and, therefore,

$$\mathbb{P}'(F)\le\mathbb{P}(F^{\varepsilon}).$$

To apply Case B, we need to approximate $\mathbb{P}'$ by a measure on a finite number of points with rationals probabilities. For large enough $n \geq 1$, let

$$\mathbb{P}''(x_k) = \frac{\lfloor n\mathbb{P}'(x_k)\rfloor}{n}.$$

Clearly, as $n \to \infty$, $\mathbb{P}''(x_k) \uparrow \mathbb{P}'(x_k)$. Notice that only a finite number of points carry non-zero weights $\mathbb{P}''(x_k) > 0$. Let x_0 be some auxiliary point outside of the sequence $\{x_k\}$ and let us assign to it the remaining probability

$$\mathbb{P}''(x_0) = 1 - \sum_{k\geq 1} \mathbb{P}''(x_k).$$

If we take n large enough so that $\mathbb{P}''(x_0) < \varepsilon/2$ then

$$\sum_{k\geq 0} |\mathbb{P}''(x_k) - \mathbb{P}'(x_k)| \leq \varepsilon.$$

All the relations above also hold true for $\mathbb{Q}, \mathbb{Q}'$ and $\mathbb{Q}''$ that are defined similarly, and we can assume that the same point x_0 plays a role of the auxiliary point for both $\mathbb{P}''$ and $\mathbb{Q}''$. Given $F \subseteq S$, we can write

$$\begin{aligned}\mathbb{P}''(F) &\leq \mathbb{P}'(F) + \varepsilon \leq \mathbb{P}(F^{\varepsilon}) + \varepsilon \leq \mathbb{Q}(F^{\varepsilon+\alpha}) + \beta + \varepsilon \\ &\leq \mathbb{Q}'(F^{\alpha+2\varepsilon}) + \beta + \varepsilon \leq \mathbb{Q}''(F^{\alpha+2\varepsilon}) + \beta + 2\varepsilon.\end{aligned}$$

By Case B, there exists a decomposition $\mu'' = \eta'' + \gamma''$ on $S \times S$ with marginals $\mathbb{P}''$ and $\mathbb{Q}''$ such that

$$\eta''\big(d(x,y) > \alpha + 3\varepsilon\big) = 0,\ \gamma''(S\times S) \leq \beta + 3\varepsilon.$$

Let us also "move" the points (x_0, x_i) and (x_i, x_0) for $i \geq 0$ into the support of γ'': since the total weight of these points is at most ε, the total weight of γ'' does not increase much:

$$\gamma''(S\times S) \leq \beta + 5\varepsilon.$$

It remains to "redistribute" these measures from the sequence $\{x_i\}_{i\geq 0}$ to the entire space S in a way that recovers marginal distributions $\mathbb{P}$ and $\mathbb{Q}$ and so that not much accuracy is lost. Define a sequence of measures on S by

$$\mathbb{P}_i(C) = \frac{\mathbb{P}(C\cap B_i)}{\mathbb{P}(B_i)} \text{ if } \mathbb{P}(B_i) > 0, \text{ and } \mathbb{P}_i(C) = 0 \text{ otherwise,}$$

and define $\mathbb{Q}_i$ similarly. The measures $\mathbb{P}_i$ and $\mathbb{Q}_i$ are concentrated on B_i. Define

$$\eta = \sum_{i,j\geq 1} \eta''(x_i, x_j)(\mathbb{P}_i \times \mathbb{Q}_j).$$

Since $\eta''(x_i,x_j)=0$ unless $d(x_i,x_j)\le\alpha+3\varepsilon$, the measure η is concentrated on the set $\{d(x,y)\le\alpha+5\varepsilon\}$ because for $x\in B_i, y\in B_j$,

$$d(x,y)\le d(x,x_i)+d(x_i,x_j)+d(x_j,y)\le\varepsilon+(\alpha+3\varepsilon)+\varepsilon=\alpha+5\varepsilon.$$

The marginals u and v of η satisfy

$$\begin{aligned}u(C):=\eta(C\times S)&\le\sum_{i,j\ge1}\eta''(x_i,x_j)\mathbb{P}_i(C)=\sum_{i\ge1}\eta''(\{x_i\}\times S)\mathbb{P}_i(C)\\&\le\sum_{i\ge1}\mathbb{P}''(x_i)\mathbb{P}_i(C)\le\sum_{i\ge1}\mathbb{P}'(x_i)\mathbb{P}_i(C)=\sum_{i\ge1}\mathbb{P}(B_i)\mathbb{P}_i(C)=\mathbb{P}(C)\end{aligned}$$

and, similarly, $v(C):=\eta(S\times C)\le\mathbb{Q}(C)$. If $u(S)=v(S)=1$ then $\eta(S\times S)=1$ and η is a probability measure with marginals $\mathbb{P}$ and $\mathbb{Q}$, so we can take $\gamma=0$. Otherwise, take $t=1-u(S)=1-v(S)$ and define

$$\gamma=\frac{1}{t}(\mathbb{P}-u)\times(\mathbb{Q}-v).$$

It is easy to check that $\mu=\eta+\gamma$ has marginals $\mathbb{P}$ and $\mathbb{Q}$. For example,

$$\begin{aligned}\mu(C\times S)&=\eta(C\times S)+\gamma(C\times S)=u(C)+\frac{1}{t}\big(\mathbb{P}(C)-u(C)\big)\times\big(\mathbb{Q}(S)-v(S)\big)\\&=u(C)+\frac{1}{t}\big(\mathbb{P}(C)-u(C)\big)\times\big(1-v(S)\big)=u(C)+\big(\mathbb{P}(C)-u(C)\big)=\mathbb{P}(C).\end{aligned}$$

Also,

$$\gamma(S\times S)=t=1-\eta(S\times S)=1-\eta''(S\times S)=\gamma''(S\times S)\le\beta+5\varepsilon.$$

Finally, by construction, μ is a finite sum of product measures. □

The following relationship between Ky Fan and Levy-Prohorov metrics is an immediate consequence of Strassen's theorem. We already saw the inequality $\rho(\mathscr{L}(X),\mathscr{L}(Y))\le\alpha(X,Y)$.

Theorem 4.12. *If (S,d) is a separable metric space and $\mathbb{P},\mathbb{Q}$ are laws on S then, for any $\varepsilon>0$, there exist random variables X and Y on the same probability space with the distributions $\mathscr{L}(X)=\mathbb{P}$ and $\mathscr{L}(Y)=\mathbb{Q}$ such that $\alpha(X,Y)\le\rho(\mathbb{P},\mathbb{Q})+\varepsilon$. If $\mathbb{P}$ and $\mathbb{Q}$ are tight, one can take $\varepsilon=0$.*

Proof. Let us take $\alpha=\beta=\rho(\mathbb{P},\mathbb{Q})$. Then, by the definition of the Levy-Prohorov metric, for any $\varepsilon>0$ and any set A,

$$\mathbb{P}(A)\le\mathbb{Q}(A^{\rho+\varepsilon})+\rho+\varepsilon.$$

By Strassen's theorem, there exists a measure μ on $S\times S$ with the marginals $\mathbb{P},\mathbb{Q}$ such that

$$\mu\big(d(x,y)>\rho+2\varepsilon\big)\le\rho+2\varepsilon. \tag{4.6}$$

Therefore, if X and Y are the coordinate projections on $S \times S$, i.e.

$$X,Y\colon S\times S\to S,\; X(x,y)=x,\; Y(x,y)=y,$$

then, by definition of the Ky Fan metric, $\alpha(X,Y)\le\rho+2\varepsilon$. If $\mathbb{P}$ and $\mathbb{Q}$ are tight then there exists compact K such that $\mathbb{P}(K),\mathbb{Q}(K)\ge 1-\delta$. For $\varepsilon=1/n$ find μ_n as in (4.6). Since μ_n has marginals $\mathbb{P}$ and $\mathbb{Q}$, $\mu_n(K\times K)\ge 1-2\delta$, which means that the sequence $(\mu_n)_{n\ge 1}$ is uniformly tight. By the Selection Theorem, there exists a converging subsequence $\mu_{n(k)}\to\mu$. Obviously, μ again has marginals $\mathbb{P}$ and $\mathbb{Q}$. Since, by construction,

$$\mu_n\Bigl(d(x,y)>\rho+\frac{2}{n}\Bigr)\le\rho+\frac{2}{n}$$

and $\{d(x,y)>\rho+\varepsilon\}$ is an open set in $S\times S$, by the Portmanteau Theorem,

$$\mu\Bigl(d(x,y)>\rho+\varepsilon\Bigr)\le\liminf_{k\to\infty}\mu_{n(k)}\Bigl(d(x,y)>\rho+\varepsilon\Bigr)\le\rho.$$

Letting $\varepsilon\downarrow 0$ we get $\mu(d(x,y)>\rho)\le\rho$. As above, $\alpha(X,Y)\le\rho$ for this measure μ. □

This also implies the relationship between the bounded Lipschitz metric β and Levy-Prohorov metric ρ.

Lemma 4.8. *If (S,d) is a separable metric space then*

$$\beta(\mathbb{P},\mathbb{Q})\le 2\rho(\mathbb{P},\mathbb{Q})\le 4\sqrt{\beta(\mathbb{P},\mathbb{Q})}.$$

Proof. We already proved the second inequality. To prove the first one, by the previous theorem, given $\varepsilon>0$, we can find random variables X and Y with the distributions $\mathbb{P}$ and $\mathbb{Q}$, such that $\alpha(X,Y)\le\rho+\varepsilon$. Consider a bounded Lipschitz function f, $\|f\|_{\mathrm{BL}}<\infty$. Then

$$\begin{aligned}\Bigl|\int f\,d\mathbb{P}-\int f\,d\mathbb{Q}\Bigr| &= |\mathbb{E}f(X)-\mathbb{E}f(Y)|\le\mathbb{E}|f(X)-f(Y)|\\ &\le\|f\|_{\mathrm{L}}(\rho+\varepsilon)+2\|f\|_\infty\mathbb{P}\bigl(d(X,Y)>\rho+\varepsilon\bigr)\\ &\le\|f\|_{\mathrm{L}}(\rho+\varepsilon)+2\|f\|_\infty(\rho+\varepsilon)\le 2\|f\|_{\mathrm{BL}}(\rho+\varepsilon).\end{aligned}$$

Thus, $\beta(\mathbb{P},\mathbb{Q})\le 2(\rho(\mathbb{P},\mathbb{Q})+\varepsilon)$ and letting $\varepsilon\downarrow 0$ finishes the proof. □

Exercise 4.3.1. Show that the inequality $\beta(\mathbb{P},\mathbb{Q})\le 2\rho(\mathbb{P},\mathbb{Q})$ is sharp in the following sense: for any $t\in(0,1)$ and $\varepsilon>0$, one can find laws $\mathbb{P}$ and $\mathbb{Q}$ on $\mathbb{R}$ such that $\rho(\mathbb{P},\mathbb{Q})=t$ and $\beta(\mathbb{P},\mathbb{Q})\ge 2t-\varepsilon$. *Hint:* let $\mathbb{P},\mathbb{Q}$ have atoms of size t at points far apart, with $\mathbb{P},\mathbb{Q}$ otherwise the same.

Exercise 4.3.2. Suppose that X, Y and U are random variables on some probability space such that $\mathbb{P}(X \in A) \leq \mathbb{P}(Y \in A^{\delta}) + \varepsilon$ for all measurable sets $A \subseteq \mathbb{R}$, and U is independent of (X,Y) and has the uniform distribution on $[0,1]$. Prove that there exists a measurable function $f\colon \mathbb{R} \times [0,1] \to \mathbb{R}$ such that $\mathbb{P}\big(|X - f(X,U)| > \delta\big) \leq \varepsilon$ and $f(X,U)$ has the same distribution as Y.

Exercise 4.3.3. Let (S,d) be a separable metric space, and let $(\mathrm{Pr}(S,d), \rho)$ be the space of probability measures on S equipped with the Levy-Prohorov metric ρ. Let $\mathrm{Pr}^{(2)}(S,d) := \mathrm{Pr}(\mathrm{Pr}(S,d))$ be the space of probability measures on $(\mathrm{Pr}(S,d), \rho)$ and denote by $\rho^{(2)}$ the Levy-Prohorov metric on $\mathrm{Pr}^{(2)}(S,d)$. (Elements of $\mathrm{Pr}^{(2)}(S,d)$ can be viewed as laws of *random* measures on S.) Given $\mathbb{P}^{(2)} \in \mathrm{Pr}^{(2)}(S,d)$, we can define a two-step process to generate a random variable on S: first, generate a probability measure $\mathbb{P} \in \mathrm{Pr}(S,d)$ from this measure $\mathbb{P}^{(2)}$ and, second, generate a random variable X from $\mathbb{P}$, so that the distribution $\mathscr{L}(X)$ of X is given by

$$\mathrm{Pr}(X \in A) = \int \mathbb{P}(A)\, d\mathbb{P}^{(2)}(\mathbb{P}).$$

Let $\mathbb{Q}^{(2)} \in \mathrm{Pr}^{(2)}(S,d)$ be another measure, and generate a random variable Y in the same way. Prove that $\rho(\mathscr{L}(X), \mathscr{L}(Y)) \leq 2\rho^{(2)}(\mathbb{P}^{(2)}, \mathbb{Q}^{(2)})$.

Exercise 4.3.4. Suppose that probability measures $\mathbb{P}, \mathbb{Q}$ on a separable metric space (S,d) are tight. Let $M(\mathbb{P},\mathbb{Q})$ be the set of measures on the product space $S \times S$ with marginals $\mathbb{P}$ and $\mathbb{Q}$. Given $\alpha > 0$, show that

$$\inf_{\mu \in M(\mathbb{P},\mathbb{Q})} \mu\Big(d(x,y) > \alpha\Big) = \sup\Big(\mathbb{P}(A) - \mathbb{Q}(A^{\alpha]})\Big),$$

where sup is over all measurable (or closed, or open) sets A. *Hint:* Let β be the supremum on the right hand side; see remark below Theorem 4.11; and notice that $\mathrm{I}(x \in A) - \mathrm{I}(y \in A^{\alpha]}) \leq \mathrm{I}(d(x,y) > \alpha)$.

4.4 Wasserstein distance and Kantorovich-Rubinstein theorem

Let (S,d) be a separable metric space. Denote by $\mathscr{P}_1(S)$ the set of all laws on S such that for some $z \in S$ (equivalently, for all $z \in S$),

$$\int_S d(x,z)\, d\mathbb{P}(x) < \infty.$$

Let us consider a set

$$M(\mathbb{P},\mathbb{Q}) = \big\{\mu \, : \, \mu \text{ is a law on } S\times S \text{ with marginals } \mathbb{P} \text{ and } \mathbb{Q}\big\}.$$

For $\mathbb{P},\mathbb{Q} \in \mathscr{P}_1(S)$, the quantity

$$W(\mathbb{P},\mathbb{Q}) = \inf\Big\{\int d(x,y)\, d\mu(x,y) \, : \, \mu \in M(\mathbb{P},\mathbb{Q})\Big\}$$

is called the *Wasserstein distance* between $\mathbb{P}$ and $\mathbb{Q}$. If $\mathbb{P}$ and $\mathbb{Q}$ are tight, this infimum is attained (see exercise below).

A measure $\mu \in M(\mathbb{P},\mathbb{Q})$ represents a *transportation* between the measures $\mathbb{P}$ and $\mathbb{Q}$. We can think of the conditional distribution $\mu(y|x)$ as a way to redistribute the mass in the neighborhood of a point x so that the distribution $\mathbb{P}$ will be redistributed to the distribution $\mathbb{Q}$. If the distance $d(x,y)$ represents the cost of moving x to y then the Wasserstein distance gives the optimal total cost of transporting $\mathbb{P}$ to $\mathbb{Q}$. Given any two laws $\mathbb{P}$ and $\mathbb{Q}$ on S, let us define

$$\gamma(\mathbb{P},\mathbb{Q}) = \sup\Big\{\Big|\int f\, d\mathbb{P} - \int f\, d\mathbb{Q}\Big| : \|f\|_{\mathrm{L}} \le 1\Big\}$$

(notice that $\beta(\mathbb{P},\mathbb{Q}) \le \gamma(\mathbb{P},\mathbb{Q})$) and

$$m_d(\mathbb{P},\mathbb{Q}) = \sup\Big\{\int f\, d\mathbb{P} + \int g\, d\mathbb{Q} \, : \, f,g \in C(S),\; f(x)+g(y) < d(x,y)\Big\}.$$

Notice that, obviously, for $\mathbb{P},\mathbb{Q} \in \mathscr{P}_1(S)$, both $\gamma(\mathbb{P},\mathbb{Q}), m_d(\mathbb{P},\mathbb{Q}) < \infty$. Let us show that these two quantities are equal.

Lemma 4.9. *We have* $\gamma(\mathbb{P},\mathbb{Q}) = m_d(\mathbb{P},\mathbb{Q})$.

Proof. Given a function f such that $\|f\|_{\mathrm{L}} \le 1$, let us take a small $\varepsilon > 0$ and set $g(y) = -f(y) - \varepsilon$. Then

$$f(x) + g(y) = f(x) - f(y) - \varepsilon \le d(x,y) - \varepsilon < d(x,y)$$

and

$$\int f\, d\mathbb{P} + \int g\, d\mathbb{Q} = \int f\, d\mathbb{P} - \int f\, d\mathbb{Q} - \varepsilon.$$

Combining with the choice of $-f(x)$ and $g(y) = f(y) - \varepsilon$ we get

$$\begin{aligned}&\Big|\int f\,d\mathbb{P}-\int f\,d\mathbb{Q}\Big|\\&\le\sup\Big\{\int f\,d\mathbb{P}+\int g\,d\mathbb{Q}\,:\,f,g\in C(S),\,f(x)+g(y)<d(x,y)\Big\}+\varepsilon\end{aligned}$$

which, of course, proves that $\gamma(\mathbb{P},\mathbb{Q})\le m_d(\mathbb{P},\mathbb{Q})$.

Let us now consider functions f,g such that $f(x)+g(y)<d(x,y)$. Define

$$e(x)=\inf_y(d(x,y)-g(y))=-\sup_y(g(y)-d(x,y))$$

Clearly, $f(x)\le e(x)\le d(x,x)-g(x)=-g(x)$ and, therefore,

$$\int f\,d\mathbb{P}+\int g\,d\mathbb{Q}\le\int e\,d\mathbb{P}-\int e\,d\mathbb{Q}.$$

The function e satisfies

$$\begin{aligned}e(x)-e(x')&=\sup_y\big(g(y)-d(x',y)\big)-\sup_y\big(g(y)-d(x,y)\big)\\&\le\sup_y\big(d(x,y)-d(x',y)\big)\le d(x,x')\end{aligned}$$

which means that $\|e\|_{\mathrm{L}}\le 1$ and, therefore, $m_d(\mathbb{P},\mathbb{Q})\le\gamma(\mathbb{P},\mathbb{Q})$. This finishes the proof. □

Below, we will need the following version of the Hahn-Banach theorem.

Theorem 4.13 (Hahn-Banach). *Let V be a normed vector space, E - a linear subspace of V and U - an open convex set in V such that $U\cap E\neq\emptyset$. If $r\colon E\to\mathbb{R}$ is a linear non-zero functional on E then there exists a linear functional $\rho\colon V\to\mathbb{R}$ such that $\rho|_E=r$ and $\sup_U\rho(x)=\sup_{U\cap E}r(x)$.*

Proof. Let $t=\sup\{r(x):x\in U\cap E\}$ and let $B=\{x\in E:r(x)>t\}$. Since B is convex and $U\cap B=\emptyset$, the Hahn-Banach separation theorem implies that there exists a linear functional $q:V\to\mathbb{R}$ that separates U and B. For any $x_0\in U\cap E$ let $F=\{x\in E:q(x)=q(x_0)\}$. Since $q(x_0)<\inf_B q(x)$, $F\cap B=\emptyset$. This means that the hyperplanes $\{x\in E:q(x)=q(x_0)\}$ and $\{x\in E:r(x)=t\}$ in the subspace E are parallel and this implies that $q(x)=\alpha r(x)$ on E for some $\alpha\neq 0$. Let $\rho=q/\alpha$. Then $r=\rho|_E$ and

$$\sup_U\rho(x)=\frac{1}{\alpha}\sup_U q(x)\le\frac{1}{\alpha}\inf_B q(x)=\inf_B r(x)=t=\sup_{U\cap E}r(x)=\sup_{U\cap E}\rho(x).$$

Since $U\supseteq U\cap E$, the inequality is actually equality, which finishes the proof. □

Using this, we will prove the following *Kantorovich-Rubinstein theorem* for compact metric spaces.

Theorem 4.14. *If S is a compact metric space then $W(\mathbb{P},\mathbb{Q}) = m_d(\mathbb{P},\mathbb{Q}) = \gamma(\mathbb{P},\mathbb{Q})$ for $\mathbb{P},\mathbb{Q} \in \mathscr{P}_1(S)$.*

Proof. We only need to show the first equality. Consider a vector space $V = C(S \times S)$ equipped with $\|\cdot\|_\infty$ norm and let

$$U = \{h \in V : h(x,y) < d(x,y)\}.$$

Obviously, U is convex and open, because $S \times S$ is compact and any continuous function on a compact achieves its maximum. Consider a linear subspace E of V defined by

$$E = \{\phi \in V : \phi(x,y) = f(x) + g(y), f,g \in C(S)\}$$

so that

$$U \cap E = \{\phi \in V : \phi(x,y) = f(x) + g(y) < d(x,y), f,g \in C(S)\}.$$

Define a linear functional r on E by

$$r(\phi) = \int f\,d\mathbb{P} + \int g\,d\mathbb{Q} \text{ if } \phi = f(x) + g(y).$$

By the above Hahn-Banach theorem, r can be extended to $\rho : V \to \mathbb{R}$ such that $\rho|_E = r$ and

$$\sup_U \rho(\phi) = \sup_{U \cap E} r(\phi) = m_d(\mathbb{P},\mathbb{Q}) < \infty.$$

Let us look at the properties of this functional. First of all, if $a(x,y) \geq 0$ then $\rho(a) \geq 0$. Indeed, for any $c \geq 0$,

$$U \ni d(x,y) - c \cdot a(x,y) - \varepsilon < d(x,y)$$

and, therefore, for all $c \geq 0$,

$$\rho(d - ca - \varepsilon) = \rho(d) - c\rho(a) - \rho(\varepsilon) \leq \sup_U \rho < \infty.$$

This can hold only if $\rho(a) \geq 0$. This implies that if $\phi_1 \leq \phi_2$ then $\rho(\phi_1) \leq \rho(\phi_2)$. For any function ϕ, both $-\phi, \phi \leq \|\phi\|_\infty \cdot 1$ and, by monotonicity of ρ,

$$|\rho(\phi)| \leq \|\phi\|_\infty \rho(1) = \|\phi\|_\infty.$$

Since $S \times S$ is compact and ρ is a continuous functional on $(C(S \times S), \|\cdot\|_\infty)$, by the Riesz representation theorem, there exists a unique measure μ on the Borel σ-algebra on $S \times S$ such that

$$\rho(h) = \int h(x,y)\,d\mu(x,y).$$

Since $\rho|_E = r$,

$$\int (f(x)+g(y))\,d\mu(x,y) = \int f\,d\mathbb{P} + \int g\,d\mathbb{Q},$$

which implies that μ has marginal $\mathbb{P}$ and $\mathbb{Q}$, i.e. $\mu \in M(\mathbb{P},\mathbb{Q})$. This proves that

$$\begin{aligned} m_d(\mathbb{P},\mathbb{Q}) = \sup_U \rho(\phi) &= \sup\Big\{ \int h(x,y)\,d\mu(x,y) \;:\; h(x,y) < d(x,y) \Big\} \\ &= \int d(x,y)\,d\mu(x,y) \geq W(\mathbb{P},\mathbb{Q}). \end{aligned}$$

The opposite inequality is easy, because for any f,g such that $f(x)+g(y) < d(x,y)$ and any $\nu \in M(\mathbb{P},\mathbb{Q})$,

$$\int f\,d\mathbb{P} + \int g\,d\mathbb{Q} = \int (f(x)+g(y))\,d\nu(x,y) \leq \int d(x,y)d\nu(x,y). \qquad (4.7)$$

This finishes the proof and, moreover, it shows that the infimum in the definition of W is attained. □

Remark 4.5. Notice that in the proof of this theorem we never used the fact that d is a metric. Theorem holds for any $d \in C(S \times S)$ under the corresponding integrability assumptions. For example, one can consider loss functions of the type $d(x,y)^p$ for $p > 1$, which are not necessarily metrics. However, in Lemma 4.9, the fact that d is a metric was essential. □

Our next goal will be to show that $W = \gamma$ on separable and not necessarily compact metric spaces. We start with the following.

Lemma 4.10. *If (S,d) is a separable metric space then W and γ are metrics on $\mathscr{P}_1(S)$.*

Proof. Since for a bounded Lipschitz metric β we have $\beta(\mathbb{P},\mathbb{Q}) \leq \gamma(\mathbb{P},\mathbb{Q})$, γ is also a metric, because if $\gamma(\mathbb{P},\mathbb{Q}) = 0$ then $\beta(\mathbb{P},\mathbb{Q}) = 0$ and, therefore, $\mathbb{P} = \mathbb{Q}$. As in (4.7), it should be obvious that $\gamma(\mathbb{P},\mathbb{Q}) = m_d(\mathbb{P},\mathbb{Q}) \leq W(\mathbb{P},\mathbb{Q})$ and if $W(\mathbb{P},\mathbb{Q}) = 0$ then $\gamma(\mathbb{P},\mathbb{Q}) = 0$ and $\mathbb{P} = \mathbb{Q}$. The symmetry of W is obvious. It remains to show that $W(\mathbb{P},\mathbb{Q})$ satisfies the triangle inequality. The idea here is very simple, and let us first explain it in the case when (S,d) is complete. Consider three laws $\mathbb{P},\mathbb{Q},\mathbb{T}$ on S and let $\mu \in M(\mathbb{P},\mathbb{Q})$ and $\nu \in M(\mathbb{Q},\mathbb{T})$ be such that

$$\int d(x,y)\,d\mu(x,y) \leq W(\mathbb{P},\mathbb{Q}) + \varepsilon \;\text{ and }\; \int d(y,z)\,d\nu(y,z) \leq W(\mathbb{Q},\mathbb{T}) + \varepsilon.$$

Let us generate a distribution γ on $S \times S \times S$ with marginals $\mathbb{P},\mathbb{Q}$ and $\mathbb{T}$ and marginals on pairs of coordinates (x,y) and (y,z) given by μ and ν by "gluing" μ and ν in the following way. We know that when (S,d) is complete and separable, there exist regular conditional distributions $\mu(dx \mid y)$ and $\nu(dz \mid y)$ of the coordinates x and z given y. Then, we define a distribution γ on $S \times S \times S$ by first generating y from the distribution $\mathbb{Q}$ and, given y, generating the pair x and z according to the conditional distributions $\mu(dx \mid y)$ and $\nu(dz \mid y)$ independently of each other, i.e. according to the product measure on $S\times S$:

$$\gamma(dxdz \mid y) = \mu(dx \mid y) \times \nu(dz \mid y).$$

This is called "conditionally independent coupling of x and z given y". Obviously, by construction, (x,y) has the distribution μ and (y,z) has the distribution ν. Therefore, the marginals of x and z are $\mathbb{P}$ and $\mathbb{T}$, which means that the pair (x,z) has the distribution $\eta\in M(\mathbb{P},\mathbb{T})$. Finally,

$$\begin{aligned} W(\mathbb{P},\mathbb{T}) &\le \int d(x,z)\,d\eta(x,z) = \int d(x,z)\,d\gamma(x,y,z) \le \int d(x,y)\,d\gamma + \int d(y,z)\,d\gamma \\ &= \int d(x,y)\,d\mu + \int d(y,z)\,d\nu \le W(\mathbb{P},\mathbb{Q}) + W(\mathbb{Q},\mathbb{T}) + 2\varepsilon. \end{aligned}$$

Letting $\varepsilon\to 0$ proves the triangle inequality for W. In the case when (S,d) is not complete, we will apply the same basic idea after we discretize the space without losing much in the transportation cost integral. This can be done as in the proof of Strassen's theorem, Case C. Given $\varepsilon > 0$, consider a partition $(S_n)_{n\ge 1}$ of S such that diameter$(S_n) < \varepsilon$ for all n. On each box $S_n\times S_m$ let

$$\mu^1_{nm}(C) = \frac{\mu((C\cap S_n)\times S_m)}{\mu(S_n\times S_m)},\ \mu^2_{nm}(C) = \frac{\mu(S_n\times (C\cap S_m))}{\mu(S_n\times S_m)}$$

be the marginal distributions of the conditional distribution of μ on $S_n\times S_m$. Define

$$\mu' = \sum_{n,m} \mu(S_n\times S_m)\ \mu^1_{nm}\times\mu^2_{nm}.$$

In this construction, locally on each small box $S_n\times S_m$, measure μ is replaced by the product measure with the same marginals. Let us compute the marginals of μ'. Given a set $C\subseteq S$,

$$\begin{aligned} \mu'(C\times S) &= \sum_{n,m} \mu(S_n\times S_m)\ \mu^1_{nm}(C)\times\mu^2_{nm}(S) \\ &= \sum_{n,m} \mu((C\cap S_n)\times S_m) = \sum_n \mu((C\cap S_n)\times S) = \sum_n \mathbb{P}(C\cap S_n) = \mathbb{P}(C). \end{aligned}$$

Similarly, $\mu'(S\times C) = \mathbb{Q}(C)$, so μ' has the same marginals as μ, $\mu'\in M(\mathbb{P},\mathbb{Q})$. It should be obvious that transportation cost integral does not change much by replacing μ with μ'. One can visualize this by looking at what happens locally on each small box $S_n\times S_m$. Let (X_n,Y_m) be a random pair with distribution μ restricted to $S_n\times S_m$ so that

$$\mathbb{E}d(X_n,Y_m) = \frac{1}{\mu(S_n\times S_m)}\int_{S_n\times S_m} d(x,y)d\mu(x,y).$$

Let Y'_m be an independent copy of Y_m, also independent of X_n. Then the joint distribution of (X_n,Y'_m) is $\mu^1_{nm}\times\mu^2_{nm}$ and

$$\mathbb{E}d(X_n, Y'_m) = \int_{S_n \times S_m} d(x,y)\, d(\mu^1_{nm} \times \mu^2_{nm})(x,y).$$

Then

$$\int d(x,y)\, d\mu(x,y) = \sum_{n,m} \mu(S_n \times S_m) \mathbb{E}d(X_n, Y_m),$$

$$\int d(x,y)\, d\mu'(x,y) = \sum_{n,m} \mu(S_n \times S_m) \mathbb{E}d(X_n, Y'_m).$$

Finally, $d(Y_m, Y'_m) \leq \mathrm{diam}(S_m) \leq \varepsilon$ and these two integrals differ by at most ε. Therefore,

$$\int d(x,y)\, d\mu'(x,y) \leq W(\mathbb{P}, \mathbb{Q}) + 2\varepsilon.$$

Similarly, we can define

$$\nu' = \sum_{n,m} \nu(S_n \times S_m)\ \nu^1_{nm} \times \nu^2_{nm}$$

such that

$$\int d(x,y)\, d\nu'(x,y) \leq W(\mathbb{Q}, \mathbb{T}) + 2\varepsilon.$$

We will now show that this special simple form of the distributions $\mu'(x,y), \nu'(y,z)$ ensures that the conditional distributions of x and z given y are well defined. Let $\mathbb{Q}_m$ be the restriction of $\mathbb{Q}$ to S_m,

$$\mathbb{Q}_m(C) = \mathbb{Q}(C \cap S_m) = \sum_n \mu(S_n \times S_m)\ \mu^2_{nm}(C).$$

Obviously, if $\mathbb{Q}_m(C) = 0$ then $\mu^2_{nm}(C) = 0$ for all n, which means that μ^2_{nm} are absolutely continuous with respect to $\mathbb{Q}_m$ and the Radon-Nikodym derivatives

$$f_{nm}(y) = \frac{d\mu^2_{nm}}{d\mathbb{Q}_m}(y) \ \text{ exist and } \sum_n \mu(S_n \times S_m) f_{nm}(y) = 1 \ \text{ a.s. for } y \in S_m.$$

Of course, we can set $f_{nm}(y) = 0$ for y outside of S_m. Let us define a conditional distribution of x given y by

$$\mu'(A \mid y) = \sum_{n,m} \mu(S_n \times S_m) f_{nm}(y) \mu^1_{nm}(A).$$

Notice that for any $A \in \mathscr{B}$, $\mu'(A \mid y)$ is measurable in y and $\mu'(A|y)$ is a probability distribution on $\mathscr{B}$, for $\mathbb{Q}$-almost all y, because

$$\mu'(S \mid y) = \sum_{n,m} \mu(S_n \times S_m) f_{nm}(y) = 1\ a.s.$$

Let us check that for Borel sets $A, B \in \mathscr{B}$,

$$\mu'(A \times B) = \int_B \mu'(A \mid y)\, d\mathbb{Q}(y).$$

Indeed, since $f_{nm}(y) = 0$ for $y \notin S_m$,

$$\begin{aligned}
\int_B \mu'(A \mid y)\, d\mathbb{Q}(y) &= \sum_{n,m} \mu(S_n \times S_m)\mu^1_{nm}(A) \int_B f_{nm}(y)\, d\mathbb{Q}(y) \\
&= \sum_{n,m} \mu(S_n \times S_m)\mu^1_{nm}(A) \int_B f_{nm}(y)\, d\mathbb{Q}_m(y) \\
&= \sum_{n,m} \mu(S_n \times S_m)\mu^1_{nm}(A)\mu^2_{nm}(B) = \mu'(A \times B).
\end{aligned}$$

Conditional distribution $\nu'(\cdot \mid y)$ can be defined similarly, and we finish the proof as in the case of the complete space above. □

Next lemma shows that on a separable metric space any law with the "first moment", i.e. $\mathbb{P} \in \mathscr{P}_1(S)$, can be approximated in metrics W and γ by laws concentrated on finite sets.

Lemma 4.11. *If (S,d) is separable and $\mathbb{P} \in \mathscr{P}_1(S)$ then there exists a sequence of laws $\mathbb{P}_n$ such that $\mathbb{P}_n(F_n) = 1$ for some finite sets F_n and $W(\mathbb{P}_n, \mathbb{P}), \gamma(\mathbb{P}_n, \mathbb{P}) \to 0$.*

Proof. For each $n \geq 1$, let $(S_{nj})_{j\geq 1}$ be a partition of S such that $\mathrm{diam}(S_{nj}) \leq 1/n$. Take a point $x_{nj} \in S_{nj}$ in each set S_{nj} and for $k \geq 1$ define a function

$$f_{nk}(x) = \begin{cases} x_{nj}, & \text{if } x \in S_{nj} \text{ for } j \leq k, \\ x_{n1}, & \text{if } x \in S_{nj} \text{ for } j > k. \end{cases}$$

We have,

$$\begin{aligned}
\int d(x, f_{nk}(x))\, d\mathbb{P}(x) &= \sum_{j\geq 1} \int_{S_{nj}} d(x, f_{nk}(x))\, d\mathbb{P}(x) \\
\leq \frac{1}{n} \sum_{j\leq k} \mathbb{P}(S_{nj}) &+ \int_{S\setminus(S_{n1}\cup\cdots\cup S_{nk})} d(x, x_{n1})\, d\mathbb{P}(x) \leq \frac{2}{n}
\end{aligned}$$

for k large enough, because $\mathbb{P} \in \mathscr{P}_1(S)$, i.e. $\int d(x, x_{n1})\, d\mathbb{P}(x) < \infty$, and the set $S \setminus (S_{n1} \cup \cdots \cup S_{nk}) \downarrow \emptyset$.

Let μ_n be the image on $S \times S$ of the measure $\mathbb{P}$ under the map $x \to (f_{nk}(x), x)$ so that $\mu_n \in M(\mathbb{P}_n, \mathbb{P})$ for some $\mathbb{P}_n$ concentrated on the set of points $\{x_{n1}, \ldots, x_{nk}\}$. Finally,

$$W(\mathbb{P}_n, \mathbb{P}) \leq \int d(x,y)\, d\mu_n(x,y) = \int d(f_{nk}(x), x)\, d\mathbb{P}(x) \leq \frac{2}{n}.$$

Since $\gamma(\mathbb{P}_n, \mathbb{P}) \leq W(\mathbb{P}_n, \mathbb{P})$, this finishes the proof. □

We are finally ready to extend Theorem 4.14 to separable metric spaces.

Theorem 4.15 (Kantorovich-Rubinstein theorem). *If (S,d) is a separable metric space then $W(\mathbb{P},\mathbb{Q}) = \gamma(\mathbb{P},\mathbb{Q})$ for any distributions $\mathbb{P},\mathbb{Q} \in \mathcal{P}_1(S)$.*

Proof. By the previous lemma, we can approximate $\mathbb{P}$ and $\mathbb{Q}$ by $\mathbb{P}_n$ and $\mathbb{Q}_n$ concentrated on finite (hence, compact) sets. By Theorem 4.14, $W(\mathbb{P}_n,\mathbb{Q}_n) = \gamma(\mathbb{P}_n,\mathbb{Q}_n)$. Finally, since both W,γ are metrics,

$$\begin{aligned} W(\mathbb{P},\mathbb{Q}) &\le W(\mathbb{P},\mathbb{P}_n) + W(\mathbb{P}_n,\mathbb{Q}_n) + W(\mathbb{Q}_n,\mathbb{Q}) \\ &\le W(\mathbb{P},\mathbb{P}_n) + \gamma(\mathbb{P}_n,\mathbb{Q}_n) + W(\mathbb{Q}_n,\mathbb{Q}) \\ &\le W(\mathbb{P},\mathbb{P}_n) + W(\mathbb{Q}_n,\mathbb{Q}) + \gamma(\mathbb{P}_n,\mathbb{P}) + \gamma(\mathbb{Q}_n,\mathbb{Q}) + \gamma(\mathbb{P},\mathbb{Q}). \end{aligned}$$

Letting $n \to \infty$ proves that $W(\mathbb{P},\mathbb{Q}) \le \gamma(\mathbb{P},\mathbb{Q})$. We saw above that the opposite inequality always holds. □

Wasserstein's distance $W_p(\mathbb{P},\mathbb{Q})$ on $\mathbb{R}^n$. We will now prove a version of the Kantorovich-Rubinstein theorem on $\mathbb{R}^n$ in some cases when $d(x,y)$ is not a metric. Given $p \ge 1$, let us define the Wasserstein distance $W_p(\mathbb{P},\mathbb{Q})$ on

$$\mathcal{P}_p(\mathbb{R}^n) = \Big\{\mathbb{P} - \text{law on } \mathbb{R}^n \, : \, \int |x|^p d\mathbb{P}(x) < \infty\Big\}$$

corresponding to the cost function $d(x,y) = |x-y|^p$ by

$$W_p(\mathbb{P},\mathbb{Q})^p = \inf\Big\{\int |x-y|^p \, d\mu(x,y) \, : \, \mu \in M(\mathbb{P},\mathbb{Q})\Big\}. \tag{4.8}$$

Even though $d(x,y)$ is not a metric for $p > 1$, W_p is still a metric on $\mathcal{P}_p(\mathbb{R}^n)$, which can be shown the same way as in Lemma 4.10. Namely, given nearly optimal $\mu \in M(\mathbb{P},\mathbb{Q})$ and $\nu \in M(\mathbb{Q},\mathbb{T})$ we can construct $(X,Y,Z) \sim M(\mathbb{P},\mathbb{Q},\mathbb{T})$ such that $(X,Y) \sim \mu$ and $(Y,Z) \sim \nu$ and, therefore,

$$\begin{aligned} W_p(\mathbb{P},\mathbb{T}) &\le (\mathbb{E}|X-Z|^p)^{\frac{1}{p}} \le (\mathbb{E}|X-Y|^p)^{\frac{1}{p}} + (\mathbb{E}|Y-Z|^p)^{\frac{1}{p}} \\ &\le (W_p^p(\mathbb{P},\mathbb{Q}) + \varepsilon)^{\frac{1}{p}} + (W_p^p(\mathbb{Q},\mathbb{T}) + \varepsilon)^{\frac{1}{p}}. \end{aligned}$$

Then, we let $\varepsilon \downarrow 0$. The following dual representation holds.

Theorem 4.16. *For any $\mathbb{P},\mathbb{Q} \in \mathcal{P}_p(\mathbb{R}^n)$,*

$$W_p(\mathbb{P},\mathbb{Q})^p = \sup\Big\{\int f d\mathbb{P} + \int g d\mathbb{Q} \, : \, f,g \in C(\mathbb{R}^n), f(x) + g(y) \le |x-y|^p\Big\}. \tag{4.9}$$

Proof. We will show below that for any continuous uniformly bounded function $d(x,y)$ on $\mathbb{R}^n \times \mathbb{R}^n$,

$$\inf\Big\{\int d(x,y) \, d\mu(x,y) \, : \, \mu \in M(\mathbb{P},\mathbb{Q})\Big\} \tag{4.10}$$

$$= \sup\Big\{\int f\,d\mathbb{P} + \int g\,d\mathbb{Q} \,:\, f,g \in C(\mathbb{R}^n), f(x)+g(y) \le d(x,y)\Big\}.$$

This will imply (4.9) as follows. Let us take $R \ge 1$ large enough so that for $K = \{|x| \le R\}$,

$$\int_{K^c} |x|^p\,d\mathbb{P} \le 2^{-p-1}\varepsilon,\ \int_{K^c} |x|^p\,d\mathbb{Q} \le 2^{-p-1}\varepsilon.$$

We can find such R, because $\mathbb{P},\mathbb{Q} \in \mathscr{P}_p(\mathbb{R}^n)$. Let $d(x,y) = |x-y|^p \wedge (2R)^p$. Then for any $x,y \in K$, we have $d(x,y) = |x-y|^p$ and for any $\mu \in M(\mathbb{P},\mathbb{Q})$,

$$\int |x-y|^p\,d\mu(x,y) \le \int d(x,y)\,d\mu(x,y) + \int_{(K\times K)^c} |x-y|^p\,d\mu(x,y).$$

Let us break the second integral into two integrals over disjoint sets $\{|x| \ge |y|\}$ and $\{|x| < |y|\}$. On the first set, $|x-y|^p \le 2^p|x|^p$ and, moreover, x can not belong to K, since in that case y must be in K^c. This means that the part of the second integral over this set in bounded by

$$\int_{K^c\times\mathbb{R}^n} 2^p|x|^p\,d\mu(x,y) = \int_{K^c} 2^p|x|^p\,d\mathbb{P}(x) \le \varepsilon/2.$$

The second integral over the set $\{|x| < |y|\}$ can be similarly bounded by $\varepsilon/2$ and, therefore,

$$\int |x-y|^p\,d\mu(x,y) \le \int d(x,y)\,d\mu(x,y) + \varepsilon.$$

Together with (4.10) this implies that

$$\begin{aligned}
W_p(\mathbb{P},\mathbb{Q})^p &= \inf\Big\{\int |x-y|^p\,d\mu(x,y) \,:\, \mu \in M(\mathbb{P},\mathbb{Q})\Big\}\\
&\le \inf\Big\{\int d(x,y)\,d\mu(x,y) \,:\, \mu \in M(\mathbb{P},\mathbb{Q})\Big\} + \varepsilon\\
&= \sup\Big\{\int f\,d\mathbb{P} + \int g\,d\mathbb{Q} \,:\, f,g \in C(\mathbb{R}^n), f(x)+g(y) \le d(x,y)\Big\} + \varepsilon\\
&\le \sup\Big\{\int f\,d\mathbb{P} + \int g\,d\mathbb{Q} \,:\, f,g \in C(\mathbb{R}^n), f(x)+g(y) \le |x-y|^p\Big\} + \varepsilon.
\end{aligned}$$

Letting $\varepsilon \downarrow 0$ proves inequality in one direction, and the opposite inequality is always true, as we have seen above, since for any $f,g \in C(\mathbb{R}^n)$ such that $f(x)+g(y) \le |x-y|^p$ and any $\mu \in M(\mathbb{P},\mathbb{Q})$,

$$\int f\,d\mathbb{P} + \int g\,d\mathbb{Q} = \int (f(x)+g(y))\,d\mu(x,y) \le \int |x-y|^p\,d\mu(x,y). \tag{4.11}$$

Therefore, it remains to prove (4.10). Notice that, by adding a constant, we can assume that $d \ge 0$.

Again, the inequality $\ge$ is obvious, so we only need to prove $\le$. We will reduce this to the compact case proved in Theorem 4.14. Let us take $R \ge 1$ large enough and let $K = \{|x| \le R\}$. Let us consider measures

$$\mathbb{P}_K(C) = \frac{\mathbb{P}(C \cap K)}{\mathbb{P}(K)},\ \mathbb{Q}_K(C) = \frac{\mathbb{Q}(C \cap K)}{\mathbb{Q}(K)}$$

and let $\mu_K \in M(\mathbb{P}_K, \mathbb{Q}_K)$ be the measure on $K \times K$ that achieves the infimum

$$\int d(x,y)\,d\mu_K(x,y) = \inf\Big\{\int d(x,y)\,d\mu(x,y) \,:\, \mu \in M(\mathbb{P}_K, \mathbb{Q}_K)\Big\}.$$

If define the measure μ_* on $\mathbb{R}^n \times \mathbb{R}^n$ by

$$\mu_* = \mathbb{P}(K)\mathbb{Q}(K)\,\mu_K + \mathbb{P} \times \mathbb{Q}\big|_{(K\times K)^c},$$

it is easy to check that $\mu_* \in M(\mathbb{P}, \mathbb{Q})$. Therefore,

$$\begin{aligned}&\inf\Big\{\int d(x,y)\,d\mu(x,y) \,:\, \mu \in M(\mathbb{P}, \mathbb{Q})\Big\} \le \int d(x,y)\,d\mu_*(x,y)\\ &\le \int d(x,y)\,d\mu_K(x,y) + \|d\|_\infty \mathbb{P} \times \mathbb{Q}((K \times K)^c).\end{aligned}$$

If we take R large enough so that the last term is smaller then ε then, using the Kantorovich-Rubinstein theorem on compacts, we get

$$\begin{aligned}&\inf\Big\{\int d(x,y)\,d\mu(x,y) \,:\, \mu \in M(\mathbb{P}, \mathbb{Q})\Big\} \le \int d(x,y)\,d\mu_K(x,y) + \varepsilon\\ &= \inf\Big\{\int d(x,y)\,d\mu(x,y) \,:\, \mu \in M(\mathbb{P}_K, \mathbb{Q}_K)\Big\} + \varepsilon\\ &= \sup\Big\{\int f\,d\mathbb{P}_K + \int g\,d\mathbb{Q}_K \,:\, f, g \in C(K), f(x) + g(y) \le d(x,y)\Big\} + \varepsilon.\\ &\le \int f\,d\mathbb{P}_K + \int g\,d\mathbb{Q}_K + 2\varepsilon, \qquad (4.12)\end{aligned}$$

for some $f, g \in C(K)$ such that $f(x) + g(y) \le d(x,y)$. To finish the proof, we would like to estimate the above sum of integrals by

$$\int \phi\,d\mathbb{P} + \int \psi\,d\mathbb{Q}$$

for some functions $\phi, \psi \in C(\mathbb{R}^n)$ such that $\phi(x) + \psi(y) \le d(x,y)$. This can be achieved by "improving" our choice of functions f and g using infimum convolutions, as we will now explain.

However, before we do this, let us make the following useful observation. Since $d \ge 0$, we can assume that the supremum in (4.12) is strictly positive (otherwise, there is nothing to prove) and

$$\int f\,d\mathbb{P}_K + \int g\,d\mathbb{Q}_K \ge 0.$$

Since we can replace f and g by $f - c$ and $g + c$ without changing the sum of integrals, we can assume that each integral is nonnegative. This implies that there exist points $x_0, y_0 \in K$ such that $f(x_0), g(y_0) \geq 0$. Since $f(x) + g(y) \leq d(x,y)$,

$$f(x) \leq d(x, y_0), \; g(y) \leq d(x_0, y) \text{ for } x, y \in K. \tag{4.13}$$

Let us now define the function $\phi(x)$ for $x \in \mathbb{R}^n$ by

$$\phi(x) = \inf_{y \in K} \big(d(x,y) - g(y)\big).$$

This is called the *infimum-convolution*. Clearly, since $f(x) \leq d(x,y) - g(y)$ for all $y \in K$, we have $f(x) \leq \phi(x)$ for $x \in K$, which implies that

$$\int f \, d\mathbb{P}_K + \int g \, d\mathbb{Q}_K \leq \int \phi \, d\mathbb{P}_K + \int g \, d\mathbb{Q}_K.$$

Notice that, by definition, $\phi(x) + g(y) \leq d(x,y)$ for all $x \in \mathbb{R}^n, y \in K$. Also, using (4.13) and the fact that $g(y_0) \geq 0$,

$$\inf_{y \in K} \big(d(x,y) - d(x_0,y)\big) \leq \phi(x) \leq d(x,y_0) - g(y_0) \leq d(x,y_0). \tag{4.14}$$

Next, we define the function $\psi(y)$ for $y \in \mathbb{R}^n$ by

$$\psi(y) = \inf_{x \in \mathbb{R}^n} \big(d(x,y) - \phi(x)\big).$$

Since $g(y) \leq d(x,y) - \phi(x)$ for all $x \in \mathbb{R}^n$, we have $g(y) \leq \psi(y)$ for $y \in K$, which implies that

$$\int f \, d\mathbb{P}_K + \int g \, d\mathbb{Q}_K \leq \int \phi \, d\mathbb{P}_K + \int g \, d\mathbb{Q}_K \leq \int \phi \, d\mathbb{P}_K + \int \psi \, d\mathbb{Q}_K.$$

By definition, $\phi(x) + \psi(y) \leq d(x,y)$ for all $x, y \in \mathbb{R}^n$. Also, using (4.14) and the fact that $\phi(x_0) \geq f(x_0) \geq 0$,

$$\inf_{x \in \mathbb{R}^n} \big(d(x,y) - d(x,y_0)\big) \leq \psi(y) \leq d(x_0,y) - \phi(x_0) \leq d(x_0,y). \tag{4.15}$$

The estimates in (4.13) and (4.14) imply that $\|\phi\|_\infty, \|\psi\|_\infty \leq \|d\|_\infty$. Therefore, if we write

$$\begin{aligned}\int \phi \, d\mathbb{P} &= \int_K \phi \, d\mathbb{P} + \int_{K^c} \phi \, d\mathbb{P} = \mathbb{P}(K) \int_K \phi \, d\mathbb{P}_K + \int_{K^c} \phi \, d\mathbb{P} \\ &= \int_K \phi \, d\mathbb{P}_K - \mathbb{P}(K^c) \int_K \phi \, d\mathbb{P}_K + \int_{K^c} \phi \, d\mathbb{P},\end{aligned}$$

each of the last two terms can be bounded in absolute value by $\|d\|_\infty \mathbb{P}(K^c)$, which shows that, for large enough $R \geq 1$,

$$\int_K \phi \, d\mathbb{P}_K \leq \int \phi \, d\mathbb{P} + \varepsilon.$$

The same estimate holds for ψ and we showed that

$$\int f\,d\mathbb{P}_K + \int g\,d\mathbb{Q}_K \le \int \phi\,d\mathbb{P} + \int \psi\,d\mathbb{Q} + 2\varepsilon.$$

The equation (4.12) implies that

$$\inf\Big\{\int d(x,y)\,d\mu(x,y)\,:\,\mu \in M(\mathbb{P},\mathbb{Q})\Big\} \le \int \phi\,d\mathbb{P} + \int \psi\,d\mathbb{Q} + 4\varepsilon$$

and, since $\phi(x)+\psi(y) \le d(x,y)$, this finishes the proof. □

Exercise 4.4.1. If $\mathbb{P}$ and $\mathbb{Q}$ are tight, show that the infimum in the definition of $W(\mathbb{P},\mathbb{Q})$ is attained.

Exercise 4.4.2. Suppose (S,d) is separable. Show that $W(\mathbb{P},\mathbb{Q}) \ge \rho(\mathbb{P},\mathbb{Q})^2$ and $W(\mathbb{P},\mathbb{Q}) \ge \beta(\mathbb{P},\mathbb{Q})$. If the diameter $D = \sup_{x,y\in S} d(x,y)$ of (S,d) is finite, show that $W(\mathbb{P},\mathbb{Q}) \le \rho(\mathbb{P},\mathbb{Q})(1+D)$.

Exercise 4.4.3. Let (S,d) be a separable metric space and, given a bounded measurable function $f\colon S^2 \to \mathbb{R}$, let

$$W_f(\mathbb{P},\mathbb{Q}) := \inf\Big\{\int f(x,y)\,d\mu(x,y)\,:\,\mu \in M(\mathbb{P},\mathbb{Q})\Big\}$$

be a Wasserstein-like 'transportation distance' between probability measures $\mathbb{P}$ and $\mathbb{Q}$ on (S,d). Prove that $W_f(\mathbb{P},\mathbb{Q})$ is convex in the pair $(\mathbb{P},\mathbb{Q})$.

Exercise 4.4.4. (a) Consider a countable set A and two probability measures $\mathbb{P}$ and $\mathbb{Q}$ on A. Prove that

$$\inf\Big\{\int \mathrm{I}(x\ne y)\,d\mu(x,y)\,:\,\mu \in M(\mathbb{P},\mathbb{Q})\Big\} = \frac{1}{2}\sum_{a\in A}\Big|\mathbb{P}(\{a\}) - \mathbb{Q}(\{a\})\Big|.$$

Hint: use the Kantorovich-Rubinstein theorem with metric $d(x,y) = \mathrm{I}(x \ne y)$.

(b) Let (S,d) be a separable metric space and $\mathscr{B}$ be the Borel σ-algebra. Given two probability measures $\mathbb{P}$ and $\mathbb{Q}$ on $\mathscr{B}$, construct a measure $\nu \in M(\mathbb{P},\mathbb{Q})$ that witnesses equality

$$\inf\Big\{\int \mathrm{I}(x\ne y)\,d\mu(x,y)\,:\,\mu \in M(\mathbb{P},\mathbb{Q})\Big\} = \sup_{A\in\mathscr{B}}|\mathbb{P}(A)-\mathbb{Q}(A)| =: \mathrm{TV}(\mathbb{P},\mathbb{Q}),$$

where $\mathrm{TV}(\mathbb{P},\mathbb{Q})$ is called the total variation distance between $\mathbb{P}$ and $\mathbb{Q}$. *Hint:* use the Hahn-Jordan decomposition.

Exercise 4.4.5. Let $\mathscr{P}_1(\mathbb{R})$ be the set of laws on $\mathbb{R}$ such that $\int |x|\,d\mathbb{P}(x) < \infty$. Let us define a map $\Phi : \mathscr{P}_1(\mathbb{R}) \to \mathscr{P}_1(\mathbb{R})$ as follows. Consider a random variable τ with values in $\mathbb{N}$ such that $\mathbb{E}\tau < \infty$ and an independent random variable ξ such that

$\mathbb{E}|\xi| < \infty$. Given a sequence (X_i) of i.i.d. random variables with the distribution $\mathbb{P} \in \mathscr{P}_1(\mathbb{R})$, independent of τ and ξ, let $\Phi(\mathbb{P})$ be the distribution of the sum $\xi + \sum_{i=1}^{\tau} X_i$, where the sum $\sum_{i=1}^{\tau} X_i$ is zero if $\tau = 0$. If $\mathbb{E}\tau \in [0,1)$, prove that Φ has a unique fixed point, $\Phi(\mathbb{P}) = \mathbb{P}$. *Hint*: use the Banach Fixed Point Theorem for the Wasserstein metric W_1. (You need to prove that the metric space $(\mathscr{P}_1(\mathbb{R}), W_1)$ is complete.)

Exercise 4.4.6. Let $\mathscr{P}$ be the set of probability laws on $[-1,1]$ and define a map $\Phi : \mathscr{P} \to \mathscr{P}$ as follows. Consider a random variable τ with values in $\mathbb{N}$ and an independent random variable ξ. Given a sequence (X_i) of i.i.d. random variables on $[-1,1]$ with the distribution $\mathbb{P} \in \mathscr{P}$, independent of τ and ξ, let $\Phi(\mathbb{P})$ be the distribution of $\cos(\xi + \sum_{i=1}^{\tau} X_i)$, where the sum $\sum_{i=1}^{\tau} X_i$ is zero if $\tau = 0$. Prove that Φ has a fixed point, $\Phi(\mathbb{P}) = \mathbb{P}$. *Hint*: use the Schauder Fixed Point Theorem for $\mathscr{P}$ equipped with the topology of weak convergence. (Notice that now a fixed point is possibly not unique.)

Exercise 4.4.7. Consider a finite set A and nonnegative real-valued functions

$$f_1, \ldots, f_n \colon A \to \mathbb{R}$$

such that for any convex combination $g = \lambda_1 f_1 + \ldots + \lambda_n f_n$ there is a point $a \in A$ such that $g(a) < 1$. Prove that there is a probability measure $\mathbb{P}$ on A such that $\int f_i \, d\mathbb{P} < 1$ for all $i \leq n$. *Hint:* use the Hahn-Banach separation theorem.

4.5 Wasserstein distance and entropy

In this section we will make several connections between the Wasserstein distance and other classical objects, with application to Gaussian concentration. Let us start with the following classical inequality.

Theorem 4.17 (Brunn-Minkowski inequality on $\mathbb{R}$). *If γ is the Lebesgue measure and A,B are two* non-empty *Borel sets on $\mathbb{R}$ then $\gamma(A+B) \geq \gamma(A)+\gamma(B)$, where $A+B=\{a+b : a\in A, b\in B\}$.*

Proof. First, suppose that A and B are open and bounded. Since the Lebesgue measure is invariant under translations, let us translate A and B so that $\sup A = \inf B = 0$. Let us check that, in this case, $A\cup B\subseteq A+B$. Since A is open, for each $a\in A$ there exists $\varepsilon>0$ such that $(a-\varepsilon, a+\varepsilon)\subseteq A$. Since $\inf B=0$, there exists $b\in B$ such that $0\leq b<\varepsilon/2$. Then $a\in(a+b-\varepsilon, a+b+\varepsilon)\subseteq A+B$, which proves that $A\subseteq A+B$. One can prove similarly that B is also a subset of $A+B$. Since A and B are disjoint, we proved that $\gamma(A)+\gamma(B)=\gamma(A\cup B)\leq\gamma(A+B)$.

Now, suppose that A and B are compact. Then, obviously, $A+B$ is also compact. For $\varepsilon>0$, let us denote by C^ε an open ε-neighborhood of the set C. Since $A^\varepsilon + B^\varepsilon\subseteq(A+B)^{2\varepsilon}$, using the previous case of the open bounded sets, we get $\gamma(A^\varepsilon)+\gamma(B^\varepsilon)\leq\gamma(A^\varepsilon+B^\varepsilon)\leq\gamma((A+B)^{2\varepsilon})$. Since A is closed, $A^\varepsilon\downarrow A$ as $\varepsilon\downarrow 0$ and, by the continuity of measure, $\gamma(A^\varepsilon)\downarrow\gamma(A)$. The same holds for B and $A+B$, and we proved the inequality for two compact sets.

Finally, consider arbitrary measurable sets A and B. We can assume that $\gamma(A)<\infty$ and $\gamma(B)<\infty$, otherwise, there is nothing to prove. By the regularity of the Lebesgue measure, we can find compacts $C\subseteq A$ and $D\subseteq B$ such that $\gamma(A\setminus C)\leq\varepsilon$ and $\gamma(B\setminus D)\leq\varepsilon$. Using the previous case of the compact sets, we can write $\gamma(A)+\gamma(B)-2\varepsilon\leq\gamma(C)+\gamma(D)\leq\gamma(C+D)\leq\gamma(A+B)$, and letting $\varepsilon\downarrow 0$ finishes the proof. □

Using this, we will prove another classical inequality.

Theorem 4.18 (Prekopa-Leindler inequality). *Consider nonnegative integrable functions $w,u,v:\mathbb{R}^n\to[0,\infty)$ such that for some $\lambda\in[0,1]$,*

$$w(\lambda x+(1-\lambda)y)\geq u(x)^\lambda v(y)^{1-\lambda}\quad \textit{for all}\ \ x,y\in\mathbb{R}^n.$$

Then,

$$\int w\,dx\geq\Big(\int u\,dx\Big)^\lambda\Big(\int v\,dx\Big)^{1-\lambda}.$$

Using the Prekopa-Leindler inequality one can prove that the Lebesgue measure γ on $\mathbb{R}^n$ satisfies the Brunn-Minkowski inequality

$$\gamma(A)^{1/n}+\gamma(B)^{1/n}\leq\gamma(A+B)^{1/n}. \tag{4.16}$$

We will leave this as an exercise. From this one can easily deduce the famous isoperimetric property of Euclidean balls with respect to the Lebesgue measure, namely, that among all sets with the same measure, a ball has the smallest surface area. If B is a unit open ball in $\mathbb{R}^n$ and $\gamma(A) = \gamma(B)$ then, by (4.16),

$$\begin{aligned}\gamma(A^\varepsilon) &= \gamma(A+\varepsilon B) \geq (\gamma(A)^{1/n} + \gamma(\varepsilon B)^{1/n})^n \\ &= (\gamma(B)^{1/n} + \gamma(\varepsilon B)^{1/n})^n = (1+\varepsilon)^n \gamma(B) = \gamma(B^\varepsilon).\end{aligned}$$

In other words, volume of A grows faster than B as we expand the sets, which means that the surface area of B is smaller.

Proof (of Theorem 4.18). The proof will proceed by induction on n. Let us first show the induction step. Suppose the statement holds for n and we would like to show it for $n+1$. By the assumption, for any $x, y \in \mathbb{R}^n$ and $a, b \in \mathbb{R}$,

$$w(\lambda x + (1-\lambda)y, \lambda a + (1-\lambda)b) \geq u(x,a)^\lambda v(y,b)^{1-\lambda}.$$

Let us fix a and b and consider functions $w_1(x) = w(x, \lambda a + (1-\lambda)b)$, $u_1(x) = u(x,a)$ and $v_1(x) = v(x,b)$ on $\mathbb{R}^n$ that satisfy $w_1(\lambda x + (1-\lambda)y) \geq u_1(x)^\lambda v_1(y)^{1-\lambda}$. By the induction assumption,

$$\int_{\mathbb{R}^n} w_1\,dx \geq \Big(\int_{\mathbb{R}^n} u_1\,dx\Big)^\lambda \Big(\int_{\mathbb{R}^n} v_1\,dx\Big)^{1-\lambda}.$$

These integrals still depend on a and b and we can define

$$w_2(\lambda a + (1-\lambda)b) = \int_{\mathbb{R}^n} w_1\,dx = \int_{\mathbb{R}^n} w(x, \lambda a + (1-\lambda)b)\,dx$$

and, similarly,

$$u_2(a) = \int_{\mathbb{R}^n} u_1(x,a)\,dx \text{ and } v_2(b) = \int_{\mathbb{R}^n} v_1(x,b)\,dx.$$

Then the above inequality can be rewritten as

$$w_2(\lambda a + (1-\lambda)b) \geq u_2(a)^\lambda v_2(b)^{1-\lambda}.$$

These functions are defined on $\mathbb{R}$ and, by the induction assumption,

$$\begin{aligned}&\int_{\mathbb{R}} w_2\,ds \geq \Big(\int_{\mathbb{R}} u_2\,ds\Big)^\lambda \Big(\int_{\mathbb{R}} v_2\,ds\Big)^{1-\lambda} \\ &\Longrightarrow \int_{\mathbb{R}^{n+1}} w\,dz \geq \Big(\int_{\mathbb{R}^{n+1}} u\,dz\Big)^\lambda \Big(\int_{\mathbb{R}^{n+1}} v\,dz\Big)^{1-\lambda},\end{aligned}$$

which finishes the proof of the induction step.

It remains to prove the case $n = 1$. We can assume that $u, v, w \colon \mathbb{R} \to [0,1]$, because both inequalities in the statement of the theorem are homogeneous to truncation and scaling. Also, we can assume that u and v are not identically zero, since there is nothing to prove in that case, and we can scale them by their $\|\cdot\|_\infty$ norm and assume that $\|u\|_\infty = \|v\|_\infty = 1$. We have the following set inclusion,

$$\{w \geq a\} \supseteq \lambda\{u \geq a\} + (1-\lambda)\{v \geq a\},$$

because if $u(x) \geq a$ and $v(y) \geq a$ then, by assumption,

$$w(\lambda x + (1-\lambda)y) \geq u(x)^{\lambda} v(y)^{1-\lambda} \geq a^{\lambda} a^{1-\lambda} = a.$$

When $a \in [0,1]$, the sets $\{u \geq a\}$ and $\{v \geq a\}$ are not empty and the Brunn-Minkowski inequality implies that

$$\gamma(w \geq a) \geq \lambda \gamma(u \geq a) + (1-\lambda)\gamma(v \geq a).$$

Finally,

$$\begin{aligned}
\int_{\mathbb{R}} w(z)\,dz &= \int_{\mathbb{R}} \int_0^1 \mathrm{I}(x \leq w(z))\,da dz = \int_0^1 \gamma(w \geq a)\,da \\
&\geq \lambda \int_0^1 \gamma(u \geq a)\,da + (1-\lambda) \int_0^1 \gamma(v \geq a)\,da \\
&= \lambda \int_{\mathbb{R}} u(z)\,dz + (1-\lambda) \int_{\mathbb{R}} v(z)\,dz \geq \Big(\int_{\mathbb{R}} u(z)\,dz\Big)^{\lambda} \Big(\int_{\mathbb{R}} v(z)\,dz\Big)^{1-\lambda},
\end{aligned}$$

where in the last step we used the arithmetic-geometric mean inequality. This finishes the proof. □

Remark 4.6. Another common proof of the last step $n = 1$ uses transportation of measure, as follows. We can assume that $\int u = \int v = 1$ by rescaling

$$u \to \frac{u}{\int u}, \quad v \to \frac{v}{\int v}, \quad w \to \frac{w}{(\int u)^{\lambda} (\int v)^{1-\lambda}}.$$

Then we need to show that $\int w \geq 1$. Consider cumulative distribution functions

$$F(x) = \int_{-\infty}^{x} u(y)\,dy \text{ and } G(x) = \int_{-\infty}^{x} v(y)\,dy,$$

and let $x(t)$ and $y(t)$ be their quantile transforms. Then, $F(x(t)) = t$ and $G(y(t)) = t$ for $0 \leq t \leq 1$. By the Lebesgue differentiation theorem, $F'(x) = u(x)$ for all $x \notin \mathscr{N}$ for some set $\mathscr{N}$ of Lebesgue measure zero. Since $x(t)$ is (strictly) monotone, the derivative $x'(t)$ exists for all $t \notin \mathscr{N}'$ for some set $\mathscr{N}'$ of Lebesgue measure zero. Also, notice that

$$x(t) \in \mathscr{N} \Longrightarrow t \in F(\mathscr{N}),$$

and, since F is absolutely continuous, $F(\mathscr{N})$ also has Lebesgue measure zero (Lusin N property). Therefore, for $t \notin F(\mathscr{N}) \cup \mathscr{N}'$, the derivative $x'(t)$ exists and $F'(x(t)) = u(x(t))$. Using the chain rule in the equation $F(x(t)) = t$, for all such t,

$u(x(t))x'(t) = 1$. Similarly, outside of some set of measure zero, $v(y(t))y'(t) = 1$. Now, consider the function $z(t) = \lambda x(t) + (1-\lambda)y(t)$. This function is strictly increasing and differentiable almost everywhere. Therefore,

$$\int_{-\infty}^{+\infty} w(z)\,dz \geq \int_0^1 w(z(t))z'(t)\,dt = \int_0^1 w\big(\lambda x(t) + (1-\lambda)y(t)\big)z'(t)\,dt.$$

By the arithmetic-geometric mean inequality

$$z'(t) = \lambda x'(t) + (1-\lambda)y'(t) \geq x'(t)^{\lambda} y'(t)^{1-\lambda}$$

and, by assumption,

$$w(\lambda x(t) + (1-\lambda)y(t)) \geq u(x(t))^{\lambda} v(y(t))^{1-\lambda}.$$

Therefore,

$$\int_{-\infty}^{\infty} w(z)dz \geq \int_0^1 \Big(u(x(t))x'(t)\Big)^{\lambda} \Big(v(y(t))y'(t)\Big)^{1-\lambda} dt = \int_0^1 1\,dt = 1.$$

This finishes the proof. □

Entropy and the Kullback-Leibler divergence. Consider a probability measure $\mathbb{P}$ on some measurable space and a nonnegative measurable function $u\colon \Omega \to \mathbb{R}_+$. We define the *entropy* of u with respect to $\mathbb{P}$ by

$$\mathrm{Ent}_{\mathbb{P}}(u) = \int u \log u\,d\mathbb{P} - \int u\,d\mathbb{P} \cdot \log \int u\,d\mathbb{P},$$

where $0 \cdot \log 0 = 0$. Notice that $\mathrm{Ent}_{\mathbb{P}}(u) \geq 0$, by Jensen's inequality, since $u \log u$ is a convex function. Entropy has the following variational representation, known in physics as the Gibbs variational principle (see exercise below).

Lemma 4.12. *The entropy can be written as*

$$\mathrm{Ent}_{\mathbb{P}}(u) = \sup\Big\{\int uv\,d\mathbb{P} \ :\ \int e^v\,d\mathbb{P} \leq 1\Big\}. \tag{4.17}$$

Proof. Take any measurable function v such that $\int e^v\,d\mathbb{P} \leq 1$. Then, for any $\lambda \geq 0$,

$$\int uv\,d\mathbb{P} \leq \int uv\,d\mathbb{P} + \lambda\Big(1 - \int e^v\,d\mathbb{P}\Big) = \lambda + \int (uv - \lambda e^v)\,d\mathbb{P}.$$

For $\lambda > 0$, the integrand $(uv - \lambda e^v)$ is strictly concave in v and can be maximized pointwise by taking v such that $u = \lambda e^v$. Therefore, for $\int e^v d\mathbb{P} \leq 1$ and $\lambda > 0$,

$$\int uv\,d\mathbb{P} \leq \lambda + \int u \log\frac{u}{\lambda}\,d\mathbb{P} - \int u\,d\mathbb{P}.$$

The right hand side is convex in λ for $\lambda \geq 0$ (since u is nonnegative), the minimum is achieved on $\lambda = \int u\, d\mathbb{P}$, and with this choice of λ the right hand side is equal to $\mathrm{Ent}_{\mathbb{P}}(u)$. On the other hand, it is easy to see that this upper bound is achieved on the function $v = \log\big(u / \int u\, d\mathbb{P}\big)$, which satisfies $\int e^v d\mathbb{P} = 1$. □

Suppose now that a law $\mathbb{Q}$ is absolutely continuous with respect to $\mathbb{P}$ and denote its Radon-Nikodym derivative by

$$u = \frac{d\mathbb{Q}}{d\mathbb{P}}.$$

Then the *Kullback-Leibler divergence of* $\mathbb{P}$ *from* $\mathbb{Q}$ is defined by

$$D(\mathbb{Q}\,\|\,\mathbb{P}) := \int \log u\, d\mathbb{Q} = \int \log \frac{d\mathbb{Q}}{d\mathbb{P}}\, d\mathbb{Q}. \tag{4.18}$$

Notice that this quantity is not symmetric in $\mathbb{P}$ and $\mathbb{Q}$, so the order is important. Clearly, $D(\mathbb{Q}\,\|\,\mathbb{P}) = \mathrm{Ent}_{\mathbb{P}}(u)$, since

$$\mathrm{Ent}_{\mathbb{P}}(u) = \int \log \frac{d\mathbb{Q}}{d\mathbb{P}} \cdot \frac{d\mathbb{Q}}{d\mathbb{P}}\, d\mathbb{P} - \int \frac{d\mathbb{Q}}{d\mathbb{P}}\, d\mathbb{P} \cdot \log \int \frac{d\mathbb{Q}}{d\mathbb{P}}\, d\mathbb{P} = \int \log \frac{d\mathbb{Q}}{d\mathbb{P}}\, d\mathbb{Q}.$$

The variational characterization (4.17) implies that

$$\text{if } \int e^v\, d\mathbb{P} \leq 1 \text{ then } \int v\, d\mathbb{Q} = \int uv\, d\mathbb{P} \leq D(\mathbb{Q}\,\|\,\mathbb{P}). \tag{4.19}$$

Talagrand's quadratic transportation cost inequality for Gaussian measures. In this subsection we consider a non-degenerate normal distribution $N(0,C)$ with the covariance matrix C such that $\det(C) \neq 0$. We know that this distribution has density $e^{-V(x)}$, where

$$V(x) = \frac{1}{2}(C^{-1}x, x) + \text{const}.$$

If we denote $A = C^{-1}/2$ then, for any $t \in [0,1]$,

$$\begin{aligned} &tV(x) + (1-t)V(y) - V(tx + (1-t)y) \\ &= t(Ax,x) + (1-t)(Ay,y) - \Big(A(tx+(1-t)y), (tx+(1-t)y)\Big) \\ &= t(1-t)\big(A(x-y),(x-y)\big) \\ &\geq \frac{1}{2\lambda_{\max}(C)} t(1-t)|x-y|^2 = Kt(1-t)|x-y|^2, \end{aligned} \tag{4.20}$$

where $\lambda_{\max}(C)$ is the largest eigenvalue of C and $K = 1/(2\lambda_{\max}(C))$. We will use this to prove the following useful inequality for the Wasserstein distance W_2 defined at the end of the previous section.

Theorem 4.19. *If* $\mathbb{P} = N(0,C)$ *and* $\mathbb{Q}$ *is absolutely continuous with respect to* $\mathbb{P}$ *with* $\int |x|^2 \, d\mathbb{Q}(x) < \infty$ *then*

$$W_2(\mathbb{Q},\mathbb{P})^2 \le 2\lambda_{\max}(C) D(\mathbb{Q}\,\|\,\mathbb{P}). \tag{4.21}$$

Proof. Using the Kantorovich-Rubinstein theorem (see (4.9)), write

$$\begin{aligned} W_2(\mathbb{Q},\mathbb{P})^2 &= \frac{1}{K}\inf\Big\{\int K|x-y|^2\, d\mu(x,y) \;:\; \mu \in M(\mathbb{P},\mathbb{Q})\Big\} \\ &= \frac{1}{K}\sup\Big\{\int f\, d\mathbb{P} + \int g\, d\mathbb{Q} \;:\; f(x)+g(y) \le K|x-y|^2\Big\}. \end{aligned}$$

Consider functions $f,g \in C(\mathbb{R}^n)$ such that $f(x)+g(y) \le K|x-y|^2$. Then, by (4.20), for any $t \in (0,1)$,

$$f(x)+g(y) \le \frac{1}{t(1-t)}\Big(tV(x)+(1-t)V(y)-V(tx+(1-t)y)\Big)$$

and

$$t(1-t)f(x)-tV(x)+t(1-t)g(y)-(1-t)V(y) \le -V(tx+(1-t)y).$$

If we introduce the functions

$$u(x) = e^{(1-t)f(x)-V(x)},\ v(y) = e^{tg(y)-V(y)} \text{ and } w(z) = e^{-V(z)}$$

then $w(tx+(1-t)y) \ge u(x)^t v(y)^{1-t}$ and, by the Prekopa-Leindler inequality,

$$\Big(\int e^{(1-t)f(x)-V(x)}\, dx\Big)^t \Big(\int e^{tg(x)-V(x)}\, dx\Big)^{1-t} \le \int e^{-V(x)}\, dx.$$

Since e^{-V} is the density of $\mathbb{P}$, we showed that

$$\Big(\int e^{(1-t)f}\, d\mathbb{P}\Big)^t \Big(\int e^{tg}\, d\mathbb{P}\Big)^{1-t} \le 1, \text{ so } \Big(\int e^{(1-t)f}\, d\mathbb{P}\Big)^{\frac{1}{1-t}} \Big(\int e^{tg}\, d\mathbb{P}\Big)^{\frac{1}{t}} \le 1.$$

It is a simple calculus exercise to show that

$$\lim_{s\downarrow 0}\Big(\int e^{sf}\, d\mathbb{P}\Big)^{\frac{1}{s}} = e^{\int f\, d\mathbb{P}},$$

and, therefore, letting $t \uparrow 1$ proves that

$$\int e^g\, d\mathbb{P} \cdot e^{\int f\, d\mathbb{P}} \le 1.$$

This inequality can be written as $\int e^v d\mathbb{P} \le 1$ for $v = g + \int f\, d\mathbb{P}$ and (4.19) implies that

$$\int v\,d\mathbb{Q} = \int f\,d\mathbb{P} + \int g\,d\mathbb{Q} \le D(\mathbb{Q}\,\|\,\mathbb{P}).$$

Together with the Kantorovich-Rubinstein representation above this implies that

$$W_2(\mathbb{Q},\mathbb{P})^2 \le \frac{1}{K} D(\mathbb{Q}\,\|\,\mathbb{P}) = 2\lambda_{\max}(C)\, D(\mathbb{Q}\,\|\,\mathbb{P}),$$

which finishes the proof. □

Concentration of Gaussian measure. Given a measurable set $A \subseteq \mathbb{R}^n$ with $\mathbb{P}(A) > 0$, define the distribution $\mathbb{P}_A$ by

$$\mathbb{P}_A(C) = \frac{\mathbb{P}(C \cap A)}{\mathbb{P}(A)}.$$

Then, obviously, the Radon-Nikodym derivative

$$\frac{d\mathbb{P}_A}{d\mathbb{P}} = \frac{1}{\mathbb{P}(A)} \mathrm{I}_A$$

and the Kullback-Leibler divergence

$$D(\mathbb{P}_A\,\|\,\mathbb{P}) = \int_A \log \frac{1}{\mathbb{P}(A)}\, d\mathbb{P}_A = \log \frac{1}{\mathbb{P}(A)}.$$

Since W_2 is a metric, for any two Borel sets A and B,

$$\begin{aligned} W_2(\mathbb{P}_A,\mathbb{P}_B) &\le W_2(\mathbb{P}_A,\mathbb{P}) + W_2(\mathbb{P}_B,\mathbb{P}) \\ &\le \sqrt{2\lambda_{\max}(C)}\Big(\log^{1/2}\frac{1}{\mathbb{P}(A)} + \log^{1/2}\frac{1}{\mathbb{P}(B)}\Big), \end{aligned}$$

using (4.21). Suppose that the sets A and B are apart from each other by a distance t, i.e. $d(A,B) \ge t > 0$. Then any two points in the support of measures $\mathbb{P}_A$ and $\mathbb{P}_B$ are at a distance at least t from each other, which implies that the transportation distance $W_2(\mathbb{P}_A,\mathbb{P}_B) \ge t$. Therefore,

$$\begin{aligned} t \le W_2(\mathbb{P}_A,\mathbb{P}_B) &\le \sqrt{2\lambda_{\max}(C)}\Big(\log^{1/2}\frac{1}{\mathbb{P}(A)} + \log^{1/2}\frac{1}{\mathbb{P}(B)}\Big) \\ &\le \sqrt{4\lambda_{\max}(C)}\log^{1/2}\frac{1}{\mathbb{P}(A)\mathbb{P}(B)}. \end{aligned}$$

Therefore,

$$\mathbb{P}(B) \le \frac{1}{\mathbb{P}(A)} \exp\Big(-\frac{t^2}{4\lambda_{\max}(C)}\Big).$$

In particular, if $B = \{x : d(x,A) \ge t\}$ then

$$\mathbb{P}\big(d(x,A) \ge t\big) \le \frac{1}{\mathbb{P}(A)} \exp\Big(-\frac{t^2}{4\lambda_{\max}(C)}\Big).$$

If the set A is not too small, e.g. $\mathbb{P}(A) \geq 1/2$, this implies that

$$\mathbb{P}\big(d(x,A) \geq t\big) \leq 2\exp\Big(-\frac{t^2}{4\lambda_{\max}(C)}\Big).$$

This shows that the Gaussian measure is exponentially concentrated near any "large enough" set. The constant $1/4$ in the exponent is not optimal and can be replaced by $1/2$; this is just an example of application of the above ideas. The optimal result is the famous Gaussian isoperimetry,

$$\text{if } \mathbb{P}(A) = \mathbb{P}(B) \text{ for some half-space } B \text{ then } \mathbb{P}(A^t) \geq \mathbb{P}(B^t).$$

Gaussian concentration via infimum-convolution. If we denote $c := 1/\lambda_{\max}(C)$ then setting $t = 1/2$ in (4.20),

$$V(x) + V(y) - 2V\Big(\frac{x+y}{2}\Big) \geq \frac{c}{4}|x-y|^2.$$

Given a function f on $\mathbb{R}^n$, let us define its *infimum-convolution* by

$$g(y) = \inf_x \Big(f(x) + \frac{c}{4}|x-y|^2\Big).$$

Then, for all x and y, we have the inequality

$$g(y) - f(x) \leq \frac{c}{4}|x-y|^2 \leq V(x) + V(y) - 2V\Big(\frac{x+y}{2}\Big). \tag{4.22}$$

If we consider the functions $u(x) = e^{-f(x)-V(x)}$, $v(y) = e^{g(y)-V(y)}$ and $w(z) = e^{-V(z)}$, then (4.22) implies that

$$w\Big(\frac{x+y}{2}\Big) \geq u(x)^{1/2} v(y)^{1/2}$$

and the Prekopa-Leindler inequality with $\lambda = 1/2$ implies that

$$\int e^g\, d\mathbb{P} \int e^{-f}\, d\mathbb{P} \leq 1. \tag{4.23}$$

Given a measurable set A, let f be equal to 0 on A and $+\infty$ on the complement of A. Then $g(y) = \frac{c}{4} d(x,A)^2$ and (4.23) implies

$$\int \exp \frac{c}{4} d(x,A)^2\, d\mathbb{P}(x) \leq \frac{1}{\mathbb{P}(A)}.$$

By Chebyshev's inequality,

$$\mathbb{P}\big(d(x,A) \geq t\big) \leq \frac{1}{\mathbb{P}(A)} \exp\Big(-\frac{ct^2}{4}\Big) = \frac{1}{\mathbb{P}(A)} \exp\Big(-\frac{t^2}{4\lambda_{\max}(C)}\Big),$$

which is the same Gaussian concentration inequality we proved above. □

Discrete metric and total variation. The *total variation distance* between two probability measures $\mathbb{P}$ and $\mathbb{Q}$ on a measurable space $(S,\mathscr{B})$ is defined by

$$\mathrm{TV}(\mathbb{P},\mathbb{Q}) = \sup_{A\in\mathscr{B}} |\mathbb{P}(A)-\mathbb{Q}(A)|.$$

Using the Hahn-Jordan decomposition, we can represent a signed measure $\mu = \mathbb{P}-\mathbb{Q}$ as $\mu = \mu^+ - \mu^-$ such that, for some set $D\in\mathscr{B}$ and for any set $E\in\mathscr{B}$,

$$\mu^+(E) = \mu(ED) \geq 0 \text{ and } \mu^-(E) = -\mu(ED^c) \geq 0.$$

Therefore, for any $A\in\mathscr{B}$,

$$\mathbb{P}(A)-\mathbb{Q}(A) = \mu^+(A)-\mu^-(A) = \mu^+(AD)-\mu^-(AD^c),$$

which makes it obvious that

$$\sup_{A\in\mathscr{B}} |\mathbb{P}(A)-\mathbb{Q}(A)| = \mu^+(D).$$

Let us describe some connections of the total variation distance to the Kullback-Leibler divergence and the Kantorovich-Rubinstein theorem. Let us start with the following simple observation.

Lemma 4.13. *If f is a measurable function on S such that $|f|\leq 1$ and $\int f\,d\mathbb{P} = 0$ then for any $\lambda\in\mathbb{R}$,*

$$\int e^{\lambda f}\,d\mathbb{P} \leq e^{\lambda^2/2}.$$

Proof. Since $(1+f)/2, (1-f)/2 \in [0,1]$ and

$$\lambda f = \frac{1+f}{2}\lambda + \frac{1-f}{2}(-\lambda),$$

by the convexity of e^x, we get

$$e^{\lambda f} \leq \frac{1+f}{2}e^{\lambda} + \frac{1-f}{2}e^{-\lambda} = \mathrm{ch}(\lambda) + f\,\mathrm{sh}(\lambda).$$

Therefore,

$$\int e^{\lambda f} d\mathbb{P} \leq \mathrm{ch}(\lambda) \leq e^{\lambda^2/2},$$

where the last inequality is easy to see by Taylor's expansion. □

Let us now consider a *discrete metric* on S given by

$$d(x,y) = \mathrm{I}(x\neq y). \tag{4.24}$$

Then a 1-Lipschitz function f with respect to the metric d, $\|f\|_L \le 1$, is defined by the condition that for all $x, y \in S$,

$$|f(x) - f(y)| \le 1. \tag{4.25}$$

Formally, the Kantorovich-Rubinstein theorem in this case would state that

$$\begin{aligned} W(\mathbb{P},\mathbb{Q}) &= \inf\Bigl\{\int \mathrm{I}(x \ne y)\, d\mu(x,y) : \mu \in M(\mathbb{P},\mathbb{Q})\Bigr\} \\ &= \sup\Bigl\{\Bigl|\int f\, d\mathbb{Q} - \int f\, d\mathbb{P}\Bigr| : \|f\|_L \le 1\Bigr\} =: \gamma(\mathbb{P},\mathbb{Q}). \end{aligned}$$

However, since any uncountable set S is not separable w.r.t. the discrete metric d, we can not apply the Kantorovich-Rubinstein theorem directly. In this case, one can use the Hahn-Jordan decomposition to show that W coincides with the total variation distance, $W(\mathbb{P},\mathbb{Q}) = \mathrm{TV}(\mathbb{P},\mathbb{Q})$ and it is easy to construct a measure $\mu \in M(\mathbb{P},\mathbb{Q})$ explicitly that witnesses the above equality. This was one of the exercises at the end of the previous section. One can also easily check directly that $\gamma(\mathbb{P},\mathbb{Q}) = \mathrm{TV}(\mathbb{P},\mathbb{Q})$. Thus, for the discrete metric d,

$$W(\mathbb{P},\mathbb{Q}) = \mathrm{TV}(\mathbb{P},\mathbb{Q}) = \gamma(\mathbb{P},\mathbb{Q}).$$

We have the following analogue of the Kullback-Leibler divergence bound in Theorem 4.19, which now holds for any measure $\mathbb{P}$, not only Gaussian.

Theorem 4.20 (Pinsker's inequality). *If $\mathbb{Q}$ is absolutely continuous with respect to $\mathbb{P}$ then* $\mathrm{TV}(\mathbb{P},\mathbb{Q}) \le \sqrt{2D(\mathbb{Q}\,\|\,\mathbb{P})}$.

Proof. Take f such that (4.25) holds. If we define $g(x) = f(x) - \int f\, d\mathbb{P}$ then, clearly, $|g| \le 1$ and $\int g\, d\mathbb{P} = 0$. The above lemma implies that for any $\lambda \in \mathbb{R}$,

$$\int e^{\lambda f - \lambda \int f\, d\mathbb{P} - \lambda^2/2}\, d\mathbb{P} \le 1.$$

The variational characterization of entropy (4.19) implies that

$$\lambda \int f\, d\mathbb{Q} - \lambda \int f\, d\mathbb{P} - \lambda^2/2 \le D(\mathbb{Q}\,\|\,\mathbb{P})$$

and for $\lambda > 0$ we get

$$\int f\, d\mathbb{Q} - \int f\, d\mathbb{P} \le \frac{\lambda}{2} + \frac{1}{\lambda} D(\mathbb{Q}\,\|\,\mathbb{P}).$$

Minimizing the right hand side over $\lambda > 0$, we get

$$\int f\, d\mathbb{Q} - \int f\, d\mathbb{P} \le \sqrt{2D(\mathbb{Q}\,\|\,\mathbb{P})}.$$

Applying this to f and $-f$ yields the result. □

Hamming metric on a product space. Let us consider a finite set A and, given integer $n \geq 1$, consider the following Hamming metric on A^n,

$$d_H(x,y) = \frac{1}{n}\sum_{i=1}^{n} \mathrm{I}(x_i \neq y_i).$$

For two measures μ and ν on A^n, consider the corresponding Wasserstein distance

$$\begin{aligned} W_H(\mu,\nu) &= \inf\Big\{ \int d_H(x,y)\,d\lambda(x,y) \, : \, \lambda \in M(\mu,\nu)\Big\} \\ &= \inf\Big\{ \frac{1}{n}\sum_{i=1}^{n} \lambda(x_i \neq y_i) \, : \, \lambda \in M(\mu,\nu)\Big\}. \end{aligned}$$

Since $d_H(x,y) \leq \mathrm{I}(x \neq y)$, by Pinsker's inequality,

$$W_H(\mu,\nu) \leq \mathrm{TV}(\mu,\nu) \leq \sqrt{2D(\mu \,\|\, \nu)}. \tag{4.26}$$

On the other hand, on the product space A^n, one is often interested to understand how different a measure μ is from some (or any) product measure $\nu = \nu_1 \times \ldots \times \nu_n$ with respect to the above Wasserstein metric. Consider the function

$$\phi_A(x) = -x\log x - (1-x)\log(1-x) + x\log \mathrm{card}(A), \; x \in [0,1].$$

This function is concave, $\phi_A(x) \geq 0$ and it is equal to zero only at $x = 0$. The following reverse analogue of the above Pinsker's inequality holds.

Lemma 4.14. *If $\mu_1,\ldots,\mu_n$ are the marginals of μ then, for any product measure $\nu = \nu_1 \times \ldots \times \nu_n$ on A^n,*

$$\phi_A(W_H(\mu,\nu)) \geq \frac{1}{2n} D(\mu \,\|\, \mu_1 \times \ldots \times \mu_n). \tag{4.27}$$

In particular,

$$\phi_A(W_H(\mu,\mu_1 \times \ldots \times \mu_n)) \geq \frac{1}{2n} D(\mu \,\|\, \mu_1 \times \ldots \times \mu_n).$$

For the proof, we refer to Lemma 4.4 in arXiv:1705.00302. One can also rewrite the KL divergence on the right hand side as

$$D(\mu \,\|\, \mu_1 \times \ldots \times \mu_n) = \sum_{i=1}^{n} H(\mu_i) - H(\mu),$$

where $H(\mathbb{P}) = -\sum_s \mathbb{P}(s)\log \mathbb{P}(s)$ is called *the entropy* of the discrete measure $\mathbb{P}$. The inequality (4.27) then shows that the KL divergence between μ and the product measure with the same marginals, $\mu_1 \times \ldots \times \mu_n$ (which is an explicit quantity that in principle should be easier to compute), can be used to control from below the Wasserstein distance $W_H(\mu,\nu)$ between μ and an arbitrary product measure ν with respect to the Hamming distance d_H. Let us also note that one can strengthen the inequality (4.27) by replacing $\phi_A(W_H(\mu,\nu))$ with

$$\inf\Big\{\frac{1}{n}\sum_{i=1}^{n}\phi_A(\lambda(x_i\neq y_i)) \,:\, \lambda\in M(\mu,\nu)\Big\}.$$

Exercise 4.5.1. Prove the Brunn-Minkowski inequality (4.16) on $\mathbb{R}^n$ using the Prekopa-Leindler inequality. *Hint*: Apply the Prekopa-Leindler inequality to sets A/λ and $B/(1-\lambda)$ and optimize over $\lambda\in(0,1)$.

Exercise 4.5.2. Using the Prekopa-Leindler inequality prove that if γ_N is the standard Gaussian measure on $\mathbb{R}^N$, A and B are Borel sets and $\lambda\in[0,1]$ then

$$\log\gamma_N(\lambda A+(1-\lambda)B)\geq\lambda\log\gamma_N(A)+(1-\lambda)\log\gamma_N(B).$$

Exercise 4.5.3 (Anderson's inequality). If C is convex and symmetric around 0, i.e. $-C=C$, then for any $z\in\mathbb{R}^N$, $\gamma_N(C)\geq\gamma_N(C+z)$. More generally, for any centred Gaussian distribution $\mathbb{P}=N(0,\Sigma)$ on $\mathbb{R}^N$, $\mathbb{P}(C)\geq\mathbb{P}(C+z)$. *Hint:* use the previous problem.

Exercise 4.5.4. If C is convex and symmetric around 0, X and Y are (centred) Gaussian on $\mathbb{R}^N$ and $\mathrm{Cov}(X)\geq\mathrm{Cov}(Y)$ (i.e. their difference is nonnegative definite) then $\mathbb{P}(X\in C)\leq\mathbb{P}(Y\in C)$. *Hint:* use the previous problem.

Exercise 4.5.5. Suppose that g is standard Gaussian on $\mathbb{R}^n$ and $f\colon\mathbb{R}^n\to\mathbb{R}$ is a Lipschitz function with $\|f\|_{\mathrm{L}}=a$. If M is a median of $f(g)$, prove that

$$\mathbb{P}\big(f(g)>M+at\big)\leq 2e^{-t^2/4}.$$

Hint: Use Gaussian concentration inequality. Median means that $\mathbb{P}(f(g)\geq M)\geq 1/2$ and $\mathbb{P}(f(g)\leq M)\geq 1/2$.

Exercise 4.5.6 (Gibbs' variational principle). Given a probability measure $\mathbb{P}$ on $\mathbb{R}^n$ and a function v such that $\int e^v\,d\mathbb{P}<\infty$, prove that

$$\log\int e^v d\mathbb{P}=\max_{\mathbb{Q}}\Big\{\int v\,d\mathbb{Q}-\int\log\frac{d\mathbb{Q}}{d\mathbb{P}}\,d\mathbb{Q}\Big\},$$

where the maximum is taken over all probability measures $\mathbb{Q}$ absolutely continuous with respect to $\mathbb{P}$ and is achieved on $d\mathbb{Q}\sim e^v d\mathbb{P}$.

Chapter 5
Martingales

5.1 Martingales. Uniform integrability

Let $(\Omega, \mathscr{B}, \mathbb{P})$ be a probability space and let $(T, \leq)$ be a linearly ordered set. We will mostly work with countable sets T, such as $\mathbb{Z}$ or subsets of $\mathbb{Z}$. Consider a family of random variables $X_t : \Omega \to \mathbb{R}$ and a family of σ-algebras $(\mathscr{B}_t)_{t\in T}$ such that $\mathscr{B}_t \subseteq \mathscr{B}_u \subseteq \mathscr{B}$ for $t \leq u$. A family $(X_t, \mathscr{B}_t)_{t\in T}$ is called a *martingale* if the following hold:

1. X_t is $\mathscr{B}_t$-measurable for all $t \in T$ (X_t is *adapted* to $\mathscr{B}_t$),
2. $\mathbb{E}|X_t| < \infty$ for all $t \in T$,
3. $\mathbb{E}(X_u|\mathscr{B}_t) = X_t$ for all $t \leq u$.

If the last equality is replaced by $\mathbb{E}(X_u|\mathscr{B}_t) \leq X_t$ then the process is called a *supermartingale* and if $\mathbb{E}(X_u|\mathscr{B}_t) \geq X_t$ then it is called a *submartingale.* If $(X_t, \mathscr{B}_t)$ is a martingale and $X_t = \mathbb{E}(X|\mathscr{B}_t)$ for some random variables X then the martingale is called *right-closable.* If some $t_0 \in T$ is an upper bound of T (i.e. $t \leq t_0$ for all $t \in T$) then the martingale is called *right-closed.* Of course, in this case $X_t = \mathbb{E}(X_{t_0}|\mathscr{B}_t)$. Clearly, a right-closable martingale can be made into a right-closed martingale by adding an additional point, say t_0, to the set T so that $t \leq t_0$ for all $t \in T$, and defining $X_{t_0} := X$. Let us give several examples of martingales.

Example 5.1.1. Consider a sequence $(X_n)_{n\geq 1}$ of independent random variables such that $\mathbb{E}X_i = 0$ and let $S_n = \sum_{i\leq n} X_i$. If $\mathscr{B}_n = \sigma(X_1, \ldots, X_n)$ then $(S_n, \mathscr{B}_n)_{n\geq 1}$ is a martingale since, for $m \geq n$,

$$\mathbb{E}(S_m|\mathscr{B}_n) = \mathbb{E}(S_n + X_{n+1} + \ldots + X_m|\mathscr{B}_n) = S_n + \mathbb{E}X_{n+1} + \ldots + \mathbb{E}X_m = S_n,$$

since $X_{n+1}, \ldots, X_m$ are independent of $\mathscr{B}_n$ and S_n is $\mathscr{B}_n$-measurable.

Example 5.1.2. Consider a sequence of σ-algebras $\ldots \subseteq \mathscr{B}_n \subseteq \mathscr{B}_{n+1} \subseteq \ldots \subseteq \mathscr{B}$ and a random variable X such that $\mathbb{E}|X| < \infty$. If we let $X_n = \mathbb{E}(X|\mathscr{B}_n)$ then $(X_n, \mathscr{B}_n)$ is a right-closable martingale, since for $n < m$,

$$\mathbb{E}(X_m|\mathscr{B}_n) = \mathbb{E}(\mathbb{E}(X|\mathscr{B}_m)|\mathscr{B}_n) = \mathbb{E}(X|\mathscr{B}_n) = X_n,$$

by the smoothing property of conditional expectation.

Example 5.1.3. As a typical example, one can consider an integrable function $X = f(\xi_1,\ldots,\xi_n)$ of some random variables $\xi_1,\ldots,\xi_n$, let $\mathscr{B}_0$ be the trivial σ-algebra and $\mathscr{B}_m = \sigma(\xi_1,\ldots,\xi_m)$ for $m \le n$. If we let $X_m = \mathbb{E}(X|\mathscr{B}_m)$ then $X_0 = \mathbb{E}X$, $X_n = X$ and the representation

$$X = \mathbb{E}X + \sum_{m=1}^{n} (X_m - X_{m-1})$$

is called the martingale-difference representation of X. This representation is particularly useful when $\xi_1,\ldots,\xi_n$ are independent, because in this case X_{m-1} is just the expectation of X_m with respect to ξ_m.

Example 5.1.4. Let $(X_i)_{i\ge 1}$ be i.i.d. and let $S_n = \sum_{i\le n} X_i$. Let $T = \{\ldots,-2,-1\}$ be the set of negative integers and, for $n \ge 1$, let us define

$$\mathscr{B}_{-n} = \sigma(S_n, S_{n+1},\ldots) = \sigma(S_n, X_{n+1}, X_{n+2},\ldots).$$

Clearly, $\mathscr{B}_{-(n+1)} \subseteq \mathscr{B}_{-n}$. For $1 \le k \le n$, by symmetry (we leave the details for the exercise at the end of the section),

$$\mathbb{E}(X_1|\mathscr{B}_{-n}) = \mathbb{E}(X_k|\mathscr{B}_{-n}).$$

Therefore,

$$S_n = \mathbb{E}(S_n|\mathscr{B}_{-n}) = \sum_{1\le k\le n} \mathbb{E}(X_k|\mathscr{B}_{-n}) = n\mathbb{E}(X_1|\mathscr{B}_{-n})$$

and $Z_{-n} := \frac{S_n}{n} = \mathbb{E}(X_1|\mathscr{B}_{-n})$. Thus, $(Z_{-n}, \mathscr{B}_{-n})_{-n\le -1}$ is a right-closed martingale.

Theorem 5.1 (Doob's decomposition). *If $(X_n,\mathscr{B}_n)_{n\ge 0}$ is a submartingale then it can be uniquely decomposed $X_n = Z_n + Y_n$, where $(Y_n,\mathscr{B}_n)$ is a martingale, $Z_0 = 0, Z_n \le Z_{n+1}$ almost surely and Z_n is $\mathscr{B}_{n-1}$-measurable.*

The sequence (Z_n) is called *predictable*, since it is a function of $X_1,\ldots,X_{n-1}$, so we know it at time $n-1$. Since an increasing sequence is always convergent, the question of convergence for submartingales is reduced to the question of convergence for martingales.

Proof. Let $D_n = X_n - X_{n-1}$ and

$$G_n = \mathbb{E}(D_n|\mathscr{B}_{n-1}) = \mathbb{E}(X_n|\mathscr{B}_{n-1}) - X_{n-1} \ge 0$$

by the definition of submartingale. Let,

$$H_n = D_n - G_n,\; Y_n = H_1 + \ldots + H_n,\; Z_n = G_1 + \cdots + G_n.$$

Since $G_n \geq 0$ a.s., $Z_n \leq Z_{n+1}$ and, by construction, Z_n is $\mathscr{B}_{n-1}$-measurable. We have,

$$\mathbb{E}(H_n|\mathscr{B}_{n-1}) = \mathbb{E}(D_n|\mathscr{B}_{n-1}) - G_n = 0$$

and, therefore, $\mathbb{E}(Y_n|\mathscr{B}_{n-1}) = Y_{n-1}$. Uniqueness follows by construction. Suppose that $X_n = Z_n + Y_n$ with all stated properties. First, since $Z_0 = 0$, $Y_0 = X_0$. By induction, given a unique decomposition up to $n-1$, we can write

$$Z_n = \mathbb{E}(Z_n|\mathscr{B}_{n-1}) = \mathbb{E}(X_n - Y_n|\mathscr{B}_{n-1}) = \mathbb{E}(X_n|\mathscr{B}_{n-1}) - Y_{n-1}$$

and $Y_n = X_n - Z_n$. This finishes the proof. □

When we study convergence properties of martingales, an important role will be played by their uniform integrability properties. We say that a collection of random variables $(X_t)_{t\in T}$ is *uniformly integrable* if

$$\sup_t \mathbb{E}|X_t|\mathrm{I}(|X_t| > M) \to 0 \text{ as } M \to \infty. \tag{5.1}$$

For example, when $|X_t| \leq Y$ for all $t \in T$ and $\mathbb{E}Y < \infty$ then, clearly, $(X_t)_{t\in T}$ is uniformly integrable. Other basic criteria of uniform integrability will be given in the exercises below. The following criterion of the L^1-convergence, which can be viewed as a strengthening of the dominated convergence theorem, will be useful.

Lemma 5.1. *Consider random variables (X_n), X, such that $\mathbb{E}|X_n| < \infty$, $\mathbb{E}|X| < \infty$. The following are equivalent:*

1. *$\mathbb{E}|X_n - X| \to 0$ as $n \to \infty$.*
2. *$(X_n)_{n\geq 1}$ is uniformly integrable and $X_n \to X$ in probability.*

Proof. 2⟹1. For any $\varepsilon > 0$ and $K > 0$, we can write,

$$\begin{aligned}
\mathbb{E}|X_n - X| &\leq \varepsilon + \mathbb{E}|X_n - X|\mathrm{I}(|X_n - X| > \varepsilon) \\
&\leq \varepsilon + 2K\mathbb{P}(|X_n - X| > \varepsilon) + 2\mathbb{E}|X_n|\mathrm{I}(|X_n| > K) + 2\mathbb{E}|X|\mathrm{I}(|X| > K) \\
&\leq \varepsilon + 2K\mathbb{P}(|X_n - X| > \varepsilon) + 2\sup_n \mathbb{E}|X_n|\mathrm{I}(|X_n| > K) + 2\mathbb{E}|X|\mathrm{I}(|X| > K).
\end{aligned}$$

Since $X_n \to X$ in probability, letting $n \to \infty$,

$$\limsup_{n\to\infty} \mathbb{E}|X_n - X| \leq \varepsilon + 2\sup_n \mathbb{E}|X_n|\mathrm{I}(|X_n| > K) + 2\mathbb{E}|X|\mathrm{I}(|X| > K).$$

Now, first letting $\varepsilon \to 0$, and then letting $K \to \infty$ and using that (X_n) is uniformly integrable proves the result.

1⟹2. By Chebyshev's inequality,

$$\mathbb{P}(|X_n - X| > \varepsilon) \leq \frac{1}{\varepsilon}\mathbb{E}|X_n - X| \to 0$$

as $n \to \infty$, so $X_n \to X$ in probability. To prove uniform integrability, let us recall that for any $\varepsilon > 0$ there exists $\delta > 0$ such that $\mathbb{P}(A) \le \delta$ implies that $\mathbb{E}|X|\mathrm{I}_A \le \varepsilon$. This is because

$$\mathbb{E}|X|\mathrm{I}_A \le K\mathbb{P}(A) + \mathbb{E}|X|\mathrm{I}(|X| \ge K),$$

and if we take K such that $\mathbb{E}|X|\mathrm{I}(|X| \ge K) \le \varepsilon/2$ then it is enough to take $\delta = \varepsilon/(2K)$. Now, given $\varepsilon > 0$, take δ as above and take $M > 0$ large enough, so that for all $n \ge 1$,

$$\mathbb{P}(|X_n| > M) \le \frac{\mathbb{E}|X_n|}{M} \le \delta.$$

We showed that $\mathbb{E}|X|\mathrm{I}(|X_n| > M) \le \varepsilon$ for such δ and, therefore,

$$\mathbb{E}|X_n|\mathrm{I}(|X_n| > M) \le \mathbb{E}|X_n - X| + \mathbb{E}|X|\mathrm{I}(|X_n| > M) \le \mathbb{E}|X_n - X| + \varepsilon.$$

For large enough $n \ge n_0$, $\mathbb{E}|X_n - X| \le \varepsilon$ and, therefore,

$$\mathbb{E}|X_n|\mathrm{I}(|X_n| > M) \le 2\varepsilon.$$

We can also choose M large enough so that $\mathbb{E}|X_n|\mathrm{I}(|X_n| > M) \le 2\varepsilon$ for $n \le n_0$ and this finishes the proof. □

Next, we will prove some uniform integrability properties for martingales and submartingales, but first let us make the following simple observation.

Lemma 5.2. *Let $f : \mathbb{R} \to \mathbb{R}$ be a convex function such that $\mathbb{E}|f(X_t)| < \infty$ for all t. Suppose that either of the following two conditions holds:*

1. *$(X_t, \mathscr{B}_t)$ is a martingale;*
2. *$(X_t, \mathscr{B}_t)$ is a submartingale and f is increasing.*

Then $(f(X_t), \mathscr{B}_t)$ is a submartingale.

Proof. Under the first condition, for $t \le u$, by conditional Jensen's inequality,

$$f(X_t) = f(\mathbb{E}(X_u|\mathscr{B}_t)) \le \mathbb{E}(f(X_u)|\mathscr{B}_t).$$

Under the second condition, since $X_t \le \mathbb{E}(X_u|\mathscr{B}_t)$ for $t \le u$ and f is increasing,

$$f(X_t) \le f(\mathbb{E}(X_u|\mathscr{B}_t)) \le \mathbb{E}(f(X_u)|\mathscr{B}_t),$$

where we again used conditional Jensen's inequality. □

Lemma 5.3. *The following holds.*

1. *If $(X_t, \mathscr{B}_t)_{t\in T}$ is a right-closable martingale then (X_t) is uniformly integrable.*
2. *If $(X_t, \mathscr{B}_t)_{t\in T}$ is a right-closable submartingale then $(\max(X_t, a))$ is uniformly integrable for any $a \in \mathbb{R}$.*

The fact that submartingale $(X_t, \mathscr{B}_t)_{t\in T}$ is right-closable means that $X_t \le \mathbb{E}(X|\mathscr{B}_t)$ for some X such that $\mathbb{E}|X| < \infty$.

Proof. 1. Since there exists an integrable random variable X such that $X_t = \mathbb{E}(X|\mathscr{B}_t)$, we have

$$|X_t| = |\mathbb{E}(X|\mathscr{B}_t)| \le \mathbb{E}\big(|X| \mid \mathscr{B}_t\big) \text{ and } \mathbb{E}|X_t| \le \mathbb{E}|X| < \infty.$$

Since $\{|X_t| > M\} \in \mathscr{B}_t$,

$$X_t \mathrm{I}(|X_t| > M) = \mathrm{I}(|X_t| > M)\mathbb{E}(X|\mathscr{B}_t) = \mathbb{E}(X\mathrm{I}(|X_t| > M)|\mathscr{B}_t)$$

and, therefore,

$$\begin{aligned}\mathbb{E}|X_t|\mathrm{I}(|X_t| > M) &\le \mathbb{E}|X|\mathrm{I}(|X_t| > M) \le K\mathbb{P}(|X_t| > M) + \mathbb{E}|X|\mathrm{I}(|X| > K)\\ &\le K\frac{\mathbb{E}|X_t|}{M} + \mathbb{E}|X|\mathrm{I}(|X| > K) \le K\frac{\mathbb{E}|X|}{M} + \mathbb{E}|X|\mathrm{I}(|X| > K).\end{aligned}$$

Letting $M \to \infty, K \to \infty$ proves that $\sup_t \mathbb{E}|X_t|\mathrm{I}(|X_t| > M) \to 0$ as $M \to \infty$.

2. Since $X_t \le \mathbb{E}(X|\mathscr{B}_t)$ and $\{X_t > M\} \in \mathscr{B}_t$, as above this implies that

$$\mathbb{E}X_t\mathrm{I}(X_t > M) \le \mathbb{E}X\mathrm{I}(X_t > M).$$

Also, since the function $\max(a,x)$ is convex and increasing in x, by the part (b) of the previous lemma,

$$\mathbb{E}\max(a,X_t) \le \mathbb{E}\max(a,X). \tag{5.2}$$

Finally, if we take $M > |a|$ then the inequality $|\max(X_t,a)| > M$ can hold only if $\max(X_t,a) = X_t > M$. Combining all these observations,

$$\begin{aligned}\mathbb{E}|\max(X_t,a)|\mathrm{I}(|\max(X_t,a)| > M) &= \mathbb{E}X_t\mathrm{I}(X_t > M) \le \mathbb{E}X\mathrm{I}(X_t > M)\\ &\le K\mathbb{P}(X_t > M) + \mathbb{E}|X|\mathrm{I}(|X| > K)\\ &\le K\frac{\mathbb{E}\max(X_t,0)}{M} + \mathbb{E}|X|\mathrm{I}(|X| > K)\\ \{\text{by (5.2)}\} &\le K\frac{\mathbb{E}\max(X,0)}{M} + \mathbb{E}|X|\mathrm{I}(|X| > K).\end{aligned}$$

Letting $M \to \infty$ and then $K \to \infty$ proves that $(\max(X_t,a))_{t\in T}$ is uniformly integrable. □

Exercise 5.1.1. In the setting of the Example 3 above, prove carefully that, for $1 \le k \le n$, $\mathbb{E}(X_1|\mathscr{B}_{-n}) = \mathbb{E}(X_k|\mathscr{B}_{-n})$.

Exercise 5.1.2. Suppose that $(S_n, \mathscr{B}_n)$ is a martingale, where $S_n = X_1 + \ldots + X_n$ and $\mathscr{B}_n = \sigma(X_1,\ldots,X_n)$. For $i < j$, if we assume that $\mathbb{E}|X_iX_j| < \infty$ then show that $\mathbb{E}X_iX_j = 0$.

Exercise 5.1.3. Let $(X_n)_{n\ge1}$ be i.i.d. random variables and, given a bounded measurable function $f : \mathbb{R}^2 \to \mathbb{R}$ such that $f(x,y) = f(y,x)$, consider

$$S_{-n} = \frac{2}{n(n-1)} \sum_{1 \le \ell < \ell' \le n} f(X_\ell, X_{\ell'})$$

(called the U-statistics) and the σ-algebras

$$\mathscr{F}_{-n} = \sigma\big(X_{(1)}, \ldots, X_{(n)}, (X_\ell)_{\ell > n}\big),$$

where $X_{(1)}, \ldots, X_{(n)}$ are the order statistics of $X_1, \ldots, X_n$. Prove that $\mathscr{F}_{-(n+1)} \subseteq \mathscr{F}_{-n}$ and $S_{-n} = \mathbb{E}(S_{-2}|\mathscr{F}_{-n})$, i.e. $(S_{-n}, \mathscr{F}_{-n})_{-n \le -1}$ is a right-closed martingale.

Exercise 5.1.4. Suppose that $(X_n)_{n\ge 1}$ are i.i.d. and $\mathbb{E}|X_1| < \infty$. If $S_n = X_1 + \ldots + X_n$, show that the sequence $(S_n/n)_{n\ge 1}$ is uniformly integrable.

Exercise 5.1.5. Show that if $\sup_{t\in T} \mathbb{E}|X_n|^{1+\delta} < \infty$ for some $\delta > 0$ then $(X_t)_{t\in T}$ is uniformly integrable.

Exercise 5.1.6. Show that $(X_t)_{t\in T}$ is uniformly integrable if and only if

(a) $\sup_t \mathbb{E}|X_t| < \infty$ and
(b) for any $\varepsilon > 0$, there exists $\delta > 0$ such that $\sup_{t\in T} \mathbb{E}|X_t| \mathrm{I}_A \le \varepsilon$ if $\mathbb{P}(A) \le \delta$.

Exercise 5.1.7. Suppose that random variables $X_n \ge 0$ for $n \ge 0$, $X_n \to X_0$ in probability and $\mathbb{E}X_n \to \mathbb{E}X_0$. Prove that the limit $\lim_{n\to\infty} \mathbb{E}|X_n - X_0| = 0$. (Hint: consider $(X_0 - X_n)^+$.)

5.2 Stopping times, optional stopping theorem

Consider a probability space $(\Omega, \mathscr{B}, \mathbb{P})$ and a sequence of σ-algebras $\mathscr{B}_n \subseteq \mathscr{B}$ for $n \geq 0$ such that $\mathscr{B}_n \subseteq \mathscr{B}_{n+1}$. An integer valued random variable $\tau \in \{0,1,2,\ldots\}$ is called a *stopping time* if $\{\tau \leq n\} \in \mathscr{B}_n$ for all n. Of course, in this case we also have

$$\{\tau = n\} = \{\tau \leq n\} \setminus \{\tau \leq n-1\} \in \mathscr{B}_n.$$

Given a stopping time τ, let us consider the σ-algebra

$$\mathscr{B}_\tau = \big\{B \in \mathscr{B} : \{\tau \leq n\} \cap B \in \mathscr{B}_n \text{ for all } n\big\}$$

consisting of all events that "depend on the data up to a stopping time τ". If a sequence of random variable (X_n) is adapted to $(\mathscr{B}_n)$, i.e. X_n is $\mathscr{B}_n$-measurable, then random variables such as X_τ or $\sum_{k=1}^{\tau} X_k$ are $\mathscr{B}_\tau$-measurable. For example,

$$\begin{aligned}\{\tau \leq n\} \cap \{X_\tau \in A\} &= \bigcup_{k \leq n} \{\tau = k\} \cap \{X_k \in A\} \\ &= \bigcup_{k \leq n} \Big(\{\tau \leq k\} \setminus \{\tau \leq k-1\} \bigcap \{X_k \in A\}\Big) \in \mathscr{B}_n.\end{aligned}$$

Let us mention several basic properties of stopping times.

1. Constant $\tau = n$ is a stopping time and $\mathscr{B}_\tau = \mathscr{B}_n$.

2. Stopping time τ is $\mathscr{B}_\tau$-measurable. To see this, we need to show that $\{\tau \leq k\} \in \mathscr{B}_\tau$ for all integer $k \geq 0$. This is true, because

$$\{\tau \leq k\} \cap \{\tau \leq n\} = \{\tau \leq k \wedge n\} \in \mathscr{B}_{k \wedge n} \subseteq \mathscr{B}_n,$$

by the definition of stopping time.

3. τ is a stopping time if and only if $\{\tau > n\} \in \mathscr{B}_n$ for all n. This is obvious, because $\{\tau > n\} = \{\tau \leq n\}^c$ and σ-algebras as closed under taking complements.

4. If $(\tau_k)_{k \geq 1}$ are all stopping times then their minimum $\wedge_{k \geq 1} \tau_k$ and maximum $\vee_{k \geq 1} \tau_k$ are also stopping times. This is true, because

$$\big\{\wedge_{k \geq 1} \tau_k > n\big\} = \bigcap_{k \geq 1} \big\{\tau_k > n\big\} \in \mathscr{B}_n, \ \big\{\vee_{k \geq 1} \tau_k \leq n\big\} = \bigcap_{k \geq 1} \big\{\tau_k \leq n\big\} \in \mathscr{B}_n.$$

5. If τ_1 and τ_2 are stopping times then the sum $\tau_1 + \tau_2$ is a stopping time. This is true, because

$$\{\tau_1 + \tau_2 \leq n\} = \bigcup_{k=0}^{n} \{\tau_1 = k\} \cap \{\tau_2 \leq n-k\} \in \mathscr{B}_n.$$

We will leave other properties, concerning two stopping times τ_1 and τ_2, as an exercise.

6. The events $\{\tau_1 < \tau_2\}$, $\{\tau_1 = \tau_2\}$ and $\{\tau_1 > \tau_2\}$ belong to $\mathscr{B}_{\tau_1}$ and $\mathscr{B}_{\tau_2}$.

7. If $B \in \mathscr{B}_{\tau_1}$ then $B \cap \{\tau_1 \le \tau_2\}$ and $B \cap \{\tau_1 < \tau_2\}$ are in $\mathscr{B}_{\tau_2}$.

8. If $\tau_1 \le \tau_2$ then $\mathscr{B}_{\tau_1} \subseteq \mathscr{B}_{\tau_2}$.

A martingale $(X_n, \mathscr{B}_n)$ often represents a fair game, so that the average value $\mathbb{E}(X_m|\mathscr{B}_n)$ at some future time $m > n$ given the data $\mathscr{B}_n$ at time n is equal to the value X_n at time n. A stopping time τ represents a strategy to stop the game based only on the information at any given time, so that the event $\{\tau \le n\}$ that we stop the game before time n depends only on the data $\mathscr{B}_n$ up to that time. A classical result that we will now prove states that, under some mild integrability conditions, the average value under any strategy is the same, so their is no "winning strategy" on average.

Theorem 5.2 (Optional Stopping). *Let $(X_n, \mathscr{B}_n)$ be a martingale and $\tau_1, \tau_2 < \infty$ be stopping times such that*

$$\mathbb{E}|X_{\tau_2}| < \infty, \quad \lim_{n \to \infty} \mathbb{E}|X_n| \mathrm{I}(n \le \tau_2) = 0. \tag{5.3}$$

Then, for any set $A \in \mathscr{B}_{\tau_1}$,

$$\mathbb{E}X_{\tau_2} \mathrm{I}_A \mathrm{I}(\tau_1 \le \tau_2) = \mathbb{E}X_{\tau_1} \mathrm{I}_A \mathrm{I}(\tau_1 \le \tau_2).$$

For example, if $\tau_1 = 0$ and $A = \Omega$ then $\mathbb{E}X_{\tau_2} = \mathbb{E}X_0$ if the condition (5.3) is satisfied. For example, if the stopping time τ_2 is bounded then (5.3) is obviously satisfied. Let us note that without some kind of integrability condition, Theorem 5.2 can not hold, as the following example shows.

Example 5.2.1. Consider independent $(X_n)_{n \ge 1}$ such that $\mathbb{P}(X_n = \pm 2^n) = 1/2$. If $\mathscr{B}_n = \sigma(X_1, \ldots, X_n)$ then $(S_n, \mathscr{B}_n)$ is a martingale. Let $\tau_1 = 1$ and $\tau_2 = \min\{k \ge 1 : S_k > 0\}$. Clearly, $S_{\tau_2} = 2$, because if $\tau_2 = k$ then

$$S_{\tau_2} = S_k = -2 - 2^2 - \ldots - 2^{k-1} + 2^k = 2.$$

However, $2 = \mathbb{E}S_{\tau_2} \ne \mathbb{E}S_1 = 0$. Notice that the second condition in (5.3) is violated, since

$$\mathbb{P}(\tau_2 = n) = 2^{-n}, \ \mathbb{P}(\tau_2 \ge n+1) = \sum_{k \ge n+1} 2^{-k} = 2^{-n}$$

and, therefore,

$$\mathbb{E}|S_n| I(n \le \tau_2) = 2\mathbb{P}(\tau_2 = n) + (2^{n+1} - 2)\mathbb{P}(n+1 \le \tau_2) = 2,$$

which does not go to zero. □

Proof (of Theorem 5.2). Consider a set $A \in \mathscr{B}_{\tau_1}$. The result is based on the following formal computation with the middle step (*) proved below,

$$\begin{aligned}\mathbb{E}X_{\tau_2}\mathrm{I}_A\mathrm{I}(\tau_1 \le \tau_2) &= \sum_{n\ge 1}\mathbb{E}X_{\tau_2}\mathrm{I}(A\cap\{\tau_1 = n\})\mathrm{I}(n\le\tau_2)\\ &\overset{(*)}{=} \sum_{n\ge 1}\mathbb{E}X_n\mathrm{I}(A\cap\{\tau_1 = n\})\mathrm{I}(n\le\tau_2) = \mathbb{E}X_{\tau_1}\mathrm{I}_A\mathrm{I}(\tau_1\le\tau_2).\end{aligned}$$

To prove (*), it is enough to show that for $A_n = A\cap\{\tau_1 = n\}\in\mathscr{B}_n$,

$$\mathbb{E}X_{\tau_2}\mathrm{I}_{A_n}\mathrm{I}(n\le\tau_2) = \mathbb{E}X_n\mathrm{I}_{A_n}\mathrm{I}(n\le\tau_2). \tag{5.4}$$

We begin by writing

$$\begin{aligned}\mathbb{E}X_n\mathrm{I}_{A_n}\mathrm{I}(n\le\tau_2) &= \mathbb{E}X_n\mathrm{I}_{A_n}\mathrm{I}(\tau_2 = n) + \mathbb{E}X_n\mathrm{I}_{A_n}\mathrm{I}(n<\tau_2)\\ &= \mathbb{E}X_{\tau_2}\mathrm{I}_{A_n}\mathrm{I}(\tau_2 = n) + \mathbb{E}X_n\mathrm{I}_{A_n}\mathrm{I}(n<\tau_2).\end{aligned}$$

Since $\{n<\tau_2\} = \{\tau_2\le n\}^c\in\mathscr{B}_n$ and $(X_n,\mathscr{B}_n)$ is a martingale, the last term

$$\mathbb{E}X_n\mathrm{I}_{A_n}\mathrm{I}(n<\tau_2) = \mathbb{E}X_{n+1}\mathrm{I}_{A_n}\mathrm{I}(n<\tau_2) = \mathbb{E}X_{n+1}\mathrm{I}_{A_n}\mathrm{I}(n+1\le\tau_2)$$

and, therefore,

$$\mathbb{E}X_n\mathrm{I}_{A_n}\mathrm{I}(n\le\tau_2) = \mathbb{E}X_{\tau_2}\mathrm{I}_{A_n}\mathrm{I}(\tau_2 = n) + \mathbb{E}X_{n+1}\mathrm{I}_{A_n}\mathrm{I}(n+1\le\tau_2).$$

We can continue the same computation and, by induction on m, we get

$$\mathbb{E}X_n\mathrm{I}_{A_n}\mathrm{I}(n\le\tau_2) = \mathbb{E}X_{\tau_2}\mathrm{I}_{A_n}\mathrm{I}(n\le\tau_2<m) + \mathbb{E}X_m\mathrm{I}_{A_n}\mathrm{I}(m\le\tau_2). \tag{5.5}$$

It remains to let $m\to\infty$ and use the assumptions in (5.3). By the second assumption, the last term

$$\big|\mathbb{E}X_m\mathrm{I}_{A_n}\mathrm{I}(m\le\tau_2)\big| \le \mathbb{E}|X_m|\mathrm{I}(m\le\tau_2)\to 0$$

as $m\to\infty$. Since

$$X_{\tau_2}\mathrm{I}_{A_n}\mathrm{I}(n\le\tau_2\le m)\to X_{\tau_2}\mathrm{I}_{A_n}\mathrm{I}(n\le\tau_2)$$

almost surely as $m\to\infty$ and $\mathbb{E}|X_{\tau_2}|<\infty$, by the dominated convergence theorem,

$$\mathbb{E}X_{\tau_2}\mathrm{I}_{A_n}\mathrm{I}(n\le\tau_2<m)\to\mathbb{E}X_{\tau_2}\mathrm{I}_{A_n}\mathrm{I}(n\le\tau_2).$$

This proves (5.4) and finishes the proof of the theorem. □

Example 5.2.2. *(Computing $\mathbb{E}X_\tau$)* If we take $\tau_1 = 0$, $n = 0$ and $A = \Omega$ then $A_n = \Omega$ and the equation (5.5) becomes

$$\mathbb{E}X_0 = \mathbb{E}X_{\tau_2}\mathrm{I}(\tau_2<m) + \mathbb{E}X_m\mathrm{I}(m\le\tau_2).$$

This implies that, for any stopping time τ,

$$\mathbb{E}X_0 = \lim_{n\to\infty} \mathbb{E}X_\tau \mathrm{I}(\tau \le n) \Longleftrightarrow \lim_{n\to\infty} \mathbb{E}X_n \mathrm{I}(\tau \ge n) = 0, \tag{5.6}$$

which gives a clean condition to check if we would like to show that $\mathbb{E}X_\tau = \mathbb{E}X_0$.

Example 5.2.3 (Hitting times of simple random walk). Given $p \in (0,1)$, consider i.i.d. random variables $(X_i)_{i\ge 1}$ such that

$$\mathbb{P}(X_i = 1) = p,\ \mathbb{P}(X_i = -1) = 1-p,$$

and consider a random walk $S_{n+1} = S_n + X_{n+1}$ starting with $S_0 = 0$. Consider two integers $a \le -1$ and $b \ge 1$ and the stopping time

$$\tau = \min\big\{k \ge 1 : S_n = a \text{ or } b\big\}.$$

If we denote $q = 1-p$ then $Y_n = (q/p)^{S_n}$ is a martingale, since

$$\mathbb{E}(Y_{n+1}|\mathscr{B}_n) = p\Big(\frac{q}{p}\Big)^{S_n+1} + q\Big(\frac{q}{p}\Big)^{S_n-1} = \Big(\frac{q}{p}\Big)^{S_n}.$$

It is easy to show that $\mathbb{P}(\tau \ge n) \to 0$, which we leave as an exercise below. Since $S_n \in [a,b]$ for $n \le \tau$ and $S_\tau = a$ or b, the equation (5.6) implies that

$$1 = \mathbb{E}Y_0 = \mathbb{E}Y_\tau = \Big(\frac{q}{p}\Big)^b \mathbb{P}(S_\tau = b) + \Big(\frac{q}{p}\Big)^a (1 - \mathbb{P}(S_\tau = b)).$$

If $p \ne 1/2$, we can solve this:

$$\mathbb{P}(S_\tau = b) = \frac{(q/p)^a - 1}{(q/p)^a - (q/p)^b}.$$

If $q = p = 1/2$ then S_n itself is a martingale and the equation (5.6) implies that

$$0 = b\mathbb{P}(S_\tau = b) + a(1 - \mathbb{P}(S_\tau = b)) \text{ and } \mathbb{P}(S_\tau = b) = \frac{a}{a-b}.$$

One can also compute $\mathbb{E}\tau$, which we will also leave as an exercise.

Example 5.2.4 (Fundamental Wald's identity). Let $(X_n)_{n\ge 1}$ be a sequence of i.i.d. random variables, $S_0 = 0$ and $S_n = X_1 + \ldots + X_n$. Suppose that the *Laplace transform* $\varphi(\lambda) = \mathbb{E}e^{\lambda X_1}$ is defined on some nontrivial interval (λ_-, λ_+) containing 0. It is obvious that

$$Y_n = \frac{e^{\lambda S_n}}{\varphi(\lambda)^n}$$

is a martingale for $\lambda \in (\lambda_-, \lambda_+)$. Let τ be a stopping time. Since $Y_\tau \ge 0$,

$$\lim_{n\to\infty}\mathbb{E}\frac{e^{\lambda S_\tau}}{\varphi(\lambda)^\tau}\mathrm{I}(\tau\le n)=\mathbb{E}\frac{e^{\lambda S_\tau}}{\varphi(\lambda)^\tau}\mathrm{I}(\tau<\infty)$$

by the monotone convergence theorem. Since $Y_0=1$, (5.6) implies in this case that

$$1=\mathbb{E}\frac{e^{\lambda S_\tau}}{\varphi(\lambda)^\tau}\mathrm{I}(\tau<\infty)\iff\lim_{n\to\infty}\mathbb{E}\frac{e^{\lambda S_n}}{\varphi(\lambda)^n}\mathrm{I}(\tau\ge n)=0. \tag{5.7}$$

The left hand side is called the *fundamental Wald's identity.* In some cases one can use this to compute the Laplace transform of a stopping time and, thus, its distribution, as we will see below.

Notice that in the case when $\tau<\infty$ almost surely and (either side of) the equation (5.7) holds, we can write

$$1=\mathbb{E}\frac{e^{\lambda S_\tau}}{\varphi(\lambda)^\tau}.$$

Computing formally the derivative in λ at $\lambda=0$, we get $\mathbb{E}S_\tau=\mathbb{E}X_1\mathbb{E}\tau$, which is the Wald's identity we proved in (2.4), so (5.7) is a generalization of that identity. Of course, this is only a formal generalization because here we make a stronger assumption on the existence of the Laplace transform in a neighbourhood of zero, while the identity in $\mathbb{E}S_\tau=\mathbb{E}X_1\mathbb{E}\tau$ was proved only under the assumption that the expectations $\mathbb{E}X_1$ and $\mathbb{E}\tau$ are well defined.

Example 5.2.5 (Symmetric random walk). Let $X_0=0$, $\mathbb{P}(X_i=\pm1)=1/2$, $S_n=\sum_{k\le n}X_k$. Given integer $z\ge1$, let

$$\tau=\min\{k:S_k=-z\text{ or }z\}.$$

Since $\varphi(\lambda)=\mathrm{ch}(\lambda)\ge1$,

$$\mathbb{E}\frac{e^{\lambda S_n}}{\varphi(\lambda)^n}\mathrm{I}(\tau\ge n)\le\frac{e^{|\lambda|z}}{\mathrm{ch}(\lambda)^n}\mathbb{P}(\tau\ge n)$$

and the right hand side of (5.7) holds. Therefore,

$$\begin{aligned}1=\mathbb{E}\frac{e^{\lambda S_\tau}}{\varphi(\lambda)^\tau}&=e^{\lambda z}\mathbb{E}\,\mathrm{ch}(\lambda)^{-\tau}\mathrm{I}(S_\tau=z)+e^{-\lambda z}\mathbb{E}\,\mathrm{ch}(\lambda)^{-\tau}\mathrm{I}(S_\tau=-z)\\&=\frac{e^{\lambda z}+e^{-\lambda z}}{2}\mathbb{E}\,\mathrm{ch}(\lambda)^{-\tau}\end{aligned}$$

by symmetry. Therefore,

$$\mathbb{E}\,\mathrm{ch}(\lambda)^{-\tau}=\frac{1}{\mathrm{ch}(\lambda z)}\quad\text{and}\quad\mathbb{E}e^{\gamma\tau}=\frac{1}{\mathrm{ch}(\mathrm{ch}^{-1}(e^{-\gamma})z)}$$

by the change of variables $e^\gamma=1/\mathrm{ch}\,\lambda$. □

For more general stopping times the condition on the right hand side of (5.7) might not be easy to check. We will now consider another point of view that may sometimes be helpful in verifying the fundamental Wald's identity. If $\mathbb{P}$ is the distribution of X_1, let $\mathbb{P}_\lambda$ be the distribution with the Radon-Nikodym derivative with respect to $\mathbb{P}$ given by

$$\frac{d\mathbb{P}_\lambda}{d\mathbb{P}} = \frac{e^{\lambda x}}{\varphi(\lambda)}.$$

This is, indeed, a density with respect to $\mathbb{P}$, since

$$\int_{\mathbb{R}} \frac{e^{\lambda x}}{\varphi(x)} d\mathbb{P} = \frac{\varphi(\lambda)}{\varphi(\lambda)} = 1.$$

For convenience of notation, we will think of (X_n) as the coordinates on the product space $(\mathbb{R}^\infty, \mathscr{B}^\infty, \mathbb{P}^{\otimes\infty})$ with the cylindrical σ-algebra. Therefore, $\{\tau = n\} \in \sigma(X_1,\ldots,X_n) \subseteq \mathscr{B}^n$ is a Borel set on $\mathbb{R}^n$. We can write,

$$\begin{aligned}
\mathbb{E}\frac{e^{\lambda S_\tau}}{\varphi(\lambda)^\tau} \mathrm{I}(\tau < \infty) &= \sum_{n=1}^\infty \mathbb{E}\frac{e^{\lambda S_n}}{\varphi(\lambda)^n}\mathrm{I}(\tau = n) \\
&= \sum_{n=1}^\infty \int_{\{\tau=n\}} \frac{e^{\lambda(x_1+\cdots+x_n)}}{\varphi(\lambda)^n} d\mathbb{P}(x_1)\ldots d\mathbb{P}(x_n) \\
&= \sum_{n=1}^\infty \int_{\{\tau=n\}} d\mathbb{P}_\lambda(x_1)\ldots d\mathbb{P}_\lambda(x_n) = \mathbb{P}_\lambda^{\otimes\infty}(\tau<\infty).
\end{aligned}$$

This means that we can think of the random variables X_n as having the distribution $\mathbb{P}_\lambda$ and, to prove Wald's identity, we need to show that $\tau < \infty$ with probability one.

Example 5.2.6 (Crossing a growing boundary). Suppose that we have a boundary given by some sequence $f\colon \mathbb{N} \to \mathbb{R}$ and a stopping time (crossing time):

$$\tau = \min\{k : S_k \geq f(k)\}.$$

To verify Wald's identity in (5.7), we need to show that $\tau < \infty$ with probability one, assuming now that random variables X_n have the distribution $\mathbb{P}_\lambda$. Under this distribution, the random variables X_n have the expectation

$$\mathbb{E}_\lambda X_i = \int x \frac{e^{\lambda x}}{\varphi(\lambda)} d\mathbb{P}(x) = \frac{\varphi'(\lambda)}{\varphi(\lambda)}.$$

By the strong law of large numbers

$$\lim_{n\to\infty} \frac{S_n}{n} = \frac{\varphi'(\lambda)}{\varphi(\lambda)}$$

and, therefore, if the growth of the crossing boundary satisfies

$$\limsup_{n\to\infty} \frac{f(n)}{n} < \frac{\varphi'(\lambda)}{\varphi(\lambda)} \tag{5.8}$$

then, obviously, the random walk S_n will cross it with probability one, in which case $\mathbb{P}_\lambda^{\otimes\infty}(\tau < \infty) = 1$. By Hölder's inequality, $\varphi(\lambda)$ is log-convex and, therefore, $\varphi'(\lambda)/\varphi(\lambda)$ is increasing, which means that if this condition holds for λ' then it holds for $\lambda > \lambda'$. Also, note that, for $\lambda \in (\lambda_-, \lambda_+)$, the derivative of $\varphi'(\lambda)/\varphi(\lambda)$ is the variance of $X \sim \mathbb{P}_\lambda$, so this function is actually strictly increasing unless the random variable $X \sim \mathbb{P}$ is constant. □

In the next two examples, we will consider a constant boundary $f(k) = z$ for some integer $z > 0$. In this case, the condition (5.8) becomes $\varphi'(\lambda) > 0$. We will consider the case when $\varphi'(0) = \mathbb{E}X_1 > 0$, which means that the random walk has a positive drift and, in particular, $\tau < \infty$ with probability one.

Example 5.2.7. Let us consider i.i.d. X_i taking integer values $\{\ldots,-2,-1,0,1\}$ with $\mathbb{E}X_1 > 0$, which of course implies that $\mathbb{P}(X_1 = 1) > 0$. These assumptions yield that $\tau = \min\{k : S_k \geq z\} < \infty$ a.s. and $S_\tau = z$ (we can not jump over the boundary). By continuity, $\varphi'(\lambda) > 0$ is a small neighbourhood of zero and, thus, by the previous example,

$$1 = \mathbb{E}\frac{e^{\lambda S_\tau}}{\varphi(\lambda)^\tau} = e^{\lambda z}\mathbb{E}\frac{1}{\varphi(\lambda)^\tau}.$$

As in the example of the symmetric walk, this determines the Laplace transform $\mathbb{E}e^{\gamma\tau}$ on some interval of γ and determines the distribution of τ. □

Example 5.2.8 (Geometric case). Let us now consider the case when X_i are i.i.d. geometric random variables and, for some $p \in (0,1)$,

$$\mathbb{P}(X_i \geq k) = (1-p)^{k-1} \text{ for } k = 1,2,3,\ldots.$$

In this case, one can show that τ and S_τ are independent and $S_\tau \stackrel{d}{=} z - 1 + X_1$. Indeed, for any $y \geq z$ and $n \geq 1$, we can write

$$\begin{aligned}\mathbb{P}(S_\tau \geq y, \tau = n) &= \sum_{x<z} \mathbb{P}(S_n \geq y, S_{n-1} = x, \tau = n)\\ &= \sum_{x<z} \mathbb{P}(X_n \geq y - x, S_{n-1} = x, \tau \geq n-1).\end{aligned}$$

Since the event $\{S_{n-1} = x, \tau \geq n-1\} \in \mathscr{B}_{n-1}$ and, thus, is independent of the geometric random variable X_n, we can continue

$$\begin{aligned}\mathbb{P}(S_\tau \geq y, \tau = n) &= \sum_{x<z} (1-p)^{y-x-1}\mathbb{P}(S_{n-1} = x, \tau \geq n-1)\\ &= (1-p)^{y-z}\sum_{x<z}(1-p)^{z-x-1}\mathbb{P}(S_{n-1} = x, \tau \geq n-1)\end{aligned}$$

$$= (1-p)^{y-z} \sum_{x<z} \mathbb{P}(X_n \geq z - x, S_{n-1} = x, \tau \geq n-1)$$
$$= (1-p)^{y-z} \mathbb{P}(S_\tau \geq z, \tau = n) = (1-p)^{y-z} \mathbb{P}(\tau = n).$$

This proves the above claim. Together with the fundamental Wald's identity this shows that

$$1 = \mathbb{E}\frac{e^{\lambda S_\tau}}{\varphi(\lambda)^\tau} = \mathbb{E}e^{\lambda S_\tau}\mathbb{E}\frac{1}{\varphi(\lambda)^\tau} = e^{\lambda(z-1)}\varphi(\lambda)\mathbb{E}\frac{1}{\varphi(\lambda)^\tau}.$$

The Laplace transform of the geometric distribution is

$$\varphi(\lambda) = \frac{pe^\lambda}{1-(1-p)e^\lambda} \text{ for } \lambda < -\log(1-p),$$

and we again determined the Laplace transform of τ on some interval. □

Exercise 5.2.1. Prove the properties 6 – 8 of the stopping times above.

Exercise 5.2.2. Let $(X_n, \mathscr{B}_n)$ be a martingale and $\tau_1 \leq \tau_2 \leq \ldots$ be a non-decreasing sequence of bounded stopping times. Show that $(X_{\tau_n}, \mathscr{B}_{\tau_n})$ is a matringale.

Exercise 5.2.3. Let $(X_n, \mathscr{B}_n)_{n\geq 0}$ be a martingale and τ a stopping time such that $\mathbb{P}(\tau < \infty) = 1$. Suppose that $\mathbb{E}|X_{\tau\wedge n}|^{1+\delta} \leq c$ for some $c, \delta > 0$ and for all n. Prove that $\mathbb{E}X_\tau = \mathbb{E}X_0$.

Exercise 5.2.4. Given $0 < p < 1$, consider i.i.d. random variables $(X_i)_{i\geq 1}$ such that $\mathbb{P}(X_i = 1) = p$, $\mathbb{P}(X_i = -1) = 1 - p$ and consider a random walk $S_0 = 0, S_{n+1} = S_n + X_{n+1}$. Consider two integers $a \leq -1$ and $b \geq 1$ and define a stopping time

$$\tau = \min\{k \geq 1 : S_k = a \text{ or } b\}.$$

(a) Show that $\mathbb{P}(\tau \geq n) \to 0$.
(b) Compute $\mathbb{E}\tau$. Hint: for $p \neq 1/2$, use that $S_n - n\mathbb{E}X_1$ is a martingale; for $p = 1/2$, use that $S_n^2 - n$ is a martingale.

Exercise 5.2.5. Let $S_n = X_1 + \ldots + X_n$ be a simple symmetric random walk with $S_0 = 0$, i.e. steps $(X_i)_{i\geq 1}$ are i.i.d. and $\mathbb{P}(X_i = \pm 1) = 1/2$. Let $\tau = \min\{n \geq 5 : S_n = S_{n-5} + 5\}$.

(a) Is τ a stopping time?
(b) Compute $\mathbb{E}\tau$. (*Hint*: If $\tau = n$ for $n \geq 6$, what must happen in the last six steps? How can one rewrite $\mathbb{P}(\tau = n)$?)

Exercise 5.2.6. Let $X_0 = 0, \mathbb{P}(X_i = \pm 1) = 1/2$ and $S_n = \sum_{k\leq n} X_k$. Let $\tau = \min\{k : S_k \geq 0.95k + 100\}$. Prove that

$$\mathbb{E}\frac{e^{\lambda S_\tau}}{\varphi(\lambda)^\tau}\mathrm{I}(\tau < \infty) = 1$$

for λ large enough.

5.3 Doob's inequalities and convergence of martingales

In this section, we will study convergence of martingales, and we begin by proving two classical inequalities.

Theorem 5.3 (Doob's inequality). *If $(X_n, \mathscr{B}_n)_{n\geq 1}$ is a submartingale and $Y_n = \max_{1\leq k\leq n} X_k$ then, for any $M > 0$,*

$$\mathbb{P}\big(Y_n \geq M\big) \leq \frac{1}{M}\mathbb{E}X_n \mathrm{I}(Y_n \geq M) \leq \frac{1}{M}\mathbb{E}X_n^+. \tag{5.9}$$

Proof. Define a stopping time

$$\tau_1 = \begin{cases} \min\{k : X_k \geq M, k \leq n\}, & \text{if such } k \text{ exists,} \\ n, & \text{otherwise.} \end{cases}$$

Let $\tau_2 = n$ so that $\tau_1 \leq \tau_2$. By the optional stopping theorem, Theorem 5.2,

$$\mathbb{E}(X_n|\mathscr{B}_{\tau_1}) = \mathbb{E}(X_{\tau_2}|\mathscr{B}_{\tau_1}) \geq X_{\tau_1}.$$

Let us average this inequality over the set $A = \{Y_n = \max_{1\leq k\leq n} X_n \geq M\}$, which belongs to $\mathscr{B}_{\tau_1}$, because

$$A \cap \{\tau_1 \leq k\} = \Big\{\max_{1\leq i\leq k\wedge n} X_i \geq M\Big\} \in \mathscr{B}_{k\wedge n} \subseteq \mathscr{B}_k.$$

On the event A, $X_{\tau_1} \geq M$ and, therefore,

$$\mathbb{E}X_n \mathrm{I}_A \geq \mathbb{E}X_{\tau_1}\mathrm{I}_A \geq M\mathbb{E}\mathrm{I}_A = M\mathbb{P}(A).$$

This is precisely the first inequality in (5.9). The second inequality is obvious. □

Example 5.3.1 (Second Kolmogorov's inequality). If (X_i) are independent and $\mathbb{E}X_i = 0$ then $S_n = \sum_{1\leq i\leq n} X_i$ is a martingale and S_n^2 is a submartingale. Therefore, by Doob's inequality,

$$\mathbb{P}\Big(\max_{1\leq k\leq n}|S_k| \geq M\Big) = \mathbb{P}\Big(\max_{1\leq k\leq n} S_k^2 \geq M^2\Big) \leq \frac{1}{M^2}\mathbb{E}S_n^2 = \frac{1}{M^2}\sum_{1\leq k\leq n}\mathrm{Var}(X_k).$$

This is a big improvement on Chebyshev's inequality, since we control the maximum $\max_{1\leq k\leq n}|S_k|$ instead of one sum $|S_n|$. □

Doob's upcrossing inequality. Let $(X_n, \mathscr{B}_n)_{n\geq 1}$ be a submartingale. Given two real numbers $a < b$, we will define an increasing sequence of stopping times $(\tau_n)_{n\geq 1}$ when X_n is crossing a downward and b upward. More specifically, we define these stopping times by

$$\tau_1 = \min\{n \geq 1 : X_n \leq a\},\ \tau_2 = \min\{n > \tau_2 : X_n \geq b\}$$

and, by induction, for $k \geq 2$,

$$\tau_{2k-1} = \min\{n > \tau_{2k-2} \,|\, X_n \leq a\},\ \tau_{2k} = \min\{n > \tau_{2k-1} : X_n \geq b\}.$$

Then, we consider the random variable

$$\nu(a,b,n) = \max\{k : \tau_{2k} \leq n\},$$

which is the number of upward crossings of the interval $[a,b]$ before time n.

Theorem 5.4 (Doob's upcrossing inequality). *If $(X_n, \mathscr{B}_n)$ is a submartingale then*

$$\mathbb{E}\nu(a,b,n) \leq \frac{\mathbb{E}(X_n - a)^+}{b-a}. \tag{5.10}$$

Proof. Since $x \to (x-a)^+$ is increasing convex function, $Z_n = (X_n - a)^+$ is also a submartingale. Clearly, the number of upcrossings of the interval $[a,b]$ by (X_n) can be expressed in terms of the number of upcrossings of $[0, b-a]$ by (Z_n),

$$\nu_X(a,b,n) = \nu_Z(0, b-a, n),$$

which means that it is enough to prove (5.10) for nonnegative submartingales. From now on we can assume that $0 \leq X_n$ and we would like to show that

$$\mathbb{E}\nu(0,b,n) \leq \frac{\mathbb{E}X_n}{b}.$$

Let us define a sequence of random variables η_j for $j \geq 1$ by

$$\eta_j = \begin{cases} 1, & \tau_{2k-1} < j \leq \tau_{2k} \text{ for some } k \\ 0, & \text{otherwise,} \end{cases}$$

i.e. η_j is the indicator of the event that at time j the process is crossing $[0,b]$ upward. Define $X_0 = 0$. Then, clearly,

$$b\nu(0,b,n) \leq \sum_{j=1}^{n} \eta_j (X_j - X_{j-1}) = \sum_{j=1}^{n} \mathrm{I}(\eta_j = 1)(X_j - X_{j-1}).$$

Since (τ_k) are stopping times, the event

$$\begin{aligned} \{\eta_j = 1\} &= \bigcup_{k \geq 1} \{\tau_{2k-1} < j \leq \tau_{2k}\} \\ &= \bigcup_{k \geq 1} \{\tau_{2k-1} \leq j-1\} \setminus \{\tau_{2k} \leq j-1\}^c \in \mathscr{B}_{j-1}, \end{aligned}$$

i.e. the fact that at time j we are crossing upward is determined completely by the sequence up to time $j-1$. Then

$$
\begin{aligned}
b\mathbb{E}\nu(0,b,n) &\le \sum_{j=1}^{n} \mathbb{E}\mathrm{I}(\eta_j = 1)(X_j - X_{j-1}) \\
&= \sum_{j=1}^{n} \mathbb{E}\mathrm{I}(\eta_j = 1)\mathbb{E}\big(X_j - X_{j-1}\big|\mathscr{B}_{j-1}\big) \\
&= \sum_{j=1}^{n} \mathbb{E}\mathrm{I}(\eta_j = 1)\big(\mathbb{E}(X_j|\mathscr{B}_{j-1}) - X_{j-1}\big).
\end{aligned}
$$

Since $(X_j, \mathscr{B}_j)$ is a submartingale, $\mathbb{E}(X_j|\mathscr{B}_{j-1}) \ge X_{j-1}$ and

$$
\mathrm{I}(\eta_j = 1)(\mathbb{E}(X_j|\mathscr{B}_{j-1}) - X_{j-1}) \le \mathbb{E}(X_j|\mathscr{B}_{j-1}) - X_{j-1}.
$$

Therefore,

$$
b\mathbb{E}\nu(0,b,n) \le \sum_{j=1}^{n} \mathbb{E}\big(\mathbb{E}(X_j|\mathscr{B}_{j-1}) - X_{j-1}\big) = \sum_{j=1}^{n} \mathbb{E}(X_j - X_{j-1}) = \mathbb{E}X_n.
$$

This finishes the proof for nonnegative submartingales, and the general case. □

Finally, we will now use Doob's upcrossing inequality to prove our main result about the convergence of submartingales, from which the convergence of martingales will follow.

Theorem 5.5. *Suppose that $(X_n, \mathscr{B}_n)_{-\infty<n<\infty}$ is a submartingale.*

1. *A.s. limit $X_{-\infty} = \lim_{n\to-\infty} X_n$ exists and $\mathbb{E}X_{-\infty}^+ < +\infty$. If $\mathscr{B}_{-\infty} = \cap_n \mathscr{B}_n$ then $(X_n, \mathscr{B}_n)_{-\infty\le n<\infty}$ is a submartingale (i.e. we can add $n = -\infty$ to our index set).*
2. *If $\sup_n \mathbb{E}X_n^+ < \infty$, then a.s. limit $X_{+\infty} = \lim_{n\to+\infty} X_n$ exists and $\mathbb{E}X_{+\infty}^+ < \infty$.*
3. *Moreover, if $(X_n^+)_{-\infty<n<\infty}$ is uniformly integrable then $(X_n, \mathscr{B}_n)_{-\infty<n\le+\infty}$ is a right-closed submartingale, where $\mathscr{B}_{+\infty} = \vee_n \mathscr{B}_n = \sigma\big(\cup_n \mathscr{B}_n\big)$.*

Proof. Let us note that X_n converges as $n \to \infty$ (here, $\infty = +\infty$ or $-\infty$) if and only if $\limsup X_n = \liminf X_n$. Therefore, X_n diverges if and only if the following event occurs

$$
\big\{\limsup X_n > \liminf X_n\big\} = \bigcup_{a<b} \big\{\limsup X_n \ge b > a \ge \liminf X_n\big\},
$$

where the union is taken over all rational numbers $a < b$. This means that the probability $\mathbb{P}(X_n \text{ diverges}) > 0$ if and only if there exist rational $a < b$ such that

$$
\mathbb{P}\big(\limsup X_n \ge b > a \ge \liminf X_n\big) > 0.
$$

If we recall that $\nu(a,b,n)$ denotes the number of upcrossings of $[a,b]$, this is equivalent to $\mathbb{P}\big(\lim_{n\to\infty} \nu(a,b,n) = \infty\big) > 0$.

Proof of 1. For each $n \geq 1$, let us define a sequence

$$Y_1 = X_{-n}, Y_2 = X_{-n+1}, \ldots, Y_n = X_{-1}.$$

By Doob's inequality,

$$\mathbb{E}\nu_Y(a,b,n) \leq \frac{\mathbb{E}(Y_n - a)^+}{b-a} = \frac{\mathbb{E}(X_{-1} - a)^+}{b-a} < \infty.$$

Since $0 \leq \nu_Y(a,b,n) \uparrow \nu(a,b)$ almost surely as $n \to +\infty$, by the monotone convergence theorem,

$$\mathbb{E}\nu(a,b) = \lim_{n \to +\infty} \mathbb{E}\nu_Y(a,b,n) < \infty.$$

Therefore, $\mathbb{P}(\nu(a,b) = \infty) = 0$, which means that the sequence $(X_n)_{n \leq 0}$ can not have infinitely many upcrossings of $[a,b]$ and, therefore, the limit

$$X_{-\infty} = \lim_{n \to -\infty} X_n$$

exists with probability one. By Fatou's lemma and the submartingale property,

$$\mathbb{E}X_{-\infty}^+ \leq \liminf_{n \to -\infty} \mathbb{E}X_{-n}^+ \leq \mathbb{E}X_{-1}^+ < \infty.$$

It remains to show that $(X_n, \mathscr{B}_n)_{-\infty \leq n < \infty}$ is a submartingale. First of all, the limit $X_{-\infty} = \lim_{n \to -\infty} X_n$ is measurable on $\mathscr{B}_n$ for all n and, therefore, measurable on $\mathscr{B}_{-\infty} = \cap \mathscr{B}_n$. Let us take a set $A \in \cap \mathscr{B}_n$. We would like to show that, for any m, $\mathbb{E}X_{-\infty}\mathrm{I}_A \leq \mathbb{E}X_m \mathrm{I}_A$. Since $(X_n, \mathscr{B}_n)_{n \leq 0}$ is a right-closed submartingale, by part 2 of Lemma 5.3 in Section 5.1, $\big(X_n \vee a\big)_{-\infty < n \leq 0}$ is uniformly integrable for any $a \in \mathbb{R}$. Since $X_{-\infty} \vee a = \lim_{n \to -\infty} X_n \vee a$ almost surely, by Section 5.1,

$$\mathbb{E}(X_{-\infty} \vee a)\mathrm{I}_A = \lim_{n \to -\infty} \mathbb{E}(X_n \vee a)\mathrm{I}_A.$$

Finally, since the function $x \to x \vee a$ is convex and non-decreasing, $(X_n \vee a, \mathscr{B}_n)_{n < \infty}$ is a submartingale by Lemma 5.2 in Section 5.1 and, therefore, for any $n \leq m$, we have $\mathbb{E}(X_n \vee a)\mathrm{I}_A \leq \mathbb{E}(X_m \vee a)\mathrm{I}_A$ since $A \in \mathscr{B}_n$. This proves that $\mathbb{E}(X_{-\infty} \vee a)\mathrm{I}_A \leq \mathbb{E}(X_m \vee a)\mathrm{I}_A$. Letting $a \downarrow -\infty$, by the monotone convergence theorem, $\mathbb{E}X_{-\infty}\mathrm{I}_A \leq \mathbb{E}X_m \mathrm{I}_A$.

Proof of 2. If $\nu(a,b,n)$ is the number of upcrossings of $[a,b]$ by the sequence $X_1, \ldots, X_n$ then, by Doob's inequality,

$$(b-a)\mathbb{E}\nu(a,b,n) \leq \mathbb{E}(X_n - a)^+ \leq |a| + \mathbb{E}X_n^+ \leq K < \infty.$$

Therefore, $\mathbb{E}\nu(a,b) < \infty$ for $\nu(a,b) = \lim_{n \to \infty} \nu(a,b,n)$ and, as above, the limit $X_{+\infty} = \lim_{n \to +\infty} X_n$ exists. Since all X_n are measurable on $\mathscr{B}_{+\infty}$, so is $X_{+\infty}$. Finally, by Fatou's lemma,

$$\mathbb{E}X_{+\infty}^+ \leq \liminf_{n \to \infty} \mathbb{E}X_n^+ < \infty.$$

Notice that a different assumption, $\sup_n \mathbb{E}|X_n| < \infty$, would similarly imply that $\mathbb{E}|X_{+\infty}| \le \liminf_{n\to\infty} \mathbb{E}|X_n| < \infty$.

Proof of 3. By 2, the limit $X_{+\infty}$ exists. We want to show that for any m and for any set $A \in \mathscr{B}_m$,

$$\mathbb{E}X_m \mathrm{I}_A \le \mathbb{E}X_{+\infty}\mathrm{I}_A.$$

We already mentioned above that $(X_n \vee a)$ is a submartingale and, therefore, $\mathbb{E}(X_m \vee a)\mathrm{I}_A \le \mathbb{E}(X_n \vee a)\mathrm{I}_A$ for $m \le n$. Clearly, if $(X_n^+)_{-\infty<n<\infty}$ is uniformly integrable then $(X_n \vee a)_{-\infty<n<\infty}$ is uniformly integrable for all $a \in \mathbb{R}$. Since $X_n \vee a \to X_{+\infty} \vee a$ almost surely, by Lemma 5.1,

$$\lim_{n\to\infty} \mathbb{E}(X_n \vee a)\mathrm{I}_A = \mathbb{E}(X_{+\infty} \vee a)\mathrm{I}_A$$

and this shows that $\mathbb{E}(X_m \vee a)\mathrm{I}_A \le \mathbb{E}(X_{+\infty} \vee a)\mathrm{I}_A$. Letting $a \to -\infty$ and using the monotone convergence theorem we get that $\mathbb{E}X_m\mathrm{I}_A \le \mathbb{E}X_{+\infty}\mathrm{I}_A$. □

Convergence of martingales. If $(X_n, \mathscr{B}_n)$ is a martingale then both $(X_n, \mathscr{B}_n)$ and $(-X_n, \mathscr{B}_n)$ are submartingales and one can apply the above theorem to both of them. For example, this implies the following.

1. If $\sup_n \mathbb{E}|X_n| < \infty$ then a.s. limit $X_{+\infty} = \lim_{n\to+\infty} X_n$ exists and $\mathbb{E}|X_{+\infty}| < \infty$.
2. If $(X_n)_{-\infty<n<\infty}$ is uniformly integrable then $(X_n, \mathscr{B}_n)_{-\infty<n\le+\infty}$ is a right-closed martingale.

In other words, a martingale is right-closable if and only if it is uniformly integrable. Of course, in this case we also conclude that $\lim_{n\to+\infty} \mathbb{E}|X_n - X_{+\infty}| = 0$. For a martingale of the type $\mathbb{E}(X|\mathscr{B}_n)$ we can identify the limit as follows.

Theorem 5.6 (Levy's convergence theorem). *Let $(\Omega, \mathscr{B}, \mathbb{P})$ be a probability space and $X\colon \Omega \to \mathbb{R}$ be a random variable such that $\mathbb{E}|X| < \infty$. Given a sequence of sigma algebras*

$$\mathscr{B}_1 \subseteq \ldots \subseteq \mathscr{B}_n \subseteq \ldots \subseteq \mathscr{B}_{+\infty} \subseteq \mathscr{B}$$

where $\mathscr{B}_{+\infty} = \sigma\big(\bigcup_{1\le n<\infty} \mathscr{B}_n\big)$, we have $X_n = \mathbb{E}(X|\mathscr{B}_n) \to \mathbb{E}(X|\mathscr{B}_{+\infty})$ a.s. .

Proof. $(X_n, \mathscr{B}_n)_{1\le n<\infty}$ is a right-closable martingale, because $X_n = \mathbb{E}(X|\mathscr{B}_n)$. Therefore, it is uniformly integrable and a.s. limit $X_{+\infty} := \lim_{n\to+\infty} \mathbb{E}(X|\mathscr{B}_n)$ exists. It remains to show that $X_{+\infty} = \mathbb{E}(X|\mathscr{B}_{+\infty})$. Consider $A \in \bigcup_n \mathscr{B}_n$. Since $A \in \mathscr{B}_m$ for some m and (X_n) is a martingale, $\mathbb{E}X_n\mathrm{I}_A = \mathbb{E}X_m\mathrm{I}_A$ for $n \ge m$. Therefore,

$$\mathbb{E}X_{+\infty}\mathrm{I}_A = \lim_{n\to\infty} \mathbb{E}X_n\mathrm{I}_A = \mathbb{E}X_m\mathrm{I}_A = \mathbb{E}(\mathbb{E}(X|\mathscr{B}_m)\mathrm{I}_A) = \mathbb{E}X\mathrm{I}_A.$$

Since $\bigcup_n \mathscr{B}_n$ is an algebra, we get $\mathbb{E}X_{+\infty}\mathrm{I}_A = \mathbb{E}X\mathrm{I}_A$ for all $A \in \mathscr{B}_{+\infty} = \sigma\big(\cup_n \mathscr{B}_n\big)$ (by Dynkin's theorem or monotone class theorem), which proves that $X_{+\infty} = \mathbb{E}(X|\mathscr{B}_{+\infty})$. □

Example 5.3.2 (Strong law of large numbers). Let $(X_n)_{n\ge1}$ be i.i.d. such that $\mathbb{E}|X_1| < \infty$, and let

$$\mathscr{B}_{-n} = \sigma(S_n, X_{n+1}, \ldots),\ S_n = X_1 + \cdots + X_n.$$

We showed before that

$$Z_{-n} = \frac{S_n}{n} = \mathbb{E}(X_1 | \mathscr{B}_{-n}),$$

so $(Z_{-n}, \mathscr{B}_{-n})$ is a reverse martingale and, by the above theorem, the almost sure limit $Z_{-\infty} = \lim_{n\to+\infty} Z_{-n}$ exists and is measurable on the σ-algebra $\mathscr{B}_{-\infty} = \bigcap_{n\geq 1} \mathscr{B}_{-n}$. Since each event in $\mathscr{B}_{-\infty}$ is obviously symmetric, i.e. invariant under the permutations of finitely many (X_n), by the Savage-Hewitt $0-1$ law, the probability of each event is 0 or 1, i.e. $\mathscr{B}_{-\infty}$ consists of $\emptyset$ and Ω up to sets of measure zero. Therefore, $Z_{-\infty}$ is a constant a.s. and, since $(Z_n)_{-\infty\leq n\leq -1}$ is also a martingale, $\mathbb{E}Z_{-\infty} = \mathbb{E}X_1$. Therefore, we proved that $S_n/n \to \mathbb{E}X_1$ almost surely.

Example 5.3.3 (Improved Kolmogorov's law of large numbers). Consider a sequence $(X_n)_{n\geq 1}$ such that

$$\mathbb{E}(X_{n+1}|X_1, \ldots, X_n) = 0.$$

We do not assume independence here. This assumption, obviously, implies that $\mathbb{E}X_k X_\ell = 0$ for $k \neq \ell$. Let us show that if a sequence (b_n) is such that

$$b_n \leq b_{n+1},\ \lim_{n\to\infty} b_n = \infty \ \text{ and } \ \sum_{n=1}^{\infty} \frac{\mathbb{E}X_n^2}{b_n^2} < \infty$$

then $S_n/b_n \to 0$ almost surely. Indeed, $Y_n = \sum_{k\leq n}(X_k/b_k)$ is a martingale and (Y_n) is uniformly integrable, since

$$\mathbb{E}|Y_n|\mathrm{I}(|Y_n| > M) \leq \frac{1}{M}\mathbb{E}|Y_n|^2 = \frac{1}{M}\sum_{k=1}^{n} \frac{\mathbb{E}X_k^2}{b_k^2}$$

(we used here that $\mathbb{E}X_k X_\ell = 0$ for $k \neq \ell$) and, therefore,

$$\sup_n \mathbb{E}|Y_n|\mathrm{I}(|Y_n| > M) \leq \frac{1}{M}\sum_{k=1}^{\infty} \frac{\mathbb{E}X_k^2}{b_k^2} \to 0$$

as $M \to \infty$. By the martingale convergence theorem, the limit $Y = \lim_{n\to\infty} Y_n$ exists almost surely, and Kronecker's lemma implies that $S_n/b_n \to 0$ almost surely. □

Exercise 5.3.1. Let $(X_n)_{n\geq 1}$ be a martingale with $\mathbb{E}X_n = 0$ and $\mathbb{E}X_n^2 < \infty$ for all n. Show that

$$\mathbb{P}\Big(\max_{1\leq k\leq n} X_k \geq M\Big) \leq \frac{\mathbb{E}X_n^2}{\mathbb{E}X_n^2 + M^2}.$$

Hint: $(X_n + c)^2$ is a submartingale for any $c > 0$.

Exercise 5.3.2. Let (g_i) be i.i.d. standard normal random variables and let $X_n = \sum_{i=1}^{n} \sigma_i^2 (g_i^2 - 1)$ for some numbers (σ_i). Prove that for $t > 0$,

$$\mathbb{P}\big(\max_{1\le i\le n}|X_i|\ge t\big)\le 2t^{-2}\sum_{i=1}^n \sigma_i^4.$$

Exercise 5.3.3. Show that for any random variable Y,

$$\mathbb{E}|Y|^p=\int_0^\infty pt^{p-1}\mathbb{P}(|Y|\ge t)dt.$$

Hint: represent $|Y|^p=\int_0^{|Y|}pt^{p-1}dt$ and switch the order of integration.

Exercise 5.3.4. Let X,Y be two non-negative random variables such that for every $t>0$,

$$\mathbb{P}(Y\ge t)\le t^{-1}\int X\mathrm{I}(Y\ge t)d\mathbb{P}.$$

For any $p>1$, $\|f\|_p=(\int|f|^p d\mathbb{P})^{1/p}$ and $1/p+1/q=1$, show that $\|Y\|_p\le q\|X\|_p$. *Hint:* Use previous exercise, switch the order of integration to integrate in t first, then use Holder's inequality and solve for $\|Y\|_p$.

Exercise 5.3.5. Given a non-negative submartingale $(X_n,\mathscr{B}_n)$, let $X_n^*:=\max_{j\le n}X_j$. Prove that for any $p>1$ and $1/p+1/q=1$, $\|X_n^*\|_p\le q\|X_n\|_p$. *Hint:* use previous exercise and Doob's maximal inequality.

Exercise 5.3.6 (Polya's urn model). Suppose we have b blue and r red balls in the urn. We pick a ball randomly and return it with c balls of the same color. Let us consider the sequence

$$Y_n=\frac{\#(\text{blue balls after } n \text{ iterations})}{\#(\text{total after } n \text{ iterations})}.$$

Prove that the limit $Y=\lim_{n\to\infty}Y_n$ exists almost surely.

Exercise 5.3.7 (Improved Kolmogorov's $0-1$ law). Consider arbitrary random variables $(X_i)_{i\ge1}$ and consider sigma algebras $\mathscr{B}_n=\sigma(X_1,\ldots,X_n)$ and $\mathscr{B}_{+\infty}=\sigma((X_n)_{n\ge1})$. Consider any set $A\in\mathscr{B}_{+\infty}$ that is independent of $\mathscr{B}_n$ for all n (also called a *tail event*). By considering conditional expectations $\mathbb{E}(I_A|\mathscr{B}_n)$ prove that $\mathbb{P}(A)=0$ or 1.

Exercise 5.3.8. Suppose that a random variable ξ takes values $0,1,2,3,\ldots$. Suppose that $\mathbb{E}\xi=\mu>1$ and $\mathrm{Var}(\xi)=\sigma^2<\infty$. Let ξ_{nk} be independent copies of ξ. Let $X_1=1$ and define recursively $X_{n+1}=\sum_{k=1}^{X_n}\xi_{nk}$. Prove that $S_n=X_n/\mu^n$ converges almost surely.

Exercise 5.3.9. Let $(X_n)_{n\ge1}$ be i.i.d. random variables and, given a bounded measurable function $f\colon\mathbb{R}^2\to\mathbb{R}$ such that $f(x,y)=f(y,x)$, consider the U-statistics

$$S_{-n}=\frac{2}{n(n-1)}\sum_{1\le\ell<\ell'\le n}f(X_\ell,X_{\ell'})$$

and the σ-algebras

$$\mathscr{F}_{-n} = \sigma\big(X_{(1)}, \ldots, X_{(n)}, (X_\ell)_{\ell > n}\big),$$

where $X_{(1)}, \ldots, X_{(n)}$ are the order statistics of $X_1, \ldots, X_n$. Prove that $\lim_{n\to\infty} S_{-n} = \mathbb{E} f(X_1, X_2)$ almost surely. Hint: notice that $\bigcap_{n\geq 1} \mathscr{F}_{-n}$ consists of symmetric events.

Exercise 5.3.10. Consider two bounded measurable functions $f(x,y), g(x,y)$ on $[0,1]^2$. Suppose that for almost all (x,y) on $[0,1]$, we have $\int_0^1 f(x,u)g(y,u)\,du \leq 1$. Prove that $\int_0^1 f(x,u)g(x,u)\,du \leq 1$ for almost all x on $[0,1]$. (For almost all in both cases means with respect to the Lebesgue measure.)

Chapter 6
Brownian Motion

6.1 Stochastic processes. Brownian motion

We have developed a general theory of convergence of laws on (separable) metric spaces and in the following two sections we will look at some specific examples of convergence on the spaces of continuous functions $(C[0,1],\|\cdot\|_\infty)$ and $(C(\mathbb{R}_+),d)$, where d is a metric metrizing uniform convergence on compacts. These examples will describe a certain central limit theorem type results on these spaces and in this section we will define the corresponding limiting Gaussian laws, namely, the Brownian motion and Brownian bridge. We will start with basic definitions and basic regularity results in the presence of continuity. Given a set T and a probability space $(\Omega,\mathscr{F},\mathbb{P})$, a *stochastic process* is a function

$$X_t(\omega) = X(t,\omega) : T\times\Omega \to \mathbb{R}$$

such that for each $t\in T, X_t : \Omega\to\mathbb{R}$ is a random variable, i.e. a measurable function. In other words, a stochastic process is a collection of random variables X_t indexed by a set T. A stochastic process is often defined by specifying finite dimensional (f.d.) distributions $\mathbb{P}_F = \mathscr{L}(\{X_t\}_{t\in F})$ for all finite subsets $F\subseteq T$. Kolmogorov's theorem then guarantees the existence of a probability space on which the process is defined, under the natural consistency condition

$$F_1\subseteq F_2 \Longrightarrow \mathbb{P}_{F_1} = \mathbb{P}_{F_2}\Big|_{\mathbb{R}^{F_1}}.$$

One can also think of a process as a function on Ω with values in $\mathbb{R}^T = \{f : T\to \mathbb{R}\}$, because for a fixed $\omega\in\Omega, X_t(\omega)\in\mathbb{R}^T$ is a (random) function of t. In Kolmogorov's theorem, given a family of consistent f.d. distributions, a process was defined on the probability space $(\mathbb{R}^T,\mathscr{B}_T)$, where $\mathscr{B}_T$ is the cylindrical σ-algebra generated by the algebra of cylinders $B\times\mathbb{R}^{T\setminus F}$ for Borel sets B in $\mathbb{R}^F$ and all finite F. When T is uncountable, some very natural sets such as

$$\left\{\omega : \sup_{t\in T} X_t > 1\right\} = \bigcup_{t\in T}\{X_t > 1\}$$

might not be measurable on $\mathscr{B}_T$. However, in our examples we will deal with continuous processes that possess additional regularity properties. If (T,d) is a metric space then a process X_t is called *sample continuous* if for all $\omega \in \Omega$, $X_t(\omega) \in C(T,d)$ – the space of continuous function on (T,d). The process X_t is called *continuous in probability* if $X_t \to X_{t_0}$ in probability whenever $t \to t_0$. Let us note that sample continuity is not a property that follows automatically if we know all finite dimensional distributions of the process.

Example 6.1.1. Let $T = [0,1]$, $(\Omega,\mathbb{P}) = ([0,1],\lambda)$ where λ is the Lebesgue measure. Let $X_t(\omega) = \mathrm{I}(t = \omega)$ and $X_t'(\omega) = 0$. Finite dimensional distributions of these processes are the same, because for any fixed $t \in [0,1]$,

$$\mathbb{P}(X_t = 0) = \mathbb{P}(X_t' = 0) = 1.$$

However, $\mathbb{P}(X_t \text{ is continuous}) = 0$, but for X_t' this probability is 1. □

Let (T,d) be a metric space. The process X_t is *measurable* if

$$X_t(\omega)\colon T\times\Omega \to \mathbb{R}$$

is jointly measurable on the product space $(T,\mathscr{B})\times(\Omega,\mathscr{F})$, where $\mathscr{B}$ is the Borel σ-algebra on T.

Lemma 6.1. *If (T,d) is a separable metric space and X_t is sample continuous then X_t is measurable.*

Proof. Let $(S_j)_{j\geq 1}$ be a measurable partition of T such that $\mathrm{diam}(S_j) \leq \frac{1}{n}$. For each non-empty S_j, let us take a point $t_j \in S_j$ and define

$$X_t^n(\omega) = X_{t_j}(\omega) \text{ for } t \in S_j.$$

$X_t^n(\omega)$ is, obviously, measurable on $T\times\Omega$, because for any Borel set A on $\mathbb{R}$,

$$\left\{(\omega,t) : X_t^n(\omega) \in A\right\} = \bigcup_{j\geq 1} S_j \times \left\{\omega : X_{t_j}(\omega) \in A\right\}.$$

Since $X_t(w)$ is sample continuous, $X_t^n(\omega) \to X_t(\omega)$ as $n \to \infty$ for all (ω,t). Hence, X_t is also measurable. □

If X_t is a sample continuous process indexed by T then we can think of X_t as an element of the metric space of continuous functions $(C(T,d),\|\cdot\|_\infty)$, rather then simply an element of $\mathbb{R}^T$. We can define measurable events on this space in two different ways. On the one hand, we have the natural Borel σ-algebra $\mathscr{B}$ on $C(T,d)$ generated by the open (or closed) balls

$$B_g(\varepsilon) = \{f \in C(T,d) : \|f - g\|_\infty < \varepsilon\}.$$

On the other hand, if we think of $C(T)$ as a subspace of $\mathbb{R}^T$, we can consider a σ-algebra

$$S_T = \{B \cap C(T,d) : B \in \mathscr{B}_T\},$$

which is the intersection of the cylindrical σ-algebra $\mathscr{B}_T$ with $C(T,d)$. (Recall that, by an exercise in Section 1.3, $C(T,d) \notin \mathscr{B}_T$ when (T,d) is uncountable.) It turns out that these two definitions coincide if (T,d) is separable. An important implication of this is that the distribution of any random element with values in $(C(T,d), \|\cdot\|_\infty)$ is completely determined by its finite dimensional distributions.

Lemma 6.2. *If (T,d) is a separable metric space then $\mathscr{B} = S_T$.*

Proof. Let us first show that $S_T \subseteq \mathscr{B}$. Any element of the cylindrical algebra that generates the cylindrical σ-algebra $\mathscr{B}_T$ is given by

$$B \times \mathbb{R}^{T \setminus F} \text{ for a finite } F \subseteq T \text{ and for some Borel set } B \subseteq \mathbb{R}^F.$$

Then

$$\big(B \times \mathbb{R}^{T\setminus F}\big) \bigcap C(T,d) = \{x \in C(T,d) : (x_t)_{t \in F} \in B\} = \{\pi_F(x) \in B\}$$

where $\pi_F : C(T,d) \to \mathbb{R}^F$ is the finite dimensional projection such that $\pi_F(x) = (x_t)_{t\in F}$. Projection π_F is, obviously, continuous in the $\|\cdot\|_\infty$ norm and, therefore, measurable on the Borel σ-algebra $\mathscr{B}$ generated by the open sets in the $\|\cdot\|_\infty$ norm. This implies that $\{\pi_F(x) \in B\} \in \mathscr{B}$ and, thus, $S_T \subseteq \mathscr{B}$. Let us now show that $\mathscr{B} \subseteq S_T$. Let T' be a countable dense subset of T. Then, by continuity, any closed ε-ball in $C(T,d)$ can be written as

$$\{f \in C(T,d) : \|f - g\|_\infty \le \varepsilon\} = \bigcap_{t \in T'} \{f \in C(T,d) : |f(t) - g(t)| \le \varepsilon\} \in S_T.$$

This finishes the proof. □

In the remainder of the section we will define two specific sample continuous stochastic processes.

Brownian motion. *Brownian motion* is defined as a sample continuous process X_t on $T = \mathbb{R}_+$ such that

(a) the distribution of X_t is a centered Gaussian for each $t \ge 0$;
(b) $X_0 = 0$ and $\mathbb{E}X_1^2 = 1$;
(c) if $t < s$ then $\mathscr{L}(X_s - X_t) = \mathscr{L}(X_{s-t})$;
(d) for any $t_1 < \ldots < t_n$, the increments $X_{t_1} - X_0, \ldots, X_{t_n} - X_{t_{n-1}}$ are independent.

If we denote $\sigma^2(t) = \mathrm{Var}(X_t)$ then these properties imply

$$\sigma^2(nt) = n\sigma^2(t),\ \sigma^2\Big(\frac{t}{m}\Big) = \frac{1}{m}\sigma^2(t) \text{ and } \sigma^2(qt) = q\sigma^2(t)$$

for all rational q. Since $\sigma^2(1) = 1$, $\sigma^2(q) = q$ for all rational q and, by the sample continuity, $\sigma^2(t) = t$ for all $t \geq 0$. For any $t_1 < \ldots < t_n$, the increments $X_{t_{j+1}} - X_{t_j}$ are independent Gaussian $N(0, t_{j+1} - t_j)$ and, therefore, jointly Gaussian. This implies that the vector $(X_{t_1}, \ldots, X_{t_n})$ also has a jointly Gaussian distribution, since it can be written as a linear map of the increments. The covariance matrix can be easily computed, since for $s < t$,

$$\mathbb{E}X_s X_t = \mathbb{E}X_s(X_s + (X_t - X_s)) = s = \min(t,s).$$

As a result, we can give an equivalent definition: *Brownian motion* is a sample continuous centered Gaussian process X_t for $t \in [0,\infty)$ with the covariance $\mathrm{Cov}(X_t, X_s) = \min(t,s)$. A process is Gaussian if all its finite dimensional distributions are Gaussian.

Without the requirement of sample continuity, the existence of such process follows from Kolmogorov's theorem, since all finite dimensional distributions are consistent by construction. However, we still need to prove that there exists a sample continuous version with these prescribed finite dimensional distributions. We start with a simple estimate.

Lemma 6.3. *The tail probability of the standard normal distribution satisfies*

$$\overline{\Phi}(c) = \frac{1}{\sqrt{2\pi}} \int_c^\infty e^{-x^2/2}\, dx \leq e^{-\frac{c^2}{2}}$$

for all $c \geq 0$.

Proof. We can write

$$\overline{\Phi}(c) = \frac{1}{\sqrt{2\pi}} \int_c^\infty e^{-x^2/2}\, dx \leq \frac{1}{\sqrt{2\pi}} \int_c^\infty \frac{x}{c} e^{-x^2/2}\, dx = \frac{1}{\sqrt{2\pi}} \frac{1}{c} e^{-c^2/2}.$$

If $c > 1/\sqrt{2\pi}$ then $\overline{\Phi}(c) \leq \exp(-c^2/2)$. If $c \leq 1/\sqrt{2\pi}$ then a simpler estimate gives the result

$$\overline{\Phi}(c) \leq \overline{\Phi}(0) = \frac{1}{2} \leq \exp\Big(-\frac{1}{2}\Big(\frac{1}{\sqrt{2\pi}}\Big)^2\Big) \leq e^{-c^2/2},$$

which finishes the proof. □

Theorem 6.1 (Existence of Brownian motion). *There exists a sample continuous Gaussian process with the covariance* $\mathrm{Cov}(X_t, X_s) = \min(t,s)$.

Proof. It is enough to construct X_t on the interval $[0,1]$. Given a process X_t that has the finite dimensional distributions of the Brownian process, but is not necessarily continuous, let us define for $n \geq 1$,

$$V_k = X_{\frac{k+1}{2^n}} - X_{\frac{k}{2^n}} \text{ for } k = 0, \ldots, 2^n - 1.$$

The variance $\mathrm{Var}(V_k) = 1/2^n$ and, by the above lemma,

$$\mathbb{P}\Big(\max_k |V_k| \geq \frac{1}{n^2}\Big) \leq 2^n \mathbb{P}\Big(|V_1| \geq \frac{1}{n^2}\Big) \leq 2^{n+1} \exp\Big(-\frac{2^{n-1}}{n^4}\Big).$$

The right hand side is summable over $n \geq 1$ and, by the Borel-Cantelli lemma,

$$\mathbb{P}\Big(\Big\{\max_k |V_k| \geq \frac{1}{n^2}\Big\} \text{ i.o.}\Big) = 0. \tag{6.1}$$

Given $t \in [0,1]$ and its dyadic expansion $t = \sum_{j=1}^{\infty} \frac{t_j}{2^j}$ for $t_j \in \{0,1\}$, let us define $t(n) = \sum_{j=1}^{n} \frac{t_j}{2^j}$ so that

$$X_{t(n)} - X_{t(n-1)} \in \{0\} \cup \{V_k : k = 0, \ldots, 2^n - 1\}.$$

Then, the sequence

$$X_{t(n)} = 0 + \sum_{1 \leq j \leq n} (X_{t(j)} - X_{t(j-1)})$$

converges almost surely to some limit Z_t, because, by (6.1), with probability one,

$$|X_{t(n)} - X_{t(n-1)}| \leq n^{-2}$$

for large enough (random) $n \geq n_0(\omega)$. By construction, $Z_t = X_t$ on the dense subset of all dyadic $t \in [0,1]$. If we can prove that Z_t is sample continuous then all f.d. distributions of Z_t and X_t will coincide, which means that Z_t is a continuous version of the Brownian motion. Take any $t, s \in [0,1]$ such that $|t - s| \leq 2^{-n}$. If $t(n) = \frac{k}{2^n}$ and $s(n) = \frac{m}{2^n}$, then $|k - m| \in \{0,1\}$. As a result, $|X_{t(n)} - X_{s(n)}|$ is either equal to 0 or one of the increments $|V_k|$ and, by (6.1), $|X_{t(n)} - X_{s(n)}| \leq n^{-2}$ for large enough n. Finally,

$$\begin{aligned} |Z_t - Z_s| &\leq |Z_t - X_{t(n)}| + |X_{t(n)} - X_{s(n)}| + |X_{s(n)} - Z_s| \\ &\leq \sum_{\ell \geq n} \frac{1}{\ell^2} + \frac{1}{n^2} + \sum_{\ell \geq n} \frac{1}{\ell^2} \leq \frac{c}{n}, \end{aligned}$$

which proves the continuity of Z_t. On the event in (6.1) of probability zero we set $Z_t = 0$. □

Definition 6.1. A sample continuous centered Gaussian process B_t for $t \in [0,1]$ is called a *Brownian bridge* if

$$\mathbb{E} B_t B_s = s(1-t) \text{ for } s < t.$$

Such process exists because if X_t is a Brownian motion then $B_t = X_t - tX_1$ is a Brownian bridge, since for $s < t$,

$$\mathbb{E}B_sB_t = \mathbb{E}(X_t - tX_1)(X_s - sX_1) = s - st - ts + st = s(1-t).$$

Notice that $B_0 = B_1 = 0$. □

Exercise 6.1.1 (Ciesielski's construction of Brownian motion). Let $(f_{k2^{-n}})$ be the Haar basis on $(L^2[0,1], dx)$, i.e. $f_0 = 1$ and for $n \geq 1$ and $k \in I_n = \{k \text{ - odd}, 1 \leq k \leq 2^n - 1\}$,

$$f_{k2^{-n}}(x) = \begin{cases} 2^{(n-1)/2}, & (k-1)2^{-n} \leq x < k2^{-n} \\ -2^{(n-1)/2}, & k2^{-n} \leq x < (k+1)2^{-n} \end{cases}$$

and 0 otherwise. If $(g_{k2^{-n}})$ are i.i.d. standard Gaussian random variables, prove that

$$W_t = g_0 t + \sum_{n\geq 1} \sum_{k \in I_n} g_{k2^{-n}} \int_0^t f_{k2^{-n}}(x)\, dx$$

is a Brownian motion on $[0,1]$. *Hint:* To show that $\mathbb{E}W_tW_s = \min(t,s)$, use Parseval's identity. To prove continuity, show that

$$e_n = \left\| \sum_{k \in I_n} g_{k2^{-n}} \int_0^t f_{k2^{-n}}(x)\, dx \right\|_\infty = 2^{-(n+1)/2} \max_{k \in I_n} |g_{k2^{-n}}|$$

satisfies $\mathbb{P}\left(e_n \geq 2\sqrt{2^{-n} \ln 2^n}\right) \leq 8^{-n}$ and use the Borel-Cantelli lemma.

Exercise 6.1.2. Let W_t be a Brownian motion. Prove that

1. (*symmetry*) $\{-W_t\}_{t\in[0,\infty)}$ is a Brownian motion;
2. (*invariance under scaling*) For every $\lambda > 0$, $\{\frac{1}{\sqrt{\lambda}} W_{\lambda t}\}_{t\in[0,\infty)}$ is a Brownian motion;
3. (*simple Markov property*) For every $s \geq 0$, $\{W_{s+t} - W_s\}_{t\in[0,\infty)}$ is a Brownian motion independent of $\sigma(\{W_r\}_{r\in[0,s]})$;
4. (*time reversal*) If $\{W_t\}_{t\in[0,T]}$ is a Brownian motion, then $\{W_T - W_{T-t}\}_{t\in[0,T]}$ is a Brownian motion.

Exercise 6.1.3. If W_t is a Brownian motion, prove that $\mathbb{P}(\max_{0\leq t\leq T} W_t \geq x) \leq e^{-x^2/(2T)}$ for $x \geq 0$. *Hint:* Apply Doob's inequality to $\exp(\lambda W_t - \lambda^2 t/2)$.

6.2 Donsker's invariance principle

In this section we will show how Brownian motion W_t arises in a classical central limit theorem on the space of continuous functions on $\mathbb{R}_+$. When working with continuous processes defined on $\mathbb{R}_+$, such as the Brownian motion, the metric $\|\cdot\|_\infty$ on $C(\mathbb{R}_+)$ is too strong. A more appropriate metric d can be defined by

$$d(f,g) = \sum_{n\geq 1} \frac{1}{2^n} \frac{d_n(f,g)}{1+d_n(f,g)}, \quad \text{where } d_n(f,g) = \sup_{0\leq t\leq n} |f(t)-g(t)|.$$

It is obvious that $d(f_j,f) \to 0$ if and only if $d_n(f_j,f) \to 0$ for all $n \geq 1$, i.e. d metrizes uniform convergence on compacts. $(C(\mathbb{R}_+),d)$ is also a complete separable space, since any sequence is Cauchy in d if and only if it is Cauchy for each d_n. When proving uniform tightness of laws on $(C(\mathbb{R}_+),d)$, we will need a characterization of compacts via the Arzela-Ascoli theorem, which in this case can be formulated as follows. For a subset $K \subseteq C(\mathbb{R}^+)$, let us define by

$$K_n = \big\{f\big|_{[0,n]} \,:\, f \in K\big\}$$

the restriction of K to the interval $[0,n]$. Then, it is clear that K is compact with respect to d if and only if each K_n is compact with respect to d_n. Therefore, using the Arzela-Ascoli theorem to characterize compacts in $C[0,n]$ we get the following. For a function $x \in C[0,T]$, let us denote its modulus of continuity by

$$m^T(x,\delta) = \sup\Big\{|x_a - x_b| \,:\, |a-b| \leq \delta,\, a,b \in [0,T]\Big\}.$$

Theorem 6.2 (Arzela-Ascoli). *A set K is compact in $(C(\mathbb{R}_+),d)$ if and only if K is closed, uniformly bounded and equicontinuous on each interval $[0,n]$. In other words,*

$$\sup_{x\in K} |x_0| < \infty \ \textit{and} \ \lim_{\delta\downarrow 0} \sup_{x\in K} m^T(x,\delta) = 0 \ \textit{for all} \ T > 0.$$

This implies the following criterion of the uniform tightness of laws on the metric space $(C(\mathbb{R}_+),d)$, which is simply a translation of the Arzela-Ascoli theorem into probabilistic language.

Theorem 6.3. *A sequence of laws $(\mathbb{P}_n)_{n\geq 1}$ on $(C(\mathbb{R}_+),d)$ is uniformly tight if and only if*

$$\lim_{\lambda\to+\infty} \sup_{n\geq 1} \mathbb{P}_n\big(|x_0| > \lambda\big) = 0 \tag{6.2}$$

and

$$\lim_{\delta\downarrow 0} \sup_{n\geq 1} \mathbb{P}_n\big(m^T(x,\delta) > \varepsilon\big) = 0 \tag{6.3}$$

for any $T > 0$ and any $\varepsilon > 0$.

Proof. ($\Longrightarrow$) Suppose that for any $\gamma > 0$, there exists a compact K such that $\mathbb{P}_n(K) > 1-\gamma$ for all $n \geq 1$. By the Arzela-Ascoli theorem, $|x_0| \leq \lambda$ for some

$\lambda > 0$ and for all $x \in K$ and, therefore,

$$\sup_n \mathbb{P}_n(|x_0| > \lambda) \le \sup_n \mathbb{P}_n(K^c) \le \gamma.$$

Also, by equicontinuity, for any $\varepsilon > 0$ there exists $\delta_0 > 0$ such that for $\delta < \delta_0$ and for all $x \in K$ we have $m^T(x,\delta) < \varepsilon$. Therefore,

$$\sup_n \mathbb{P}_n\big(m^T(x,\delta) > \varepsilon\big) \le \sup_n \mathbb{P}_n(K^c) \le \gamma.$$

($\Longleftarrow$) Fix any $\gamma > 0$. For each integer $T \ge 1$, find $\lambda_T > 0$ such that

$$\sup_n \mathbb{P}_n\big(|x_0| > \lambda_T\big) \le \frac{\gamma}{2^{T+1}}.$$

For each integer $T \ge 1$ and $k \ge 1$, find $\delta_{T,k} > 0$ such that

$$\sup_n \mathbb{P}_n\Big(m^T(x,\delta_{T,k}) > \frac{1}{k}\Big) \le \frac{\gamma}{2^{T+k+1}}.$$

Consider the set

$$A_T = \Big\{x \in C(\mathbb{R}^+) \,:\, |x_0| \le \lambda_T, m^T(x,\delta_{T,k}) \le \frac{1}{k} \text{ for all } k \ge 1\Big\}.$$

Then,

$$\sup_n \mathbb{P}_n\big((A_T)^c\big) \le \frac{\gamma}{2^{T+1}} + \sum_{k\ge 1} \frac{\gamma}{2^{T+k+1}} = \frac{\gamma}{2^T}$$

and their intersection $A = \bigcap_{T\ge 1} A_T$ satisfies

$$\sup_n \mathbb{P}_n(A^c) = \sup_n \mathbb{P}_n\Big(\bigcup_{T\ge 1}(A_T)^c\Big) \le \sum_{T\ge 1}\frac{\gamma}{2^T} = \gamma.$$

By construction, each set A_T is closed, uniformly bounded and equicontinuous on $[0,T]$. Therefore, by the Arzela-Ascoli theorem, their intersection A is compact on $(C(\mathbb{R}_+),d)$. Since $\gamma > 0$ was arbitrary, this proves that the sequence $(\mathbb{P}_n)$ is uniformly tight. □

Remark 6.1. Of course, for the uniform tightness on $(C[0,1],\|\cdot\|_\infty)$ we only need the second condition (6.3) for $T = 1$. Also, it will be convenient to slightly relax (6.3) and replace it with *asymptotic equicontinuity* condition

$$\lim_{\delta\downarrow 0}\limsup_{n\to\infty} \mathbb{P}_n\big(m^T(x,\delta) > \varepsilon\big) = 0. \tag{6.4}$$

If this holds then, given $\gamma > 0$, we can find $\delta_0 > 0$ and $n_0 \ge 1$ such that for all $n > n_0$,

$$\mathbb{P}_n\big(m^T(x,\delta_0) > \varepsilon\big) < \gamma.$$

Since $m^T(x,\delta) \downarrow 0$ as $\delta \downarrow 0$ for all $x \in C(\mathbb{R}_+)$, for each $n \le n_0$, we can find $\delta_n > 0$ such that

$$\mathbb{P}_n\big(m^T(x,\delta_n) > \varepsilon\big) < \gamma,$$

Since $m^T(x,\delta)$ is decreasing in δ, this implies that

$$\text{if } \delta < \min(\delta_0, \delta_1, \ldots, \delta_{n_0}) \text{ then } \mathbb{P}_n\big(m^T(x,\delta) > \varepsilon\big) < \gamma$$

for all $n \ge 1$. We showed that (6.4) implies (6.3). □

Using this characterization of uniform tightness, we will now give our first example of convergence on $(C(\mathbb{R}_+),d)$ to the Brownian motion W_t. Consider a sequence $(X_i)_{i\ge 1}$ of i.i.d. random variables such that $\mathbb{E}X_i = 0$ and $\sigma^2 = \mathbb{E}X_i^2 < \infty$. Let us consider a continuous partial sum process on $[0,\infty)$ defined by

$$W_t^n = \frac{1}{\sqrt{n}\sigma}\sum_{i\le \lfloor nt\rfloor} X_i + (nt - \lfloor nt\rfloor)\frac{X_{\lfloor nt\rfloor+1}}{\sqrt{n}\sigma}, \tag{6.5}$$

where $\lfloor nt\rfloor$ is the integer part of nt, $\lfloor nt\rfloor \le nt < \lfloor nt\rfloor + 1$.

Theorem 6.4 (Donsker Invariance Principle). *The processes W_t^n converge in distribution to the Brownian motion W_t on the space $(C(\mathbb{R}_+),d)$.*

This implies for example that any continuous functionals of these processes on $(C(\mathbb{R}_+),d)$ converges in distribution, for example,

$$\sup_{t\le 1} W_t^n \to \sup_{t\le 1} W_t$$

in distribution.

Proof. Since the last term in W_t^n is of order $n^{-1/2}$, for simplicity of notations, we will simply write

$$W_t^n = \frac{1}{\sqrt{n}\sigma}\sum_{i\le nt} X_i$$

and treat nt as an integer. By the central limit theorem,

$$\frac{1}{\sqrt{n}\sigma}\sum_{i\le nt} X_i = \sqrt{t}\frac{1}{\sqrt{nt}\sigma}\sum_{i\le nt} X_i$$

converges in distribution to normal distribution $N(0,t)$. Given $t < s$, we can represent

$$W_s^n = W_t^n + \frac{1}{\sqrt{n}\sigma}\sum_{nt<i\le ns} X_i$$

and since W_t^n and $W_s^n - W_t^n$ are independent, it should be obvious that finite dimensional distributions of W_t^n converge to finite dimensional distributions of the Brownian motion W_t. By Lemma 6.2 in the previous section, this identifies W_t as the unique possible limit of W_t^n and, if we can show that the sequence of laws $(\mathscr{L}(W_t^n))_{n\geq 1}$ is uniformly tight on $\big(C[0,\infty),d\big)$, the Selection Theorem will imply that $W_t^n \to W_t$ weakly. Since $W_0^n = 0$, we only need to prove the asymptotic equicontinuity (6.4). First,

$$m^T(W^n,\delta) = \sup_{t,s\in[0,T],|t-s|\leq\delta}\Big|\frac{1}{\sqrt{n}\sigma}\sum_{ns<i\leq nt} X_i\Big| \leq \max_{0\leq k\leq nT, 0<j\leq n\delta}\Big|\frac{1}{\sqrt{n}\sigma}\sum_{k<i\leq k+j} X_i\Big|.$$

If instead of maximizing over all $0\leq k\leq nT$, we maximize in increments of $n\delta$, i.e. over indices k of the type $k=\ell n\delta$ for $0\leq \ell\leq m-1$, where $m:=T/\delta$, then it is easy to check that the maximum will decrease by at most a factor of 3,

$$\max_{0\leq k\leq nT, 0<j\leq n\delta}\Big|\frac{1}{\sqrt{n}\sigma}\sum_{k<i\leq k+j} X_i\Big| \leq 3 \max_{0\leq \ell\leq m-1, 0<j\leq n\delta}\Big|\frac{1}{\sqrt{n}\sigma}\sum_{\ell n\delta<i\leq \ell n\delta+j} X_i\Big|,$$

because the second maximum over $0<j\leq n\delta$ is taken over intervals of the same size $n\delta$. As a consequence, if $m^T(W^n,\delta)>\varepsilon$ then one of the events

$$\Big\{\max_{0<j\leq n\delta}\Big|\frac{1}{\sqrt{n}\sigma}\sum_{\ell n\delta<i\leq \ell n\delta+j} X_i\Big| > \frac{\varepsilon}{3}\Big\}$$

must occur for some $0\leq \ell\leq m-1$. Since the number of these events is $m=T/\delta$,

$$\mathbb{P}\Big(m^T(W^n,\delta)>\varepsilon\Big) \leq m\,\mathbb{P}\Big(\max_{0<j\leq n\delta}\Big|\frac{1}{\sqrt{n}\sigma}\sum_{0<i\leq j} X_i\Big| > \frac{\varepsilon}{3}\Big). \tag{6.6}$$

Kolmogorov's inequality, Theorem 2.10, implies that if $S_n = X_1+\cdots+X_n$ and $\max_{0<j\leq n}\mathbb{P}(|S_n-S_j|>\alpha)\leq p<1$ then

$$\mathbb{P}\Big(\max_{0<j\leq n}|S_j|>2\alpha\Big) \leq \frac{1}{1-p}\mathbb{P}(|S_n|>\alpha).$$

If we take $\alpha=\varepsilon\sqrt{n}\sigma/6$ then, by Chebyshev's inequality,

$$\mathbb{P}\Big(\Big|\sum_{j<i\leq n\delta} X_i\Big| > \frac{\varepsilon}{6}\sqrt{n}\sigma\Big) \leq \frac{6^2\delta n\sigma^2}{\varepsilon^2 n\sigma^2} = 36\delta\varepsilon^{-2}$$

and, therefore, if $36\delta\varepsilon^{-2}<1$,

$$\mathbb{P}\Big(\max_{0<j\leq n\delta}\Big|\sum_{0<i\leq j} X_i\Big| > \frac{\varepsilon}{3}\sqrt{n}\sigma\Big) \leq (1-36\delta\varepsilon^{-2})^{-1}\,\mathbb{P}\Big(\Big|\sum_{0<i\leq n\delta} X_i\Big| > \frac{\varepsilon}{6}\sqrt{n}\sigma\Big).$$

Finally, using (6.6) and the central limit theorem,

$$\limsup_{n\to\infty}\mathbb{P}\Big(m^T(W_t^n,\delta)>\varepsilon\Big) \leq m(1-36\delta\varepsilon^{-2})^{-1}\limsup_{n\to\infty}\mathbb{P}\Big(\Big|\sum_{0<i\leq n\delta} X_i\Big| > \frac{\varepsilon}{6}\sqrt{n}\sigma\Big)$$

$$
\begin{aligned}
&= m\big(1-36\delta\varepsilon^{-2}\big)^{-1} 2N(0,1)\Big(\frac{\varepsilon}{6\sqrt{\delta}},\infty\Big) \\
&\le 2T\delta^{-1}\big(1-36\delta\varepsilon^{-2}\big)^{-1}\exp\Big(-\frac{1}{2}\frac{\varepsilon^2}{6^2\delta}\Big).
\end{aligned}
$$

This, obviously, goes to zero as $\delta \downarrow 0$, which proves that

$$
\lim_{\delta\downarrow 0}\limsup_{n\to\infty}\mathbb{P}\big(m^T(W^n,\delta)>\varepsilon\big)=0,
$$

for all $T>0$ and $\varepsilon>0$. This proves that $W_t^n \to W_t$ weakly in $(C[0,\infty),d)$. □

Exercise 6.2.1. For $n\ge 1$, consider independent random variables $X_{n,i}$ for $1\le i\le n$ such that

$$
\mathbb{E}X_{n,i}=0, \sigma_{n,i}^2=\mathbb{E}X_{n,i}^2, \text{ and } \sum_{i=1}^n \sigma_{n,i}^2 = 1.
$$

Suppose that Lindeberg's condition holds: for all $\varepsilon>0$,

$$
\lim_{n\to\infty}\sum_{i=1}^n \mathbb{E}X_{n,i}^2 \mathrm{I}(|X_{n,i}|>\varepsilon)=0.
$$

Define $t_{n,j}=\sum_{i\le j}\sigma_{n,i}^2$ for $1\le j\le n$, and define a random process $X^n(t)$ on the interval $[0,1]$ by setting $X^n(0)=0$, $X^n(t_{n,j})=\sum_{i\le j}X_{n,i}$ for $1\le j\le n$ and interpolating linearly in between. Prove that $X^n(t)$ converges in distribution to the Brownian motion W_t on $(C[0,1],\|\cdot\|_\infty)$.

Exercise 6.2.2. Let $(\varepsilon_i)_{1\le i\le n}$ be i.i.d. Rademacher random variables, i.e. $\mathbb{P}(\varepsilon_i=\pm 1)=1/2$. For $\delta\in(0,1)$, let F_δ be the subset of vectors in $\{0,1\}^n$ with at most δn coordinates equal to 1, and such that these coordinates are consecutive. Show that, for large enough absolute constant $c>0$,

$$
\mathbb{P}\Big(\max_{f\in F_\delta}\Big|\frac{1}{\sqrt{n}}\sum_{i=1}^n \varepsilon_i f_i\Big| \ge c(\sqrt{\delta}+t)\Big)\le \frac{c}{\delta}e^{-\frac{t^2}{2\delta}}
$$

for all $t>0$ and, as a consequence,

$$
\mathbb{E}\max_{f\in F_\delta}\Big|\frac{1}{\sqrt{n}}\sum_{i=1}^n \varepsilon_i f_i\Big| \le c\Big(\sqrt{\delta\log\frac{1}{\delta}}+\sqrt{\delta}\Big)
$$

for some large enough $c>0$. *Hint:* Use discretization and Kolmogorov's inequality as in the proof of Donsker's theorem, and then apply Hoeffding's inequality instead of the CLT.

6.3 Convergence of empirical process to Brownian bridge

Empirical process and the Kolmogorov-Smirnov test. In this sections we show how the Brownian bridge B_t arises in another central limit theorem on the space of continuous functions on $[0,1]$. Let us start with a motivating example from statistics. Suppose that $x_1,\ldots,x_n$ are i.i.d. uniform random variables on $[0,1]$. By the law of large numbers, for any $t\in[0,1]$, the *empirical* c.d.f. $n^{-1}\sum_{i=1}^n \mathrm{I}(x_i\le t)$ converges to the true c.d.f. $\mathbb{P}(x_1\le t)=t$ almost surely and, moreover, by the CLT,

$$X_t^n := \sqrt{n}\Bigl(\frac{1}{n}\sum_{i=1}^n \mathrm{I}(x_i\le t)-t\Bigr)\xrightarrow{d} N\bigl(0,t(1-t)\bigr).$$

The stochastic process X_t^n is called the *empirical process*. The covariance of this process, for $s\le t$,

$$\mathbb{E}X_t^nX_s^n=\mathbb{E}(\mathrm{I}(x_1\le t)-t)(\mathrm{I}(x_1\le s)-s)=s-ts-ts+ts=s(1-t),$$

is the same as the covariance of the Brownian bridge and, by the multivariate CLT, all finite dimensional distributions of the empirical process converge to the f.d. distributions of the Brownian bridge,

$$\mathscr{L}\bigl((X_t^n)_{t\in F}\bigr)\to\mathscr{L}\bigl((B_t)_{t\in F}\bigr). \tag{6.7}$$

However, we would like to show the convergence of X_t^n to B_t in some stronger sense that would imply weak convergence of continuous functions of the process on the space $(C[0,1],\|\cdot\|_\infty)$. The Kolmogorov-Smirnov test in statistics provides some motivation. Suppose that i.i.d. $(X_i)_{i\ge1}$ have continuous distribution with c.d.f. $F(t)=\mathbb{P}(X_1\le t)$. Let $F_n(t)=n^{-1}\sum_{i=1}^n \mathrm{I}(X_i\le t)$ be the empirical c.d.f.. Since F is continuous and $F(\mathbb{R})=[0,1]$,

$$\begin{aligned}\sup_{t\in\mathbb{R}}\sqrt{n}|F_n(t)-F(t)| &= \sup_{t\in\mathbb{R}}\sqrt{n}\Bigl|\frac{1}{n}\sum_{i=1}^n \mathrm{I}(X_i\le t)-F(t)\Bigr|\\ &=\sup_{t\in\mathbb{R}}\sqrt{n}\Bigl|\frac{1}{n}\sum_{i=1}^n \mathrm{I}\bigl(F(X_i)\le F(t)\bigr)-F(t)\Bigr|\\ &=\sup_{t\in[0,1]}\sqrt{n}\Bigl|\frac{1}{n}\sum_{i=1}^n \mathrm{I}(F(X_i)\le t)-t\Bigr|\stackrel{d}{=}\sup_{t\in[0,1]}|X_t^n|,\end{aligned}$$

because $(F(X_i))$ are i.i.d. and have the uniform distribution on $[0,1]$. This means that the distribution of the left hand side does not depend on F and, in order to infer whether the sample $(X_i)_{1\le i\le n}$ comes from the distribution with the c.d.f. F, statisticians need to know only the distribution of the supremum of the empirical process or, as an approximation, the distribution of its limit. Equation (6.7) suggests that

$$\mathscr{L}\bigl(\sup_t|X_t^n|\bigr)\to\mathscr{L}\bigl(\sup_t|B_t|\bigr), \tag{6.8}$$

and the right hand side is called the Kolmogorov-Smirnov distribution that will be computed in the next section. Since B_t is sample continuous, its distribution is the law on the metric space $(C[0,1], \|\cdot\|_\infty)$. Even though X_t^n is not continuous, its jumps are equal to $n^{-1/2}$, so it can be approximated by a continuous process Y_t^n uniformly within $n^{-1/2}$. Since $\|\cdot\|_\infty$ is a continuous functional on $(C[0,1], \|\cdot\|_\infty)$, (6.8) would hold if we can prove weak convergence $\mathscr{L}(Y_t^n) \to \mathscr{L}(B_t)$. We only need to prove uniform tightness of $\mathscr{L}(Y_t^n)$ because, by Lemma 6.2, (6.7) already identifies the law of the Brownian motion as the unique possible limit. Thus, we need to address the question of uniform tightness of $(\mathscr{L}(Y_t^n))$ on the complete separable space $(C[0,1], \|\cdot\|_\infty)$ or, equivalently, by the result of the previous section, the asymptotic equicontinuity of Y_t^n,

$$\lim_{\delta \downarrow 0} \limsup_{n\to\infty} \mathbb{P}\big(m(Y^n,\delta) > \varepsilon\big) = 0.$$

Obviously, this is again equivalent to

$$\lim_{\delta \downarrow 0} \limsup_{n\to\infty} \mathbb{P}\big(m(X^n,\delta) > \varepsilon\big) = 0,$$

so we can deal with the process X_t^n, even though it is not continuous. By Chebyshev's inequality,

$$\mathbb{P}\big(m(X^n,\delta) > \varepsilon\big) \le \frac{1}{\varepsilon}\mathbb{E} m(X^n,\delta)$$

and we need to learn how to control $\mathbb{E} m(X^n,\delta)$. The modulus of continuity of X^n can be written as

$$\begin{aligned} m(X^n,\delta) &= \sup_{|t-s|\le\delta} |X_t^n - X_s^n| = \sqrt{n} \sup_{|t-s|\le\delta} \Big|\frac{1}{n}\sum_{i=1}^n \mathrm{I}(s < x_i \le t) - (t-s)\Big| \\ &= \sqrt{n} \sup_{f\in\mathscr{F}} \Big|\frac{1}{n}\sum_{i=1}^n f(x_i) - \mathbb{E} f\Big|, \end{aligned} \tag{6.9}$$

where we introduced the class of functions

$$\mathscr{F} = \Big\{ f(x) = \mathrm{I}(s < x \le t) \,:\, t,s \in [0,1], |t-s| < \delta \Big\}. \tag{6.10}$$

We will now develop an approach to control the expectation of (6.9) for general classes of functions $\mathscr{F}$ and we will only use the specific definition (6.10) at the very end. This will be done in several steps.

Symmetrization. As the first step, we will replace the empirical process (6.9) by a symmetrized version, called the *Rademacher process*, that will be easier to control. Let $x_1',\ldots,x_n'$ be independent copies of $x_1,\ldots,x_n$ and let $\varepsilon_1,\ldots,\varepsilon_n$ be i.i.d. *Rademacher* random variables, such that $\mathbb{P}(\varepsilon_i = 1) = \mathbb{P}(\varepsilon_i = -1) = 1/2$. Let us define

$$\mathbb{P}_n f = \frac{1}{n}\sum_{i=1}^n f(x_i) \quad\text{and}\quad \mathbb{P}_n' f = \frac{1}{n}\sum_{i=1}^n f(x_i').$$

Notice that $\mathbb{E}\mathbb{P}'_n f = \mathbb{E}f$. Consider the random variables

$$Z = \sup_{f\in\mathscr{F}} \big|\mathbb{P}_n f - \mathbb{E}f\big| \ \text{ and } \ R = \sup_{f\in\mathscr{F}} \Big|\frac{1}{n}\sum_{i=1}^n \varepsilon_i f(x_i)\Big|.$$

Then, using Jensen's inequality, the symmetry, and then triangle inequality, we can write

$$\begin{aligned}
\mathbb{E}Z = \mathbb{E}\sup_{f\in\mathscr{F}}\Big|\mathbb{P}_n f - \mathbb{E}f\Big| &= \mathbb{E}\sup_{f\in\mathscr{F}}\Big|\mathbb{P}_n f - \mathbb{E}\mathbb{P}'_n f\Big| \\
\le \mathbb{E}\sup_{f\in\mathscr{F}}\Big|\frac{1}{n}\sum_{i=1}^n (f(x_i)-f(x'_i))\Big| &= \mathbb{E}\sup_{f\in\mathscr{F}}\Big|\frac{1}{n}\sum_{i=1}^n \varepsilon_i(f(x_i)-f(x'_i))\Big| \\
\le \mathbb{E}\sup_{f\in\mathscr{F}}\Big|\frac{1}{n}\sum_{i=1}^n \varepsilon_i f(x_i)\Big| + \mathbb{E}\sup_{f\in\mathscr{F}}\Big|\frac{1}{n}\sum_{i=1}^n \varepsilon_i f(x'_i)\Big| &= 2\mathbb{E}R.
\end{aligned}$$

Equality in the second line holds because switching $x_i \leftrightarrow x'_i$ arbitrarily does not change the expectation, so the equality holds for any fixed (ε_i) and, therefore, for any random (ε_i). □

Given this symmetrization bound, for the specific class of functions in (6.10), one can finish the proof of asymptotic equicontinuity using the last exercise in the previous section together with a simple observation (6.13) below. However, instead, we will describe a much more general approach to this problem.

Covering numbers, Kolmogorov's chaining and Dudley's entropy integral. To control $\mathbb{E}R$ for general classes of functions $\mathscr{F}$, we will need to use some measures of complexity of $\mathscr{F}$. First, we will show how to control the Rademacher process R conditionally on $x_1,\ldots,x_n$. Suppose that (F,d) is a totally bounded metric space. For any $u>0$, a *u-packing number* of F with respect to d is defined by

$$D(F,u,d) = \max\operatorname{card}\big\{F_u \subseteq F : d(f,g) > u \text{ for all } f,g\in F_u\big\}$$

and a *u-covering number* is defined by

$$N(D,u,d) = \min\operatorname{card}\big\{F_u \subseteq F : \forall f\in F\ \exists\, g\in F_u \text{ such that } d(f,g)\le u\big\}.$$

Both packing and covering numbers measure how many points are needed to approximate any element in the set F within distance u. It is a simple exercise (at the end of the section) to show that

$$N(F,u,d) \le D(F,u,d) \le N(F,u/2,d)$$

and, in this sense, packing and covering numbers are closely related. Let F be a subset of the cube $[-1,1]^n$ equipped with a rescaled Euclidean metric

$$d(f,g)=\Big(\frac{1}{n}\sum_{i=1}^{n}(f_i-g_i)^2\Big)^{1/2}.$$

Consider the following Rademacher process on F,

$$\mathscr{R}(f)=\frac{1}{\sqrt{n}}\sum_{i=1}^{n}\varepsilon_i f_i.$$

Then we have the following version of the classical Kolmogorov chaining lemma. We will assume that $0\in F$ by increasing $D(F,\varepsilon,d)$ by 1 if necessary.

Theorem 6.5 (Kolmogorov's chaining). *For any* $u>0$,

$$\mathbb{P}\Big(\forall f\in F,\ \mathscr{R}(f)\le 2^{9/2}\int_0^{2d(0,f)}\log^{1/2}D(F,\varepsilon,d)\,d\varepsilon+2^{9/2}d(0,f)\sqrt{u}\Big)\ge 1-e^{-u}.$$

Proof. Define a sequence of subsets

$$\{0\}=F_0\subseteq F_1\subseteq\ldots\subseteq F_j\subseteq\ldots\subseteq F$$

such that F_j satisfies

1. $\forall f,g\in F_j,\ d(f,g)>2^{-j}$,
2. $\forall f\in F$ we can find $g\in F_j$ such that $d(f,g)\le 2^{-j}$.

F_0 obviously satisfies these properties for $j=0$. To construct F_{j+1} given F_j:

- Start with $F_{j+1}:=F_j$.
- If possible, find $f\in F$ such that $d(f,g)>2^{-(j+1)}$ for all $g\in F_{j+1}$.
- Let $F_{j+1}:=F_{j+1}\cup\{f\}$ and repeat until you cannot find such f.

Define projection $\pi_j:F\to F_j$ as follows:

for $f\in F$ find closest $g\in F_j$ with $d(f,g)\le 2^{-j}$ and set $\pi_j(f)=g$,

where we break the tie arbitrarily. Then any $f\in F$ can be decomposed into the telescopic series

$$f=\sum_{j=1}^{\infty}(\pi_j(f)-\pi_{j-1}(f)).$$

Note that, by the above definition, if $F_{j-1}=F_j$ then $\pi_{j-1}(f)=\pi_j(f)$, because in this case there is a unique $g\in F_j$ with $d(f,g)\le 2^{-j}$. In this case, for any f, the term $\pi_j(f)-\pi_{j-1}(f)$ belongs to the set $L_{j-1,j}:=\{0\}$. Otherwise, if $F_{j-1}\ne F_j$,

$$\begin{aligned}d(\pi_{j-1}(f),\pi_j(f))&\le d(\pi_{j-1}(f),f)+d(f,\pi_j(f))\\&\le 2^{-(j-1)}+2^{-j}=3\cdot 2^{-j}\le 2^{-j+2}.\end{aligned}$$

In this case, for any f, the term $\pi_j(f) - \pi_{j-1}(f)$ belongs to the set

$$L_{j-1,j} = \{f - g : f \in F_j, g \in F_{j-1}, d(f,g) \le 2^{-j+2}\}.$$

We will call elements of these sets *links* (as in a chain). Since $\mathscr{R}(f)$ is linear,

$$\mathscr{R}(f) = \sum_{j=1}^{\infty} \mathscr{R}\big(\pi_j(f) - \pi_{j-1}(f)\big).$$

We first show how to control $\mathscr{R}$ on the set of all non-trivial links. Assume that $0 \neq \ell \in L_{j-1,j}$. By Hoeffding's inequality,

$$\mathbb{P}\Big(\mathscr{R}(\ell) = \frac{1}{\sqrt{n}} \sum_{i=1}^{n} \varepsilon_i \ell_i \ge t\Big) \le \exp\Big(-\frac{t^2}{2n^{-1}\sum_{i=1}^{n} \ell_i^2}\Big) \le \exp\Big(-\frac{t^2}{2 \cdot 2^{-2j+4}}\Big).$$

If $|F|$ denotes the cardinality of the set F then $|L_{j-1,j}| \le |F_{j-1}| \cdot |F_j| \le |F_j|^2$ and, therefore, by the union bound,

$$\mathbb{P}\big(\forall \ell \in L_{j-1,j},\ \mathscr{R}(\ell) \le t\big) \ge 1 - |F_j|^2 \exp\Big(-\frac{t^2}{2^{-2j+5}}\Big) = 1 - \frac{1}{|F_j|^2} e^{-u},$$

after making the change of variables

$$t = \Big(2^{-2j+5}(4 \log |F_j| + u)\Big)^{1/2} \le 2^{7/2} 2^{-j} \log^{1/2} |F_j| + 2^{5/2} 2^{-j} \sqrt{u}.$$

Hence,

$$\mathbb{P}\Big(\forall \ell \in L_{j-1,j},\ \mathscr{R}(\ell) \le 2^{7/2} 2^{-j} \log^{1/2} |F_j| + 2^{5/2} 2^{-j} \sqrt{u}\Big) \ge 1 - \frac{1}{|F_j|^2} e^{-u}.$$

If $F_{j-1} = F_j$ then $L_{j-1,j} = \{0\}$, so the probability on the left hand side is 1 and, by the union bound,

$$\begin{aligned}
&\mathbb{P}\Big(\forall j \ge 1\ \forall \ell \in L_{j-1,j},\ \mathscr{R}(\ell) \le 2^{7/2} 2^{-j} \log^{1/2} |F_j| + 2^{5/2} 2^{-j} \sqrt{u}\Big) \\
&\ge 1 - \sum_{j=1}^{\infty} \mathrm{I}(|F_{j-1}| < |F_j|) \frac{1}{|F_j|^2} e^{-u} \ge 1 - \sum_{j=1}^{\infty} \frac{1}{(j+1)^2} e^{-u} \\
&= 1 - (\pi^2/6 - 1) e^{-u} \ge 1 - e^{-u}.
\end{aligned}$$

Given $f \in F$, let integer k be such that $2^{-(k+1)} < d(0,f) \le 2^{-k}$. Then in the above construction $\pi_0(f) = \ldots = \pi_{k-1}(f) = 0$ and, with probability at least $1 - e^{-u}$,

$$\sum_{j=k}^{\infty} \mathscr{R}\big(\pi_j(f) - \pi_{j-1}(f)\big) \le \sum_{j=k}^{\infty} \Big(2^{7/2} 2^{-j} \log^{1/2} |F_j| + 2^{5/2} 2^{-j} \sqrt{u}\Big)$$

$$\leq 2^{7/2}\sum_{j=k}^{\infty} 2^{-j}\log^{1/2} D(F,2^{-j},d) + 2^{7/2}2^{-k}\sqrt{u}.$$

Note that $2^{-k} < 2d(f,0)$ and $2^{7/2}2^{-k} < 2^{9/2}d(f,0)$. Finally, since the packing numbers $D(F,\varepsilon,d)$ are decreasing in ε, we can write

$$2^{9/2}\sum_{j=k}^{\infty} 2^{-(j+1)}\log^{1/2} D(F,2^{-j},d) \tag{6.11}$$

$$\leq 2^{9/2}\int_0^{2^{-k}} \log^{1/2} D(F,\varepsilon,d)\,d\varepsilon \leq 2^{9/2}\int_0^{2d(0,f)} \log^{1/2} D(F,\varepsilon,d)\,d\varepsilon.$$

This finishes the proof. □

The integral in (6.11) is called *Dudley's entropy integral.* We would like to apply the bound of the above theorem to

$$\sqrt{n}R = \sup_{f\in\mathscr{F}}\left|\frac{1}{\sqrt{n}}\sum_{i=1}^{n}\varepsilon_i f(x_i)\right|$$

for a class of functions $\mathscr{F}$ in (6.10). Suppose that $x_1,\ldots,x_n \in [0,1]$ are fixed and let

$$F = \Big\{(f_i)_{1\leq i\leq n} = \Big(\mathrm{I}(s < x_i \leq t)\Big)_{1\leq i\leq n} : t,s\in[0,1], |t-s|\leq\delta\Big\} \subseteq \{0,1\}^n.$$

Then we can control the covering numbers of this set as follows.

Lemma 6.4. $N(F,u,d) \leq Ku^{-4}$ *for some absolute* $K > 0$ *independent of the points* $x_1,\ldots,x_n$.

This kind of estimate is called *uniform covering numbers*, because the bound does not depend on the points $x_1,\ldots,x_n$ that generate the class F from $\mathscr{F}$.

Proof. We can assume that $x_1 \leq \ldots \leq x_n$. Then the class F consists of all vectors of the type

$$(0\ldots1\ldots1\ldots0),$$

i.e. the coordinates equal to 1 come in blocks. Given u, let F_u be a subset of such vectors with blocks of 1's starting and ending at the coordinates $k\lfloor nu\rfloor$. Given any vector $f\in F$, let us approximate it by a vector in $f'\in F_u$ by choosing the closest starting and ending coordinates for the blocks of 1's. The number of different coordinates will be bounded by $2\lfloor nu\rfloor$ and, therefore, the distance between f and f' will be bounded by

$$d(f,f') \leq \sqrt{2n^{-1}\lfloor nu\rfloor} \leq \sqrt{2u}.$$

The cardinality of F_u is, obviously, of order u^{-2} and this proves that $N(F,\sqrt{2u},d) \leq Ku^{-2}$. Making the change of variables $\sqrt{2u}\to u$ proves the result. □

To apply Kolmogorov's chaining bound to this class F, let us make a simple observation that if a random variable $X \geq 0$ satisfies $\mathbb{P}(X \geq a+bt) \leq Ke^{-t^2}$ for all $t \geq 0$ then

$$\begin{aligned}\mathbb{E}X = \int_0^\infty \mathbb{P}(X \geq t)\,dt \leq a + \int_0^\infty \mathbb{P}(X \geq a+t)\,dt \\ \leq a + K\int_0^\infty e^{-\frac{t^2}{b^2}}\,dt \leq a + Kb \leq K(a+b).\end{aligned}$$

Theorem 6.5 then implies that

$$\mathbb{E}_\varepsilon \sup_F \Big|\frac{1}{\sqrt{n}}\sum_{i=1}^n \varepsilon_i f_i\Big| \leq K\Big(\int_0^{D_n} \sqrt{\log\frac{K}{u}}\,du + D_n\Big) \tag{6.12}$$

where $\mathbb{E}_\varepsilon$ is the expectation with respect to (ε_i) only and

$$\begin{aligned}D_n^2 = \sup_F d(0,f)^2 = \sup_{\mathscr{F}} \frac{1}{n}\sum_{i=1}^n f(x_i)^2 = \sup_{|t-s|\leq\delta} \frac{1}{n}\sum_{i=1}^n \mathrm{I}(s < x_i \leq t) \\ = \sup_{|t-s|\leq\delta} \Big|\frac{1}{n}\sum_{i=1}^n \mathrm{I}(x_i \leq t) - \frac{1}{n}\sum_{i=1}^n \mathrm{I}(x_i \leq s)\Big|.\end{aligned}$$

Since the integral on the right hand side of (6.12) is concave in D_n, by Jensen's inequality,

$$\mathbb{E}\sup_{\mathscr{F}} \Big|\frac{1}{\sqrt{n}}\sum_{i=1}^n \varepsilon_i f(x_i)\Big| \leq K\Big(\int_0^{\mathbb{E}D_n} \sqrt{\log\frac{K}{u}}\,du + \mathbb{E}D_n\Big).$$

By the symmetrization inequality, this finally proves that

$$\mathbb{E}m(X^n,\delta) \leq K\Big(\int_0^{\mathbb{E}D_n} \sqrt{\log\frac{K}{u}}\,du + \mathbb{E}D_n\Big).$$

The strong law of large numbers easily implies (exercise) that

$$\sup_{t\in[0,1]} \Big|\frac{1}{n}\sum_{i=1}^n \mathrm{I}(x_i \leq t) - t\Big| \to 0$$

almost surely and, therefore,

$$D_n^2 \to \delta \text{ a.s. and } \mathbb{E}D_n \to \sqrt{\delta}. \tag{6.13}$$

This implies that

$$\limsup_{n\to\infty} \mathbb{E}m(X_t^n,\delta) \leq K\Big(\int_0^{\sqrt{\delta}} \sqrt{\log\frac{K}{u}}\,du + \sqrt{\delta}\Big).$$

The right-hand side goes to zero as $\delta \to 0$ and this finishes the proof of asymptotic equicontinuity of X^n. As a result, for any continuous function Φ on $(C[0,1], \|\cdot\|_\infty)$ the distibution of $\Phi(X_t^n)$ converges to the distribution of $\Phi(B_t)$. For example,

$$\sqrt{n} \sup_{0\le t\le 1} \Big|\frac{1}{n}\sum \mathrm{I}(x_i \le t) - t\Big| \to \sup_{0\le t\le 1} |B_t|$$

in distribution. We will find the distribution of the right hand side in the next section. Notice that the methods we used to prove equicontinuity were quite general and the main step where we used the specific class of function $\mathscr{F}$ was to control the covering numbers.

Exercise 6.3.1. Show that $N(F,u,d) \le D(F,u,d) \le N(F,u/2,d)$.

Exercise 6.3.2. If $(x_i)_{i\ge 1}$ are i.i.d. uniform on $[0,1]$, prove that

$$\sup_{t\in[0,1]} \Big|\frac{1}{n}\sum_{i=1}^n \mathrm{I}(x_i \le t) - t\Big| \to 0 \text{ a.s.}.$$

Exercise 6.3.3. If ε_i are i.i.d. Rademacher, g_i are i.i.d. standard Gaussian, and F is a bounded subset of $\mathbb{R}^n$, show that

$$\mathbb{E}\sup_{f\in F}\Big|\frac{1}{\sqrt{n}}\sum_{i=1}^n \varepsilon_i f_i\Big| \le \frac{1}{\mathbb{E}|g_1|}\mathbb{E}\sup_{f\in F}\Big|\frac{1}{\sqrt{n}}\sum_{i=1}^n g_i f_i\Big|.$$

Exercise 6.3.4. Consider a set $\Lambda = \{\lambda = (\lambda_1,\ldots,\lambda_d) : \sum_{i=1}^d |\lambda_i| \le 1\} \subset \mathbb{R}^d$ and the distance $d(\lambda^1,\lambda^2) = \sum_{i=1}^n |\lambda_i^1 - \lambda_i^2|$. Prove that $D(\Lambda,\varepsilon,d) \le (4/\varepsilon)^d$.

Exercise 6.3.5. Suppose that a family $\mathscr{F}$ of measurable functions $f\colon \Omega \to [0,1]$ is such that for some $V \ge 1$ and for any $\omega_1,\ldots,\omega_n \in \Omega$,

$$\log D(\mathscr{F},\varepsilon,d) \le V \log\frac{2}{\varepsilon}, \text{ where } d(f,g) = \Big(\frac{1}{n}\sum_{i=1}^n \big(f(\omega_i) - g(\omega_i)\big)^2\Big)^{1/2}.$$

If $(X_n)_{n\ge 1}$ are i.i.d. random elements with values in Ω, prove that

$$\mathbb{E}\sup_{f\in\mathscr{F}}\Big|\frac{1}{n}\sum_{i=1}^n f(X_i) - \mathbb{E}f(X_1)\Big| \le K\Big(\frac{V}{n}\Big)^{1/2}$$

for some absolute constant K. (Assume, for example, that $\mathscr{F}$ is countable to avoid any measureability issues.)

6.4 Reflection principles for Brownian motion

We showed that the empirical process converges to the Brownian bridge on $(C([0,1]),\|\cdot\|_\infty)$. As a result, the distribution of a continuous function of the process will also converge, for example,

$$\sup_{0\le t\le 1}|X_t^n| \to \sup_{0\le t\le 1}|B_t|$$

in distribution. We will compute the distribution of this supremum in Theorem 6.9 below, but first we will prove the so called strong Markov property of the Brownian motion.

Given a Brownian motion W_t on some probability space $(\Omega,\mathscr{F},\mathbb{P})$ let $\mathscr{B}_t = \sigma\big((W_s)_{s\le t}\big)\subseteq\mathscr{F}$ be the σ-algebra generated by the process up to time t. Let us consider a family of σ-algebras $\mathscr{F}_t$ for $t\ge 0$ such that

(i) $\mathscr{F}_t\subseteq\mathscr{F}_s$ for all $t\le s$;
(ii) $\mathscr{B}_t\subseteq\mathscr{F}_t$ for all $t\ge 0$;
(iii) the future increments $(W_{t+s}-W_t)_{s\ge 0}$ are independent of $\mathscr{F}_t$ for all $t\ge 0$.

For example, if we consider some σ-algebra $\mathscr{A}$ independent of the process W_t then $\mathscr{F}_t=\sigma(\mathscr{B}_t\cup\mathscr{A})$ satisfy these properties. A random variable $\tau\ge 0$ is called a *stopping time* if

$$\{\tau\le t\}\in\mathscr{F}_t \text{ for all } t\ge 0.$$

For example, a hitting time $\tau_c=\inf\{t>0, W_t=c\}$ for $c>0$ is a stopping time because, by the sample continuity, $\{\tau_c\le t\}=\bigcap_{q<c}\bigcup_{r<t}\{W_r>q\}$, where both the intersection and union are over rational numbers q,r.

Consider the σ-algebra

$$\mathscr{F}_\tau=\Big\{B\in\mathscr{F}\,:\,B\cap\{\tau\le t\}\in\mathscr{F}_t \text{ for all } t\ge 0\Big\} \tag{6.14}$$

generated by the data up to the stopping time τ. Let us denote by $\mathscr{B}(C(\mathbb{R}_+),d)$ the Borel σ-algebra on $(C(\mathbb{R}_+),d)$, where a metric d metrizes uniform convergence on compacts. The following is very similar to the property of stopping times for sums of i.i.d. random variables in Section 2.4.

Theorem 6.6 (Strong Markov Property). *On the event $\{\tau<\infty\}$, the increments $W_t':=W_{\tau+t}-W_\tau$ of the process after the stopping time are independent of $\mathscr{F}_\tau$ and, moreover, the distribution of W_t' is that of a Brownian motion. More precisely, for any $B\in\mathscr{F}_\tau$ and $A\in\mathscr{B}(C(\mathbb{R}_+),d)$,*

$$\mathbb{E}\mathrm{I}(W'\in A)\mathrm{I}_B\mathrm{I}(\tau<\infty)=\mathbb{E}\mathrm{I}(W\in A)\mathbb{E}\mathrm{I}_B\mathrm{I}(\tau<\infty). \tag{6.15}$$

Proof. Since $\mathscr{B}(C(\mathbb{R}_+),d)=\mathscr{A}\cap C(\mathbb{R}_+)$ where $\mathscr{A}$ is the cylindrical σ-algebra on $\mathbb{R}^{[0,\infty)}$, it is enough to prove (6.15) for A in the cylindrical algebra, i.e. cylinders of the form $C \times \mathbb{R}^{[0,\infty)\setminus F}$ for $C \in \mathscr{B}(\mathbb{R}^F)$ for a finite $F = \{t_1, \ldots, t_d\}$ for some integer $d \geq 1$ and $0 \leq t_1 < \ldots < t_d$. By Dynkin's theorem, it is enough to consider only open sets C and, finally, approximating open sets by bounded Lipshitz functions from below, it is enough to prove that

$$\mathbb{E}e(\tau)\mathrm{I}_B\mathrm{I}(\tau < \infty) = \mathbb{E}e(0)\mathbb{E}\mathrm{I}_B\mathrm{I}(\tau < \infty),$$

where $e(\tau) := f(W_{\tau+t_1} - W_\tau, \ldots, W_{\tau+t_d} - W_\tau)$ for some $f \in C_b(\mathbb{R}^d)$.

The main tool in the proof is the approximation of an arbitrary stopping time τ by the dyadic stopping time

$$\tau_n = \frac{\lfloor 2^n \tau \rfloor + 1}{2^n} \geq \tau.$$

First of all, let us check that τ_n is a stopping time. Indeed, if

$$\frac{k}{2^n} \leq \tau < \frac{k+1}{2^n} \quad \text{then} \quad \tau_n = \frac{k+1}{2^n}$$

and, therefore, for any $t \geq 0$, if $\ell/2^n \leq t < (\ell+1)/2^n$ then

$$\{\tau_n \leq t\} = \Big\{\tau_n \leq \frac{\ell}{2^n}\Big\} = \Big\{\tau < \frac{\ell}{2^n}\Big\} = \bigcup_{q < \ell/2^n} \{\tau \leq q\} \in \mathscr{F}_t.$$

By construction, $\tau_n \downarrow \tau$ and, by the continuity of the process, $W_{\tau_n} \to W_\tau$ almost surely. Take a set $B \in \mathscr{F}_\tau$. For any integer $k \geq 1$, the event

$$B \cap \{\tau_n = k2^{-n}\} = B \cap \big\{(k-1)2^{-n} \leq \tau < k2^{-n}\big\} \in \mathscr{F}_{k2^{-n}}$$

is independent of $e(k2^{-n})$ and, therefore,

$$\begin{aligned}
\mathbb{E}e(\tau_n)\mathrm{I}_B\mathrm{I}(\tau < \infty) &= \sum_{k \geq 0} \mathbb{E}e(k2^{-n})\mathrm{I}\big(B \cap \{\tau_n = k2^{-n}\}\big) \\
&= \sum_{k \geq 0} \mathbb{E}e(k2^{-n})\mathbb{E}\mathrm{I}\big(B \cap \{\tau_n = k2^{-n}\}\big) \\
&= \mathbb{E}e(0) \sum_{k \geq 0} \mathbb{E}\mathrm{I}\big(B \cap \{\tau_n = k2^{-n}\}\big) = \mathbb{E}e(0)\mathbb{E}\mathrm{I}_B\mathrm{I}(\tau < \infty).
\end{aligned}$$

In the second line we used the homogeneity of the Brownian motion to write $\mathbb{E}e(k2^{-n}) = \mathbb{E}e(0)$. Since $e(\tau_n) \to e(\tau)$ almost surely, this finishes the proof. □

Remark 6.2. A random variable $\tau \geq 0$ is called a *Markov time* if

$$\{\tau < t\} \in \mathscr{F}_t \quad \text{for all } t \geq 0.$$

The σ-algebra generated by the process up to a Markov time τ is defined by

$$\mathscr{F}_{\tau^+} = \big\{B \in \mathscr{F} \;:\; B \cap \{\tau < t\} \in \mathscr{F}_t \text{ for all } t \geq 0\big\}. \tag{6.16}$$

One can show that a stopping time is always Markov time and $\mathscr{F}_\tau \subseteq \mathscr{F}_{\tau+}$. One can check that the above Strong Markov property also holds for Markov times τ and σ-algebras $\mathscr{F}_{\tau+}$ (exercise below). □

We will use the strong Markov property in the above form in the next section, but in the examples in this section the events A will be allowed to depend on the stopping time τ. For this purpose, it will be convenient to have the following version of the strong Markov property. Given time $s \geq 0$, consider the following two modifications on the Brownian motion:

$$\begin{aligned} W^i(t;s) &:= W(t)\,\mathrm{I}(t \leq s) + (W_s + \tilde{W}_{t-s})\,\mathrm{I}(t > s), \\ W^r(t;s) &:= W(t)\,\mathrm{I}(t \leq s) + (W_s - (W_t - W_s))\,\mathrm{I}(t > s), \end{aligned} \tag{6.17}$$

where the increment $W_t - W_s$ of the original process W_t after time s was replaced in the first case by an *independent* Brownian motion $(\tilde{W}_h)_{h\geq 0}$ at time $t-s$ and in the second case by $-(W_t - W_s)$, i.e. the original increment *reflected* around zero. (To accommodate the Brownian motion $(\tilde{W}_h)_{h\geq 0}$ independent of all σ-algebras $\mathscr{F}_t$ we can extend the probability space Ω for example, to the product space $\Omega \times C(\mathbb{R}_+)$ if necessary.) Of course, both modifications satisfy of the properties of a Brownian motion, so

$$(W^i(t;s))_{t\geq 0} \stackrel{d}{=} (W^r(t;s))_{t\geq 0} \stackrel{d}{=} (W(t))_{t\geq 0}.$$

Our next result shows that this still holds if we replace a fixed time s by a stopping time τ.

Theorem 6.7 (Reflexion principle). *If τ is a stopping time and and $a \in \{i,r\}$ then*

$$(W^a(t;\tau))_{t\geq 0} \stackrel{d}{=} (W_t)_{t\geq 0}$$

on the event $\tau < \infty$. More precisely, for any $A \in \mathscr{B}(C(\mathbb{R}_+),d)$,

$$\mathbb{P}(W^a \in A, \tau < \infty) = \mathbb{P}(W \in A, \tau < \infty). \tag{6.18}$$

Proof. By the same reduction as in the previous theorem, it is enough to show that, for any $d \geq 1$, any $0 < t_1 < \ldots < t_d$, and any $f \in C_b(\mathbb{R}^d)$,

$$\mathbb{E}f(W^a(t_1;\tau),\ldots,W^a(t_d;\tau))\,\mathrm{I}(\tau < \infty) = \mathbb{E}f(W_{t_1},\ldots,W_{t_d})\,\mathrm{I}(\tau < \infty).$$

Let us denote $e(s) := f(W^a(t_1;s),\ldots,W^a(t_d;s))$ and note that it is continuous is s because the processes $W^a(t;s)$ for $a \in \{i,r\}$ are continuous in (t,s) and $f \in C_b(\mathbb{R}^d)$. Therefore, for the stopping time τ_n defined as in the proof of the strong Markov property above, $e(\tau_n) \to e(\tau)$ on the event $\{\tau < \infty\}$. Finally, write

$$\mathbb{E}e(\tau_n)\mathrm{I}(\tau < \infty) = \sum_{k\geq 0} \mathbb{E}e(k2^{-n})\mathrm{I}(\tau_n = k2^{-n})$$

and note that $\{\tau_n = k2^{-n}\} \in \mathscr{F}_{k2^{-n}}$, while for $s = k2^{-n}$ the only difference between W_t and $W^a(t;s)$ is how the increments after time $k2^{-n}$ were defined (all independent of $\mathscr{F}_{k2^{-n}}$). As a result the sum on the right hand side above is equal to

$$\sum_{k\geq 0} \mathbb{E}f(W_{t_1},\ldots,W_{t_d})\mathrm{I}(\tau_n = k2^{-n}) = \mathbb{E}f(W_{t_1},\ldots,W_{t_d})\mathrm{I}(\tau < \infty),$$

which finishes the proof. □

Example 6.4.1. As a first example of application of the reflexion principle, let us compute the following probability,

$$\mathbb{P}\Big(\sup_{t\leq b} W_t \geq c\Big) = \mathbb{P}(\tau_c \leq b),$$

for any $c > 0$ and the hitting time τ_c. Let us write

$$\mathbb{P}(\tau_c \leq b) = \mathbb{P}(\tau_c \leq b, W_b - W_{\tau_c} > 0) + \mathbb{P}(\tau_c \leq b, W_b - W_{\tau_c} < 0),$$

where we can write strict inequalities because

$$\mathbb{P}(\tau_c \leq b, W_b - W_{\tau_c} = 0) = \mathbb{P}(\tau_c \leq b, W_b = c) \leq \mathbb{P}(W_b = c) = 0.$$

By the reflexion principle above applied to $W^r(t,\tau_c)$, the two probabilities above are equal, so

$$\mathbb{P}(\tau_c \leq b) = 2\mathbb{P}(\tau_c \leq b, W_b - W_{\tau_c} > 0) = \mathbb{P}(\tau_c \leq b, W_b > c) = \mathbb{P}(W_b > c),$$

where we could omit $\tau_c \leq b$ because, by continuity of the process, the event $W_b > c$ automatically implies that $\tau_c \leq b$. Hence,

$$\mathbb{P}\Big(\sup_{t\leq b} W_t \geq c\Big) = \mathbb{P}(\tau_c \leq b) = 2\mathbb{P}(W_b > c) = 2\int_{c/\sqrt{b}}^{\infty} \frac{1}{\sqrt{2\pi}} e^{-\frac{x^2}{2}}\,dx. \tag{6.19}$$

This implies that the density of of τ_c is given by

$$f_{\tau_c}(b) = \frac{1}{\sqrt{2\pi}} e^{-\frac{c^2}{2b}} \cdot \frac{c}{b^{3/2}},$$

and since it is of order $O(b^{-3/2})$ as $b \to +\infty$, $\mathbb{E}\tau_c = \infty$. □

Next, we compute various probabilities for the suprema of the Brownian bridge. In addition to the reflexion principle, we will need the following observation that the Brownian bridge can be viewed as a Brownian motion conditioned to be equal to zero at time $t = 1$ *(a.k.a. pinned down Brownian motion).*

Lemma 6.5 (BB as pinned down BM)). *Conditional distribution of* $(W_t)_{t\in[0,1]}$ *given* $|W_1| < \varepsilon$ *converges to the law of* $(B_t)_{t\in[0,1]}$,

$$\mathscr{L}\big(W \,\big|\, |W_1| < \varepsilon\big) \to \mathscr{L}(B) \text{ as } \varepsilon \downarrow 0.$$

Proof. If W_t is a Brownian motion then $B_t = W_t - tW_1$ is the Brownian bridge for $t \in [0,1]$. Notice that B_t is independent of W_1, because their covariance

$$\mathbb{E}B_t W_1 = \mathbb{E}W_t W_1 - t\mathbb{E}W_1^2 = t - t = 0.$$

Therefore, the Brownian motion can be written as a sum $W_t = B_t + tW_1$ of the Brownian bridge and independent process tW_1 and, if we define a random variable η_ε with distribution $\mathscr{L}(\eta_\varepsilon) = \mathscr{L}(W_1 \mid |W_1| < \varepsilon)$ independent of B_t then

$$\mathscr{L}(W \mid |W_1| < \varepsilon) = \mathscr{L}(B + t\eta_\varepsilon) \to \mathscr{L}(B).$$

as $\varepsilon \downarrow 0$. □

Using this lemma we will first prove the analogue of the above Example 6.4.1 for the Brownian bridge.

Theorem 6.8. *If B_t is a Brownian bridge then, for all $b > 0$,*

$$\mathbb{P}\Big(\sup_{t\in[0,1]} B_t \geq b\Big) = e^{-2b^2}.$$

Proof. Since $B_t = W_t - tW_1$ and W_1 are independent, we can write

$$\begin{aligned}\mathbb{P}(\exists t : B_t = b) &= \frac{\mathbb{P}(\exists t : W_t - tW_1 = b, |W_1| < \varepsilon)}{\mathbb{P}(|W_1| < \varepsilon)} \\ &= \frac{\mathbb{P}(\exists t : W_t = b + tW_1, |W_1| < \varepsilon)}{\mathbb{P}(|W_1| < \varepsilon)}.\end{aligned}$$

We can estimate the numerator from below and above by

$$\begin{aligned}\mathbb{P}\big(\exists t : W_t > b + \varepsilon, |W_1| < \varepsilon\big) &\leq \mathbb{P}\big(\exists t : W_t = b + tW_1, |W_1| < \varepsilon\big) \\ &\leq \mathbb{P}\big(\exists t : W_t \geq b - \varepsilon, |W_1| < \varepsilon\big).\end{aligned}$$

Let us first analyze the upper bound. If we define a hitting time $\tau = \inf\{t : W_t = b - \varepsilon\}$ then $W_\tau = b - \varepsilon$ and

$$\begin{aligned}&\mathbb{P}\big(\exists t : W_t \geq b - \varepsilon, |W_1| < \varepsilon\big) = \mathbb{P}\big(\tau \leq 1, |W_1| < \varepsilon\big) \\ &= \mathbb{P}\big(\tau \leq 1, W_1 - W_\tau \in (-b, -b + 2\varepsilon)\big).\end{aligned}$$

By the reflexion principle,

$$\begin{aligned}&\mathbb{P}\big(\tau \leq 1, W_1 - W_\tau \in (-b, -b+2\varepsilon)\big) = \mathbb{P}\big(\tau \leq 1, -(W_1 - W_\tau) \in (-b, -b+2\varepsilon)\big) \\ &= \mathbb{P}\big(\tau \leq 1, W_1 - W_\tau \in (b - 2\varepsilon, b)\big) = \mathbb{P}\big(W_1 \in (2b - 3\varepsilon, 2b - \varepsilon)\big),\end{aligned}$$

because the fact that $W_1 \in (2b - 3\varepsilon, 2b - \varepsilon)$ automatically implies that $\tau \leq 1$ for $b > 0$ and ε small enough. Therefore, we proved that

$$\mathbb{P}(\exists t : B_t = b) \leq \frac{\mathbb{P}(W_1 \in (2b-3\varepsilon, 2b-\varepsilon))}{\mathbb{P}(W_1 \in (-\varepsilon, \varepsilon))} \to e^{-2b^2}$$

as $\varepsilon \downarrow 0$. The lower bound can be analyzed similarly. □

Theorem 6.9 (Kolmogorov-Smirnov). *If B is a Brownian bridge then, for $b > 0$,*

$$\mathbb{P}\Big(\sup_{0\leq t\leq 1} |B_t| \geq b\Big) = 2\sum_{n\geq 1} (-1)^{n-1} e^{-2n^2b^2}.$$

Proof. For $n \geq 1$, consider an event

$$A_n = \big\{\exists t_1 < \cdots < t_n \leq 1 \,:\, B_{t_j} = (-1)^{j-1} b\big\}$$

and let τ_b and τ_{-b} be the hitting times of b and $-b$. By symmetry of the distribution of the process B_t,

$$\mathbb{P}\Big(\sup_{0\leq t\leq 1} |B_t| \geq b\Big) = \mathbb{P}\big(\tau_b \text{ or } \tau_{-b} \leq 1\big) = 2\mathbb{P}(A_1, \tau_b < \tau_{-b}).$$

Again, by symmetry,

$$\mathbb{P}(A_n, \tau_b < \tau_{-b}) = \mathbb{P}(A_n) - \mathbb{P}(A_n, \tau_{-b} < \tau_b) = \mathbb{P}(A_n) - \mathbb{P}(A_{n+1}, \tau_b < \tau_{-b})$$

and, by induction,

$$\mathbb{P}(A_1, \tau_b < \tau_{-b}) = \mathbb{P}(A_1) - \mathbb{P}(A_2) + \ldots + (-1)^{n-1}\mathbb{P}(A_n, \tau_b < \tau_{-b}).$$

As in Theorem 6.8, reflecting the Brownian motion each time we hit b or $-b$, one can show that

$$\mathbb{P}(A_n) = \lim_{\varepsilon\downarrow 0} \frac{\mathbb{P}\big(W_1 \in (2nb-\varepsilon, 2nb+\varepsilon)\big)}{\mathbb{P}\big(W_1 \in (-\varepsilon, \varepsilon)\big)} = e^{-\frac{1}{2}(2nb)^2} = e^{-2n^2b^2}$$

and this finishes the proof. □

Given $a, b > 0$, let us compute the probability that a Brownian bridge crosses one of the levels $-a$ or b.

Theorem 6.10 (Two-sided boundary). *If $a, b > 0$ then*

$$\mathbb{P}\big(\exists t : B_t = -a \text{ or } b\big) = \sum_{n\geq 0} \big(e^{-2(na+(n+1)b)^2} + e^{-2((n+1)a+nb)^2}\big) - \sum_{n\geq 1} 2e^{-2n^2(a+b)^2}. \tag{6.20}$$

Proof. We have

$$\mathbb{P}(\exists t : B_t = -a \text{ or } b) = \mathbb{P}(\exists t : B_t = -a, \tau_{-a} < \tau_b) + \mathbb{P}(\exists t : B_t = b, \tau_b < \tau_{-a}).$$

If we introduce the events

$$C_n = \{\exists t_1 < \ldots < t_n : B_{t_1} = b, B_{t_2} = -a, \ldots\}$$

and

$$A_n = \{\exists t_1 < \ldots < t_n : B_{t_1} = -a, B_{t_2} = b, \ldots\}$$

then, as in the previous theorem,

$$\mathbb{P}(C_n, \tau_b < \tau_{-a}) = \mathbb{P}(C_n) - \mathbb{P}(C_n, \tau_{-a} < \tau_b) = \mathbb{P}(C_n) - \mathbb{P}(A_{n+1}, \tau_{-a} < \tau_b)$$

and, similarly,

$$\mathbb{P}(A_n, \tau_{-a} < \tau_b) = \mathbb{P}(A_n) - \mathbb{P}(C_{n+1}, \tau_b < \tau_{-a}).$$

By induction,

$$\mathbb{P}(\exists t : B_t = -a \text{ or } b) = \sum_{n=1}^{\infty} (-1)^{n-1} (\mathbb{P}(A_n) + \mathbb{P}(C_n)).$$

Probabilities of the events A_n and C_n can be computed using the reflection principle as above, $\mathbb{P}(A_{2n}) = \mathbb{P}(C_{2n}) = e^{-2n^2(a+b)^2}$, $\mathbb{P}(C_{2n+1}) = e^{-2(na+(n+1)b)^2}$, and $\mathbb{P}(A_{2n+1}) = e^{-2((n+1)a+nb)^2}$, which finishes the proof. □

If $X = -\inf B_t$ and $Y = \sup B_t$ then *the spread* of the process B_t is $\xi = X + Y$.

Theorem 6.11 (Distribution of the spread). *For any* $t > 0$,

$$\mathbb{P}(\xi \le t) = 1 - \sum_{n \ge 1} (8n^2t^2 - 2)e^{-2n^2t^2}.$$

Proof. First of all, (6.20) gives the joint c.d.f. of (X,Y) because

$$\begin{aligned} F(a,b) = \mathbb{P}(X < a, Y < b) &= \mathbb{P}(-a < \inf B_t, \sup B_t < b) \\ &= 1 - \mathbb{P}(\exists t : B_t = -a \text{ or } b). \end{aligned}$$

If $f(a,b) = \frac{\partial^2 F}{\partial a \partial b}$ is the joint p.d.f. of (X,Y) then the c.d.f of the spread $X+Y$ is

$$\mathbb{P}\big(Y + X \le t\big) = \int_0^t \int_0^{t-a} f(a,b)\, db\, da.$$

The inner integral is

$$\int_0^{t-a} f(a,b)\, db = \frac{\partial F}{\partial a}(a, t-a) - \frac{\partial F}{\partial a}(a,0).$$

Since

$$\frac{\partial F}{\partial a}(a,b) = \sum_{n \ge 0} 4n\big(na + (n+1)b\big)\, e^{-2(na+(n+1)b)^2}$$

$$+\sum_{n\geq 0} 4(n+1)\big((n+1)a+nb\big)\,e^{-2((n+1)a+nb)^2}$$
$$-\sum_{n\geq 1} 8n^2(a+b)\,e^{-2n^2(a+b)^2},$$

plugging in the values $b=t-a$ and $b=0$ gives

$$\int_0^{t-a} f(a,b)\,db = \sum_{n\geq 0} 4n((n+1)t-a)e^{-2((n+1)t-a)^2}$$
$$+\sum_{n\geq 0} 4(n+1)(nt+a)e^{-2(nt+a)^2} - \sum_{n\geq 1} 8n^2te^{-2n^2t^2}.$$

Integrating over $a \in [0,t]$,

$$\mathbb{P}(Y+X \leq t) = \sum_{n\geq 0}(2n+1)\big(e^{-2n^2t^2} - e^{-2(n+1)^2t^2}\big) - \sum_{n\geq 1} 8n^2t^2e^{-2n^2t^2}$$
$$= 1+2\sum_{n\geq 1} e^{-2n^2t^2} - \sum_{n\geq 1} 8n^2t^2e^{-2n^2t^2},$$

and this finishes the proof. □

Exercise 6.4.1. Prove the Strong Markov Property for a Markov time τ and σ-algebra $\mathscr{F}_{\tau+}$ in (6.16).

In the next four exercises, $(\mathscr{F}_t)$ is a nested family of σ-algebras, $\mathscr{F}_t \subseteq \mathscr{F}_s$ for $t \leq s$.

Exercise 6.4.2. Show that for $\tau = t$, $\mathscr{F}_{\tau+} = \mathscr{F}_{t+} = \cap_{s>t}\mathscr{F}_s$.

Exercise 6.4.3. A family $(\mathscr{F}_t)$ is called right-continuous if $\cap_{s>t}\mathscr{F}_s = \mathscr{F}_t$ for all $t \geq 0$. If $(\mathscr{F}_t)$ is right-continuous, show that a Markov time τ is a stopping time and $\mathscr{F}_{\tau+} = \mathscr{F}_\tau$.

Exercise 6.4.4. If we define $\overline{\mathscr{F}}_t := \cap_{s>t}\mathscr{F}_s$, show that $(\overline{\mathscr{F}}_t)$ is right-continuous.

Exercise 6.4.5. If $(\mathscr{F}_t)$ satisfies conditions (i) – (iii) at the beginning of this section, prove that $(\overline{\mathscr{F}}_t)$ satisfies them too.

6.5 Skorohod's imbedding and laws of the iterated logarithm

In this section we will prove another classical limit theorem in Probability, called the *laws of the iterated logarithm*, using the method of Skorohod's imbedding, which we will describe first. Throughout this section, $(W_t)_{t\geq 0}$ will be a Brownian motion and $\mathscr{F}_t = \sigma((W_s)_{s\leq t})$. We will need the following preliminary technical result.

Theorem 6.12. *If* $\tau < \infty$ *is a stopping time such that* $\mathbb{E}\tau < \infty$ *then* $\mathbb{E}W_\tau = 0$ *and* $\mathbb{E}W_\tau^2 = \mathbb{E}\tau$.

Proof. Let us start with the case when a stopping time τ takes a finite number of values, $\tau \in \{t_1, \ldots, t_n\}$. Note that $(W_{t_j}, \mathscr{F}_{t_j})$ is a martingale, since

$$\mathbb{E}(W_{t_j}|\mathscr{F}_{t_{j-1}}) = \mathbb{E}(W_{t_j} - W_{t_{j-1}} + W_{t_{j-1}}|\mathscr{F}_{t_{j-1}}) = W_{t_{j-1}}.$$

By the optional stopping theorem for martingales, $\mathbb{E}W_\tau = \mathbb{E}W_{t_1} = 0$. Next, let us prove that $\mathbb{E}W_\tau^2 = \mathbb{E}\tau$ by induction on n. If $n = 1$ then $\tau = t_1$ and

$$\mathbb{E}W_\tau^2 = \mathbb{E}W_{t_1}^2 = t_1 = \mathbb{E}\tau.$$

To make an induction step from $n-1$ to n, define a stopping time $\alpha = \tau \wedge t_{n-1}$ and write

$$\mathbb{E}W_\tau^2 = \mathbb{E}(W_\alpha + W_\tau - W_\alpha)^2 = \mathbb{E}W_\alpha^2 + \mathbb{E}(W_\tau - W_\alpha)^2 + 2\mathbb{E}W_\alpha(W_\tau - W_\alpha).$$

First of all, by the induction assumption, $\mathbb{E}W_\alpha^2 = \mathbb{E}\alpha$. Moreover, $\tau \neq \alpha$ only if $\tau = t_n$, in which case $\alpha = t_{n-1}$. The event $\{\tau = t_n\} = \{\tau \leq t_{n-1}\}^c \in \mathscr{F}_{t_{n-1}}$, so

$$\mathbb{E}W_\alpha(W_\tau - W_\alpha) = \mathbb{E}W_{t_{n-1}}(W_{t_n} - W_{t_{n-1}})\mathrm{I}(\tau = t_n) = 0.$$

Similarly,

$$\mathbb{E}(W_\tau - W_\alpha)^2 = \mathbb{E}\mathbb{E}(\mathrm{I}(\tau = t_n)(W_{t_n} - W_{t_{n-1}})^2|\mathscr{F}_{t_{n-1}}) = (t_n - t_{n-1})\mathbb{P}(\tau = t_n).$$

Therefore,

$$\mathbb{E}W_\tau^2 = \mathbb{E}\alpha + (t_n - t_{n-1})\mathbb{P}(\tau = t_n) = \mathbb{E}\tau$$

and this finishes the proof of the induction step.

Next, let us consider the case of a uniformly bounded stopping time $\tau \leq M < \infty$. In the previous lecture we defined a dyadic approximation

$$\tau_n = \frac{\lfloor 2^n \tau \rfloor + 1}{2^n},$$

which is also a stopping time, $\tau_n \downarrow \tau$, and by the sample continuity $W_{\tau_n} \to W_\tau$ almost surely. Since (τ_n) are uniformly bounded, $\mathbb{E}\tau_n \to \mathbb{E}\tau$. To prove that $\mathbb{E}W_{\tau_n}^2 \to \mathbb{E}W_\tau^2$, we need to show that the sequence $(W_{\tau_n}^2)$ is uniformly integrable. Notice that

$\tau_n < 2M$ and, therefore, τ_n takes possible values of the type $k/2^n$ for $k \le k_0 = \lfloor 2^n(2M) \rfloor$. Since the sequence $(W_{1/2^n}, \ldots, W_{k_0/2^n}, W_{2M})$ is a martingale, adapted to a corresponding sequence of $\mathscr{F}_t$, and τ_n and $2M$ are two stopping times such that $\tau_n < 2M$, by the optional stopping theorem, Theorem 5.2, $W_{\tau_n} = \mathbb{E}(W_{2M}|\mathscr{F}_{\tau_n})$. By Jensen's inequality,

$$W_{\tau_n}^4 \le \mathbb{E}(W_{2M}^4|\mathscr{F}_{\tau_n}),\ \mathbb{E}W_{\tau_n}^4 \le \mathbb{E}W_{2M}^4 = 6M,$$

and the uniform integrability follows,

$$\mathbb{E}W_{\tau_n}^2 \mathrm{I}(|W_{\tau_n}| > N) \le \frac{\mathbb{E}W_{\tau_n}^4}{N^2} \le \frac{6M}{N^2}.$$

This proves that $\mathbb{E}W_{\tau_n} \to \mathbb{E}W_\tau$ and $\mathbb{E}W_{\tau_n}^2 \to \mathbb{E}W_\tau^2$. Since τ_n takes a finite number of values, $\mathbb{E}W_{\tau_n} = 0$ and $\mathbb{E}W_{\tau_n}^2 = \mathbb{E}\tau_n$, and letting $n \to \infty$ proves

$$\mathbb{E}W_\tau = 0,\ \mathbb{E}W_\tau^2 = \mathbb{E}\tau. \tag{6.21}$$

Before we consider the general case, let us notice that for two bounded stopping times $\tau \le \rho \le M$ one can similarly show that

$$\mathbb{E}(W_\rho - W_\tau)W_\tau = 0. \tag{6.22}$$

because, by the optional stopping, $(W_{\tau_n}, \mathscr{F}_{\tau_n})$, $(W_{\rho_n}, \mathscr{F}_{\rho_n})$ is a martingale and

$$\mathbb{E}(W_{\rho_n} - W_{\tau_n})W_{\tau_n} = \mathbb{E}W_{\tau_n}\big(\mathbb{E}(W_{\rho_n}|\mathscr{F}_{\tau_n}) - W_{\tau_n}\big) = 0.$$

Finally, we consider the general case. Let us define $\tau(n) = \min(\tau, n)$. Since $\mathbb{E}W_{\tau(n)}^2 = \mathbb{E}\tau(n) \le \mathbb{E}\tau < \infty$, the sequence $W_{\tau(n)}$ is uniformly integrable and, thus, $0 = \mathbb{E}W_{\tau(n)} \to \mathbb{E}W_\tau$, which proves that $\mathbb{E}W_\tau = 0$. Next, for $m \le n$,

$$\begin{aligned}
\mathbb{E}(W_{\tau(n)} - W_{\tau(m)})^2 &= \mathbb{E}W_{\tau(n)}^2 - \mathbb{E}W_{\tau(m)}^2 - 2\mathbb{E}W_{\tau(m)}(W_{\tau(n)} - W_{\tau(m)}) \\
&= \mathbb{E}\tau(n) - \mathbb{E}\tau(m),
\end{aligned}$$

using (6.21), (6.22) and the fact that $\tau(n), \tau(m)$ are bounded stopping times. Since $\tau(n) \uparrow \tau$, Fatou's lemma and the monotone convergence theorem imply

$$\mathbb{E}(W_\tau - W_{\tau(m)})^2 \le \liminf_{n\to\infty}\big(\mathbb{E}\tau(n) - \mathbb{E}\tau(m)\big) = \mathbb{E}\tau - \mathbb{E}\tau(m).$$

Letting $m \to \infty$ shows that

$$\lim_{m\to\infty} \mathbb{E}(W_\tau - W_{\tau(m)})^2 = 0$$

which means that $\mathbb{E}W_{\tau(m)}^2 \to \mathbb{E}W_\tau^2$. Since $\mathbb{E}W_{\tau(m)}^2 = \mathbb{E}\tau(m)$ and $\mathbb{E}\tau(m) \to \mathbb{E}\tau$ by the monotone convergence theorem, this implies that $\mathbb{E}W_\tau^2 = \mathbb{E}\tau$. □

We will now prove the Skorohod's imbedding theorem.

Theorem 6.13 (Skorohod's imbedding). *Let Y be a random variable such that $\mathbb{E}Y = 0$, $\mathbb{E}Y^2 < \infty$. There exists a stopping time $\tau < \infty$ such that $\mathbb{E}\tau < \infty$ and $\mathscr{L}(W_\tau) = \mathscr{L}(Y)$.*

Proof. Let us start with the simplest case when Y takes only two values, $Y \in \{-a, b\}$ for $a, b > 0$. The condition $\mathbb{E}Y = 0$ determines the distribution of Y,

$$pb + (1-p)(-a) = 0 \text{ and } p = \frac{a}{a+b}. \tag{6.23}$$

Let $\tau = \inf\{t > 0 \mid W_t = -a \text{ or } b\}$ be a hitting time of the two-sided boundary $-a$, b. The tail probability of τ can be bounded by

$$\begin{aligned} \mathbb{P}(\tau > n) &\le \mathbb{P}\big(|W_{j+1} - W_j| < a+b, 0 \le j \le n-1\big) \\ &= \mathbb{P}(|W_1| < a+b)^n = \gamma^n. \end{aligned} \tag{6.24}$$

Therefore, $\mathbb{E}\tau < \infty$ and, by the previous theorem, $\mathbb{E}W_\tau = 0$. Since $W_\tau \in \{-a, b\}$ we must have $\mathscr{L}(W_\tau) = \mathscr{L}(Y)$. Let us now consider the general case. If μ is the law of Y, let us define Y by the identity $Y = Y(x) = x$ on its sample probability space $(\mathbb{R}, \mathscr{B}, \mu)$. Let us construct a sequence of σ-algebras

$$\mathscr{B}_1 \subseteq \mathscr{B}_2 \subseteq \ldots \subseteq \mathscr{B}$$

as follows. Let $\mathscr{B}_1$ be generated by the set $(-\infty, 0)$, i.e.

$$\mathscr{B}_1 = \big\{\emptyset, \mathbb{R}, (-\infty, 0), [0, +\infty)\big\}.$$

Given $\mathscr{B}_j$, let us define $\mathscr{B}_{j+1}$ by splitting each finite interval $[c,d) \in \mathscr{B}_j$ into two intervals $[c, (c+d)/2)$ and $[(c+d)/2, d)$, and splitting infinite interval $(-\infty, -j)$ into $(-\infty, -(j+1))$ and $[-(j+1), -j)$ and, similarly, splitting $[j, +\infty)$ into $[j, j+1)$ and $[j+1, \infty)$. Consider a right-closed martingale

$$Y_j = \mathbb{E}(Y|\mathscr{B}_j). \tag{6.25}$$

It is almost obvious that the Borel σ-algebra $\mathscr{B}$ on $\mathbb{R}$ is generated by all these σ-algebras, $\mathscr{B} = \vee_{j\ge 1}\mathscr{B}_j$. Then, by Levy's martingale convergence, Lemma 5.6, $Y_j \to \mathbb{E}(Y|\mathscr{B}) = Y$ almost surely. Since Y_j is measurable on $\mathscr{B}_j$, it must be constant on each simple set $[c,d) \in \mathscr{B}_j$. If $Y_j(x) = y$ for $x \in [c,d)$ then, since $Y_j = \mathbb{E}(Y|\mathscr{B}_j)$,

$$y\mu([c,d)) = \mathbb{E}Y_j \mathrm{I}_{[c,d)} = \mathbb{E}Y\mathrm{I}_{[c,d)} = \int_{[c,d)} x\, d\mu(x)$$

and

$$y = \frac{1}{\mu([c,d))} \int_{[c,d)} x\, d\mu(x). \tag{6.26}$$

If $\mu([c,d)) = 0$ we pick any $y \in (c,d)$ as a value of Y_j on $[c,d)$. Since in the σ-algebra $\mathscr{B}_{j+1}$ the interval $[c,d)$ is split into two intervals, the random variable Y_{j+1} can take only two values on the interval $[c,d)$, say $c \le y_1 < y_2 < d$,

$$\begin{aligned} \mathbb{P}(Y_{j+1} = y_2) &= \mathbb{P}(Y_{j+1} - Y_j = y_2 - y, Y_j = y) \\ &= \mathbb{P}(Y_{j+1} - Y_j = y_2 - y \mid Y_j = y)\mathbb{P}(Y_j = y) = \frac{y - y_1}{y_2 - y_1}\mathbb{P}(Y_j = y), \end{aligned} \tag{6.27}$$

where in the last equation we used (6.23) together with $\mathbb{E}(Y_{j+1} - Y_j | \mathscr{B}_j) = 0$.

We will define stopping times τ_n such that $\mathscr{L}(W_{\tau_n}) = \mathscr{L}(Y_n)$ as follows. Since Y_1 takes only two values $-a$ and b, if $\tau_1 = \inf\{t > 0 \mid W_t = -a \text{ or } b\}$ then we proved above that $\mathscr{L}(W_{\tau_1}) = \mathscr{L}(Y_1)$. Given τ_j, define τ_{j+1} by:

$$\text{if } W_{\tau_j} = y \text{ for } y \text{ in (6.26), let } \tau_{j+1} = \inf\{t > \tau_j \mid W_t = y_1 \text{ or } y_2\}.$$

Since, by definition, the events $\{W_{\tau_{j+1}} = y_j\}$ for $j = 1,2$ assume that $W_{\tau_j} = y$, we can rewrite, for example,

$$\{W_{\tau_{j+1}} = y_2\} = \{(W_{\tau_j + h} - W_{\tau_j})_{h \ge 0} \text{ hits } y_2 - y \text{ before } y_1 - y\} \bigcap \{W_{\tau_j} = y\}.$$

Note that $\{W_{\tau_j} = y\} \in \mathscr{F}_{\tau_j}$ and, by the strong Markov property in Theorem 6.6,

$$\begin{aligned} \mathbb{P}(W_{\tau_{j+1}} = y_2) &= \mathbb{P}\big((W_h)_{h \ge 0} \text{ hits } y_2 - y \text{ before } y_1 - y\big)\mathbb{P}(W_{\tau_j} = y) \\ &= \frac{y - y_1}{y_2 - y_1}\mathbb{P}(W_{\tau_j} = y), \end{aligned}$$

where in the last equation we again used (6.23). Since Y_j's above satisfy the same recursive equations (6.27), this proves that $\mathscr{L}(W_{\tau_n}) = \mathscr{L}(Y_n)$.

Also, $\tau_n \le \tau = \inf\{t > 0 \mid W_t = -a \text{ or } b\}$ if we take $-a$ and b to be the smallest and largest among all values y that Y_n can take so, by (6.24), all $\mathbb{E}\tau_n < \infty$. Moreover, by the previous theorem and (6.25),

$$\mathbb{E}\tau_n = \mathbb{E}W_{\tau_n}^2 = \mathbb{E}Y_n^2 \le \mathbb{E}Y^2 < \infty.$$

The sequence τ_n is monotone, so it converges to some stopping time $\tau_n \uparrow \tau$. and, therefore, $\mathbb{E}\tau = \lim \mathbb{E}\tau_n \le \mathbb{E}Y^2 < \infty$. By sample continuity, $W_{\tau_n} \to W_\tau$ a.s. and, hence, $\mathscr{L}(W_{\tau_n}) = \mathscr{L}(Y_n) \to \mathscr{L}(W_\tau) = \mathscr{L}(Y)$. □

The Skorohod's imbedding will be used below to reduce the case of sums of i.i.d. random variables to the following law of the iterated logarithm for the Brownian motion, which we prove first.

Theorem 6.14. *If* $u(t) = \sqrt{2t\ell(t)}$ *where* $\ell(t) = \log\log(t)$ *then*

$$\limsup_{t \to +\infty} \frac{W_t}{u(t)} = 1 \ \ a.s.$$

Let us start with the following result which shows that the Brownian motion does not fluctuate much between times s and $(1+\varepsilon)s$ for large s.

Lemma 6.6. *For any* $\varepsilon \in (0,1)$,

$$\limsup_{s\to\infty} \sup\Big\{\frac{|W_t - W_s|}{u(s)} : s \le t \le (1+\varepsilon)s\Big\} \le 8\sqrt{\varepsilon} \ a.s.$$

Proof. Let $\varepsilon > 0$, $t_k = (1+\varepsilon)^k$ and $M_k = \sqrt{2\varepsilon}u(t_k)$. By symmetry, the equation (6.19) and the Gaussian tail estimate in Lemma 6.3,

$$\begin{aligned}
\mathbb{P}\Big(\sup_{t_k\le t\le t_{k+1}} |W_t - W_{t_k}| \ge M_k\Big) &\le 2\mathbb{P}\Big(\sup_{0\le t\le t_{k+1}-t_k} W_t \ge M_k\Big) \\
&= 4N(0,t_{k+1}-t_k)(M_k,\infty) \le 4\exp\Big(-\frac{1}{2}\frac{M_k^2}{(t_{k+1}-t_k)}\Big) \\
&= 4\exp\Big(-\frac{4\varepsilon t_k \ell(t_k)}{2\varepsilon t_k}\Big) = 4\Big(\frac{1}{k\log(1+\varepsilon)}\Big)^2.
\end{aligned}$$

The sum of these probabilities converges and, by the Borel-Cantelli lemma, for large enough k,

$$\sup_{t_k\le t\le t_{k+1}} |W_t - W_{t_k}| \le \sqrt{2\varepsilon}u(t_k).$$

If k is such that $t_k \le s \le t_{k+1}$ and $s \le t \le (1+\varepsilon)s$ then, clearly, $t_k \le s \le t \le t_{k+2}$ and, therefore, for large enough k,

$$|W_t - W_s| \le 2\sqrt{2\varepsilon}u(t_k) + \sqrt{2\varepsilon}u(t_{k+1}) \le 8\sqrt{\varepsilon}u(s).$$

This finishes the proof. □

Proof (of Theorem 6.14). For $\varepsilon > 0, a > 1$ and $t_k = a^k$ (recall $u(t) = \sqrt{2t\ell(t)}$),

$$\begin{aligned}
\mathbb{P}\big(W_{t_k} \ge (1+\varepsilon)u(t_k)\big) &= \mathbb{P}\Big(\frac{W_{t_k}}{\sqrt{t_k}} \ge (1+\varepsilon)\sqrt{2\ell(t_k)}\Big) \\
&\sim \frac{1}{\sqrt{2\pi}}\frac{1}{(1+\varepsilon)\sqrt{2\ell(t_k)}}e^{-(1+\varepsilon)^2\ell(t_k)} \\
&\sim \frac{1}{\sqrt{2\pi}}\frac{1}{(1+\varepsilon)\sqrt{2\ell(a^k)}}\Big(\frac{1}{\log a^k}\Big)^{(1+\varepsilon)^2}.
\end{aligned} \tag{6.28}$$

This series converges so, by the Borel-Cantelli lemma, $W_{t_k} \le (1+\varepsilon)u(a^k)$ for large enough k. If we take $a = 1+\varepsilon$, together with Lemma 6.6 this gives that, for large enough k, if $t_k \le t < t_{k+1}$ then

$$\frac{W_t}{u(t)} = \frac{W_{t_k}}{u(t_k)}\frac{u(t_k)}{u(t)} + \frac{W_t - W_{t_k}}{u(t_k)}\frac{u(t_k)}{u(t)} \le (1+\varepsilon) + 8\sqrt{\varepsilon}.$$

Letting $\varepsilon \downarrow 0$ over some sequence proves that, with probability one,

$$\limsup_{t\to\infty} \frac{W_t}{u(t)} \le 1.$$

To prove that the upper limit is equal to 1, we will use the Borel-Cantelli lemma for the independent increments $W_{a^k} - W_{a^{k-1}}$ for large values of the parameter $a > 1$. If $0 < \varepsilon < 1$ then, similarly to (6.28),

$$\begin{aligned} &\mathbb{P}\big(W_{a^k} - W_{a^{k-1}} \ge (1-\varepsilon)u(a^k - a^{k-1})\big) \\ &\quad \sim \frac{1}{\sqrt{2\pi}} \frac{1}{(1-\varepsilon)\sqrt{2\ell(a^k - a^{k-1})}} \Big(\frac{1}{\log(a^k - a^{k-1})}\Big)^{(1-\varepsilon)^2}. \end{aligned}$$

The series diverges and, since these events are independent, by Borel-Cantelli lemma they occur infinitely often with probability one. We already proved in (6.28) that, for $\varepsilon > 0$, $W_{a^k} \le (1+\varepsilon)u(a^k)$ for large enough $k \ge 1$ or, by symmetry, $W_{a^k} \ge -(1+\varepsilon)u(a^k)$ for large enough $k \ge 1$. Together with $W_{a^k} - W_{a^{k-1}} \ge (1-\varepsilon)u(a^k - a^{k-1})$ this gives

$$\begin{aligned} \frac{W_{a^k}}{u(a^k)} &\ge (1-\varepsilon)\frac{u(a^k - a^{k-1})}{u(a^k)} + \frac{W_{a^{k-1}}}{u(a^k)} \\ &\ge (1-\varepsilon)\frac{u(a^k - a^{k-1})}{u(a^k)} - (1+\varepsilon)\frac{u(a^{k-1})}{u(a^k)} \\ &= (1-\varepsilon)\sqrt{\frac{(a^k - a^{k-1})\ell(a^k - a^{k-1})}{a^k \ell(a^k)}} - (1+\varepsilon)\sqrt{\frac{a^{k-1}\ell(a^{k-1})}{a^k \ell(a^k)}} \end{aligned}$$

and

$$\limsup_{t\to\infty} \frac{W_t}{u(t)} \ge \limsup_{k\to\infty} \frac{W_{a^k}}{u(a^k)} \ge (1-\varepsilon)\sqrt{\Big(1 - \frac{1}{a}\Big)} - (1+\varepsilon)\sqrt{\frac{1}{a}}.$$

Letting $\varepsilon \downarrow 0$ and $a \to \infty$ over some sequences proves that the upper limit is equal to one. □

The LIL for Brownian motion implies the LIL for sums of independent random variables via Skorohod's imbedding.

Theorem 6.15. *Suppose that $Y_1, \ldots, Y_n$ are i.i.d. and $\mathbb{E}Y_i = 0, \mathbb{E}Y_i^2 = 1$. If $S_n = Y_1 + \ldots + Y_n$ then*

$$\limsup_{n\to\infty} \frac{S_n}{\sqrt{2n\log\log n}} = 1 \ \textit{a.s.}$$

Proof. Let us define a stopping time $\tau(1)$ such that $W_{\tau(1)} \stackrel{d}{=} Y_1$. By the strong Markov property, the increment of the process after stopping time is independent of the process before stopping time and has the law of the Brownian motion. Therefore, we can define $\tau(2)$ such that $W_{\tau(1)+\tau(2)} - W_{\tau(1)} \stackrel{d}{=} Y_2$ and, by independence, $W_{\tau(1)+\tau(2)} \stackrel{d}{=} Y_1 + Y_2$ and $\tau(1), \tau(2)$ are i.i.d. By induction, we can define i.i.d. $\tau(1), \ldots, \tau(n)$ such that, for $T(n) = \tau(1) + \ldots + \tau(n)$,

$$\left(\frac{S_n}{u(n)}\right)_{n\geq 1} \stackrel{d}{=} \left(\frac{W_{T(n)}}{u(n)}\right)_{n\geq 1}.$$

By the Exercise 2.3.6, it is enough to prove the almost sure convergence of the right hand side. Let us write

$$\frac{W_{T(n)}}{u(n)} = \frac{W_n}{u(n)} + \frac{W_{T(n)} - W_n}{u(n)}.$$

By the LIL for Brownian motion,

$$\limsup_{n\to\infty} \frac{W_n}{u(n)} = 1.$$

By the strong law of large numbers, $T(n)/n \to \mathbb{E}\tau(1) = \mathbb{E}Y_1^2 = 1$ almost surely. This means that, for any $\varepsilon > 0$, $n/\sqrt{1+\varepsilon} \leq T(n) \leq n\sqrt{1+\varepsilon}$ for large enough n. Using Lemma 6.6 with $s = n/\sqrt{1+\varepsilon}$ implies that

$$\limsup_{n\to\infty} \frac{|W_{T(n)} - W_n|}{u(n)} \leq 16\sqrt{\varepsilon},$$

and letting $\varepsilon \downarrow 0$ finishes the proof. □

The LIL for Brownian motion also implies the *local LIL*:

$$\limsup_{t\to 0} \frac{W_t}{\sqrt{2t\ell(1/t)}} = 1.$$

It is easy to check that if W_t is a Brownian motion then $tW_{1/t}$ is also the Brownian motion and the result follows by a change of variable $t \to 1/t$. To check that $tW_{1/t}$ is a Brownian motion notice that for $t < s$,

$$\mathbb{E}tW_{1/t}\big(sW_{1/s} - tW_{1/t}\big) = st\frac{1}{s} - t^2\frac{1}{t} = t - t = 0$$

and $\mathbb{E}\big(tW_{1/t} - sW_{1/s}\big)^2 = t + s - 2t = s - t.$

Exercise 6.5.1. Show that the condition $\mathbb{E}\tau < \infty$ in Theorem 6.12 above can not be replaced with the condition $\mathbb{E}W_\tau^2 < \infty$. *Hint:* see previous section.

Exercise 6.5.2. Consider the sequence $T(n) = \tau(1) + \ldots + \tau(n)$ as in the proof of Theorem 6.15. Prove that, for any $\varepsilon > 0$,

$$\lim_{n\to\infty} \mathbb{P}\left(\max_{j\leq n}\left|\frac{W_{T(j)} - W_j}{\sqrt{n}}\right| \geq \varepsilon\right) = 0.$$

Hint: First prove that, by the SLLN, for any $\delta > 0$,

$$\lim_{n\to\infty} \mathbb{P}\Big(\max_{j\le n}\Big|\frac{T(j)-j}{n}\Big| \ge \delta\Big) = 0,$$

and then use the fact that $(\sqrt{c}W_t)_{t\ge 0} \stackrel{d}{=} (W_{ct})_{t\ge 0}$.

Exercise 6.5.3. For $t \le 1$, define the process $X_t^n := W_{T(nt)}/\sqrt{n}$ when nt is an integer, and by linear interpolation in between, as in Donsker's theorem. Let K be a closed set in $C([0,1], \|\cdot\|_\infty)$. Show that, for any $\varepsilon \ge 0$,

$$\limsup_{n\to\infty} \mathbb{P}\big((X_t^n)_{t\le 1} \in K\big) \le \mathbb{P}\big((W_t)_{t\le 1} \in K^\varepsilon\big).$$

Hint: Use previous exercise for $\varepsilon > 0$ and the fact that $(\sqrt{c}W_t)_{t\ge 0} \stackrel{d}{=} (W_{ct})_{t\ge 0}$.

Exercise 6.5.4 (Donsker's theorem). In the setting of previous exercise, prove that $(X_t^n)_{t\le 1}$ converges in distribution to $(W_t)_{t\le 1}$ in $C([0,1], \|\cdot\|_\infty)$. *Hint:* Use the Portmanteau theorem.

Chapter 7
Poisson Processes

7.1 Overview of Poisson processes

In this section we will introduce and review several important properties of Poisson processes on a measurable space $(S,\mathscr{S})$. In applications, $(S,\mathscr{S})$ is usually some nice space, such as a Euclidean space $\mathbb{R}^n$ with the Borel σ-algebra. However, in general, it is enough to require that the diagonal $\{(s,s') : s = s'\}$ is measurable on the product space $S \times S$ which, in particular, implies that every singleton set $\{s\}$ in S is measurable as a section of the diagonal. This condition is needed to be able to write $\mathbb{P}(X = Y)$ for a pair (X,Y) of random variables defined on the product space. From now on, every time we consider a measurable space we will assume that it satisfies this condition. Let us also notice that a product $S \times S'$ of two such spaces will also satisfy this condition since $\{(s_1,s_1') = (s_2,s_2')\} = \{s_1 = s_2\} \cap \{s_1' = s_2'\}$. Let μ and μ_n for $n \geq 1$ be some non-atomic (not having any atoms, i.e. points of positive measure) measures on $\mathscr{S}$ such that

$$\mu = \sum_{n \geq 1} \mu_n, \; \mu_n(S) < \infty. \tag{7.1}$$

For each $n \geq 1$, let N_n be a random variable with the Poisson distribution $\Pi(\mu_n(S))$ with the mean $\mu_n(S)$ and let $(X_{n\ell})_{\ell \geq 1}$ be i.i.d. random variables, also independent of N_n, with the distribution

$$p_n(B) = \frac{\mu_n(B)}{\mu_n(S)}. \tag{7.2}$$

We assume that all these random variables are independent for different $n \geq 1$. The condition that μ is non-atomic implies that $\mathbb{P}(X_{n\ell} = X_{mj}) = 0$ if $n \neq m$ or $\ell \neq j$. Let us consider random sets

$$\Pi_n = \{X_{n1},\ldots,X_{nN_n}\} \text{ and } \Pi = \bigcup\nolimits_{n \geq 1} \Pi_n. \tag{7.3}$$

The set Π will be called a *Poisson process* on S with the *mean measure* μ. Let us point out a simple observation that will be used many times below that if, first, we are given the means $\mu_n(S)$ and distributions (7.2) that were used to generate the set (7.3) then the measure μ in (7.1) can be written as

$$\mu = \sum_{n\geq 1} \mu_n(S) p_n. \tag{7.4}$$

We will show in Theorem 7.4 below that when the measure μ is σ-finite then, in some sense, this definition of a Poisson process Π is independent of the particular representation (7.1). However, for several reasons it is convenient to think of the above construction as the definition of a Poisson process. First of all, it allows us to avoid any discussion about what "a random set" means and, moreover, many important properties of Poisson processes follow from it rather directly. In any case, we will show in Theorem 7.5 below that all such processes satisfy the traditional definition of a Poisson process.

One important immediate consequence of the definition in (7.3) is the Mapping Theorem. Given a Poisson process Π on S with the mean measure μ and a measurable map $f : S \to S'$ into another measurable space $(S', \mathscr{S}')$, let us consider the image set

$$f(\Pi) = \bigcup_{n\geq 1} \{f(X_{n1}), \ldots, f(X_{nN_n})\}.$$

Since random variables $f(X_{n\ell})$ have the distribution $p_n \circ f^{-1}$ on $\mathscr{S}'$, the set $f(\Pi)$ resembles the definition (7.3) corresponding to the measure

$$\sum_{n\geq 1} \mu_n(S)\, p_n \circ f^{-1} = \sum_{n\geq 1} \mu_n \circ f^{-1} = \mu \circ f^{-1}.$$

Therefore, to conclude that $f(\Pi)$ is a Poisson process on S' we only need to require that this image measure is non-atomic.

Theorem 7.1 (Mapping Theorem). *If Π is a Poisson process on S with the mean measure μ and $f : S \to S'$ is such that $\mu \circ f^{-1}$ is non-atomic then $f(\Pi)$ is a Poisson process on S' with the mean measure $\mu \circ f^{-1}$.*

Next, let us consider a sequence (λ_m) of measures that satisfy (7.1), $\lambda_m = \sum_{n\geq 1} \lambda_{mn}$ and $\lambda_{mn}(S) < \infty$. Let $\Pi_m = \cup_{n\geq 1} \Pi_{mn}$ be a Poisson process with the mean measure λ_m defined as in (7.3) and suppose that all these Poisson processes are generated independently over $m \geq 1$. Since

$$\lambda := \sum_{m\geq 1} \lambda_m = \sum_{m\geq 1} \sum_{n\geq 1} \lambda_{mn} = \sum_{\ell\geq 1} \lambda_{m(\ell)n(\ell)}$$

and $\Pi := \cup_{m\geq 1} \Pi_m = \cup_{\ell\geq 1} \Pi_{m(\ell)n(\ell)}$ for any enumeration of the pairs (m,n) by the indices $\ell \in \mathbb{N}$, we immediately get the following.

Theorem 7.2 (Superposition Theorem). *If Π_m for $m \geq 1$ are independent Poisson processes with the mean measures λ_m then their superposition $\Pi = \cup_{m\geq 1} \Pi_m$ is a Poisson process with the mean measure $\lambda = \sum_{m\geq 1} \lambda_m$.*

Another important property of Poisson processes is the Marking Theorem. Consider another measurable space $(S', \mathscr{S}')$ and let $K : S \times \mathscr{S}' \to [0,1]$ be a transition function (a probability kernel), which means that for each $s \in S$, $K(s, \cdot)$ is a probability measure on $\mathscr{S}'$ and for each $A \in \mathscr{S}'$, $K(\cdot, A)$ is a measurable function on $(S, \mathscr{S})$. For each point $s \in \Pi$, let us generate a point $m(s) \in S'$ from the distribution $K(s, \cdot)$, independently for different points s. The point $m(s)$ is called a *marking* of s and a random subset of $S \times S'$,

$$\Pi^* = \big\{(s, m(s)) : s \in \Pi\big\}, \tag{7.5}$$

is called a marked Poisson process. In other words, $\Pi^* = \cup_{n \geq 1} \Pi_n^*$, where

$$\Pi_n^* = \big\{(X_{n1}, m(X_{n1})), \ldots, (X_{nN_n}, m(X_{nN_n}))\big\},$$

and each point $(X_{n\ell}, m(X_{n\ell})) \in S \times S'$ is generated according to the distribution

$$p_n^*(C) = \iint_C K(s, ds')\, p_n(ds).$$

Since p_n^* is obviously non-atomic, by definition, this means that Π^* is a Poisson process on $S \times S'$ with the mean measure

$$\sum_{n \geq 1} \mu_n(S)\, p_n^*(C) = \sum_{n \geq 1} \iint_C K(s, ds') \mu_n(ds) = \iint_C K(s, ds') \mu(ds).$$

Therefore, the following holds.

Theorem 7.3 (Marking Theorem). *The random subset Π^* in (7.5) is a Poisson process on $S \times S'$ with the mean measure*

$$\mu^*(C) = \iint_C K(s, ds') \mu(ds). \tag{7.6}$$

The above results give us several ways to generate a new Poisson process from the old one. However, in each case, the new process is generated in a way that depends on a particular representation of its mean measure. We will now show that, when the measure μ is σ-finite, the random set Π can be generated in a way that, in some sense, does not depend on the particular representation (7.1). This point is very important, because, often, given a Poisson process with one representation of its mean measure, we study its properties using another, more convenient, representation. Suppose that S is equal to a disjoint union $\cup_{m \geq 1} S_m$ of sets such that $0 < \mu(S_m) < \infty$, in which case

$$\mu = \sum_{m \geq 1} \mu|_{S_m} \tag{7.7}$$

is another representation of the type (7.1), where $\mu|_{S_m}$ is the restriction of μ to the set S_m. The following holds.

Theorem 7.4 (Equivalence Theorem). *The process Π in (7.3) generated according to the representation (7.1) is statistically indistinguishable from a process generated using the representation (7.7).*

More precisely, we will show the following. Let us denote

$$p_{S_m}(B) = \frac{\mu|_{S_m}(B)}{\mu|_{S_m}(S)} = \frac{\mu(B\cap S_m)}{\mu(S_m)}. \tag{7.8}$$

We will show that, given the Poisson process Π in (7.3) generated according to the representation (7.1), the cardinalities $N(S_m) = |\Pi\cap S_m|$ are independent random variables with the distributions $\Pi(\mu(S_m))$ and, conditionally on $(N(S_m))_{m\geq 1}$, each set $\Pi\cap S_m$ "looks like" an i.i.d. sample of size $N(S_m)$ from the distribution p_{S_m}. This means that, if we arrange the points in $\Pi\cap S_m$ in a random order, the resulting vector has the same distribution as an i.i.d. sample from the measure p_{S_m}. Moreover, conditionally on $(N(S_m))_{m\geq 1}$, these vectors are independent over $m\geq 1$. This is exactly how one would generate a Poisson process using the representation (7.7). The proof of Theorem 7.4 will be based on the following two properties that usually appear as the definition of a Poisson process on S with the mean measure μ:

(i) for any $A\in\mathscr{S}$, the cardinality $N(A) = |\Pi\cap A|$ has the Poisson distribution $\Pi(\mu(A))$ with the mean $\mu(A)$;
(ii) the cardinalities $N(A_1),\ldots,N(A_k)$ are independent, for any $k\geq 1$ and any disjoint sets $A_1,\ldots,A_k\in\mathscr{S}$.

When $\mu(A)=\infty$, it is understood that $\Pi\cap A$ is countably infinite. Usually, one starts with (i) and (ii) as the definition of a Poisson process and the set Π constructed in (7.3) is used to demonstrate the existence of such processes, which explains the name of the following theorem.

Theorem 7.5 (Existence Theorem). *The process Π in (7.3) satisfies the properties (i) and (ii).*

Proof. Given $x\geq 0$, let us denote the weights of the Poisson distribution $\Pi(x)$ with the mean x by

$$\pi_j(x) = \Pi(x)(\{j\}) = \frac{x^j}{j!}e^{-x}.$$

Consider disjoint sets $A_1,\ldots,A_k$ and let $A_0 = (\cup_{i\leq k}A_i)^c$ be the complement of their union. Fix $m_1,\ldots,m_k\geq 0$ and let $m = m_1+\cdots+m_k$. Given any set A, denote $N_n(A) = |\Pi_n\cap A|$. With this notation, let us compute the probability of the event

$$\Omega = \big\{N_n(A_1) = m_1,\ldots,N_n(A_k) = m_k\big\}.$$

Recall that the random variables $(X_{n\ell})_{\ell\geq 1}$ are i.i.d. with the distribution p_n defined in (7.2) and, therefore, conditionally on N_n in (7.3), the cardinalities $(N_n(A_\ell))_{0\leq\ell\leq k}$ have multinomial distribution and we can write

$$
\begin{aligned}
\mathbb{P}(\Omega) &= \sum_{j\geq 0} \mathbb{P}\big(\Omega \,\big|\, N_n = m+j\big)\mathbb{P}\big(N_n = m+j\big) \\
&= \sum_{j\geq 0} \mathbb{P}\big(\{N_n(A_0) = j\} \cap \Omega \,\big|\, N_n = m+j\big)\pi_{m+j}\big(\mu_n(S)\big) \\
&= \sum_{j\geq 0} \frac{(m+j)!}{j!m_1!\cdots m_k!} p_n(A_0)^j p_n(A_1)^{m_1}\cdots p_n(A_k)^{m_k} \frac{\mu_n(S)^{m+j}}{(m+j)!} e^{-\mu_n(S)} \\
&= \sum_{j\geq 0} \frac{1}{j!m_1!\cdots m_k!} \mu_n(A_0)^j \mu_n(A_1)^{m_1}\cdots \mu_n(A_k)^{m_k} e^{-\mu_n(S)} \\
&= \frac{\mu_n(A_1)^{m_1}}{m_1!} e^{-\mu_n(A_1)} \cdots \frac{\mu_n(A_k)^{m_k}}{m_k!} e^{-\mu_n(A_k)}.
\end{aligned}
$$

This means that the cardinalities $N_n(A_\ell)$ for $1 \leq \ell \leq k$ are independent random variables with the distributions $\Pi(\mu_n(A_\ell))$. Since all measures p_n are non-atomic, $\mathbb{P}(X_{nj} = X_{mk}) = 0$ for any $(n,j) \neq (m,k)$. Therefore, the sets Π_n are all disjoint with probability one and

$$
N(A) = |\Pi \cap A| = \sum_{n\geq 1} |\Pi_n \cap A| = \sum_{n\geq 1} N_n(A).
$$

First of all, the cardinalities $N(A_1),\ldots,N(A_k)$ are independent, since we showed that $N_n(A_1),\ldots,N_n(A_k)$ are independent for each $n \geq 1$, and it remains to show that $N(A)$ has the Poisson distribution with the mean $\mu(A) = \sum_{n\geq 1}\mu_n(A)$. The partial sum $S_m = \sum_{n\leq m} N_n(A)$ has the Poisson distribution with the mean $\sum_{n\leq m}\mu_n(A)$ and, since $S_m \uparrow N(A)$, for any integer $r \geq 0$, the probability $\mathbb{P}(N(A) \leq r)$ equals

$$
\lim_{m\to\infty} \mathbb{P}(S_m \leq r) = \lim_{m\to\infty} \sum_{j\leq r} \pi_j\Big(\sum_{n\leq m} \mu_n(A)\Big) = \sum_{j\leq r} \pi_j\big(\mu(A)\big).
$$

If $\mu(A) < \infty$, this shows that $N(A)$ has the distribution $\Pi(\mu(A))$. If $\mu(A) = \infty$ then $\mathbb{P}(N(A) \leq r) = 0$ for all $r \geq 0$, which implies that $N(A)$ is countably infinite. □

Proof (of Theorem 7.4). First, suppose that $\mu(S) < \infty$ and suppose that the Poisson process Π was generated as in (7.3) using some sequence of measures (μ_n). By Theorem 7.5, the cardinality $N = N(S)$ has the distribution $\Pi(\mu(S))$. Let us show that, conditionally on the event $\{N = n\}$, the set Π "looks like" an i.i.d. sample of size n from the distribution $p(\cdot) = \mu(\cdot)/\mu(S)$ or, more precisely, if we randomly assign labels $\{1,\ldots,n\}$ to the points in Π, the resulting vector $(X_1,\ldots,X_n)$ has the same distribution as an i.i.d. sample from p. Let us consider some measurable sets $B_1,\ldots,B_n \in \mathscr{S}$ and compute

$$
\mathbb{P}_n\big(X_1 \in B_1,\ldots,X_n \in B_n\big) = \mathbb{P}\big(X_1 \in B_1,\ldots,X_n \in B_n \,\big|\, N = n\big), \tag{7.9}
$$

where we denoted by $\mathbb{P}_n$ the conditional probability given the event $\{N = n\}$. To use Theorem 7.5, we will reduce this case to the case of disjoint sets as follows. Let $A_1,\ldots,A_k$ be the partition of the space S generated by the sets $B_1,\ldots,B_n$, obtained by taking intersections of the sets B_ℓ or their complements over $\ell \le n$. Then, for each $\ell \le n$, we can write

$$\mathrm{I}(x \in B_\ell) = \sum_{j \le k} \delta_{\ell j} \mathrm{I}(x \in A_j),$$

where $\delta_{\ell j} = 1$ if $B_\ell \supseteq A_j$ and $\delta_{\ell j} = 0$ otherwise. In terms of these sets, (7.9) can be rewritten as

$$\begin{aligned} \mathbb{P}_n\big(X_1 \in B_1, \ldots, X_n \in B_n\big) &= \mathbb{E}_n \prod_{\ell \le n} \sum_{j \le k} \delta_{\ell j} \mathrm{I}(X_\ell \in A_j) \\ &= \sum_{j_1, \ldots, j_n \le k} \delta_{1 j_1} \cdots \delta_{n j_n} \mathbb{P}_n\big(X_1 \in A_{j_1}, \ldots, X_n \in A_{j_n}\big). \end{aligned} \tag{7.10}$$

If we fix $j_1, \ldots, j_n$ and for $j \le k$ let $I_j = \{\ell \le n : j_\ell = j\}$ then

$$\mathbb{P}_n\big(X_1 \in A_{j_1}, \ldots, X_n \in A_{j_n}\big) = \mathbb{P}_n\big(X_\ell \in A_j \text{ for all } j \le k, \ell \in I_j\big).$$

If $n_j = |I_j|$ then the last event can be expressed in words by saying that, for each $j \le k$, we observe n_j points of the random set Π in the set A_j and then assign labels in I_j to the points $\Pi \cap A_j$. By Theorem 7.5, the probability to observe n_j points in each set A_j, given that $N = n$, equals

$$\begin{aligned} \mathbb{P}_n(N(A_j) = n_j, j \le k) &= \frac{\mathbb{P}(N(A_j) = n_j, j \le k)}{\mathbb{P}(N = n)} \\ &= \prod_{j \le k} \frac{\mu(A_j)^{n_j}}{n_j!} e^{-\mu(A_j)} \Big/ \frac{\mu(S)^n}{n!} e^{-\mu(S)}, \end{aligned}$$

while the probability to randomly assign the labels in I_j to the points in A_j for all $j \le k$ is equal to $\prod_{j \le k} n_j!/n!$. Therefore,

$$\mathbb{P}_n\big(X_1 \in A_{j_1}, \ldots, X_n \in A_{j_n}\big) = \prod_{j \le k} \Big(\frac{\mu(A_j)}{\mu(S)}\Big)^{n_j} = \prod_{\ell \le n} p(A_{j_\ell}). \tag{7.11}$$

If we plug this into (7.10), we get

$$\begin{aligned} \mathbb{P}_n\big(X_1 \in B_1, \ldots, X_n \in B_n\big) &= \sum_{j_1 \le k} \delta_{1 j_1} p(A_{j_1}) \cdots \sum_{j_n \le k} \delta_{n j_n} p(A_{j_n}) \\ &= p(B_1) \cdots p(B_n). \end{aligned}$$

Therefore, conditionally on $\{N = n\}$, $X_1, \ldots, X_n$ are i.i.d. with the distribution p.

Now, suppose that μ is σ-finite and (7.7) holds. Consider the random sets in (7.3) and define $N(S_m) = |\Pi \cap S_m|$. By Theorem 7.5, the random variables $(N(S_m))_{m \ge 1}$ are independent and have the Poisson distributions $\Pi(\mu(S_m))$, and we would like to show that, conditionally on $(N(S_m))_{m \ge 1}$, each set $\Pi \cap S_m$ can be generated as a sample of size $N(S_m)$ from the distribution $p_{S_m} = \mu|_{S_m}(\cdot)/\mu(S_m)$, independently over m. This can be done by considering finitely many m at a time and using exactly the same computation leading to (7.11) based on the properties proved in Theorem 7.5, with each subset $A \subseteq S_m$ producing a factor $p_{S_m}(A)$. $\square$

Let us now consider one example of a Poisson process that will be useful below when talk about α-stable subordinators. Let us consider a parameter $\zeta \in (0,1)$ and let Π be a Poisson process on $(0,\infty)$ with the mean measure

$$\mu(dx) = \zeta x^{-1-\zeta} dx. \tag{7.12}$$

Clearly, the measure μ is σ-finite, which can be seen, for example, by considering the partition $(0,\infty) = \cup_{m\geq 1} S_m$ with $S_1 = [1,\infty)$ and $S_m = [1/m, 1/(m-1))$ for $m \geq 2$, since

$$\int_1^\infty \zeta x^{-1-\zeta} dx = 1 < \infty.$$

By Theorem 7.4, each cardinality $N(S_m) = |\Pi \cap S_m|$ has the Poisson distribution with the mean $\mu(S_m)$ and, conditionally on $N(S_m)$, the points in $\Pi \cap S_m$ are i.i.d. with the distribution $\mu|_{S_m}/\mu(S_m)$. Therefore, by Wald's identity,

$$\mathbb{E} \sum_{x \in \Pi \cap S_m} x = \mathbb{E} N(S_m) \int_{S_m} \frac{x\mu(dx)}{\mu(S_m)} = \int_{S_m} x\mu(dx)$$

and, therefore,

$$\mathbb{E} \sum_{x\in\Pi} x \mathrm{I}(x<1) = \int_0^1 x\mu(dx) = \int_0^1 \zeta x^{-\zeta} dx = \frac{\zeta}{1-\zeta} < \infty. \tag{7.13}$$

This means that the sum $\sum_{x\in\Pi} x\mathrm{I}(x<1)$ is finite with probability one and, since there are finitely many points in the set $\Pi \cap [1,\infty)$, the sum $\sum_{x\in\Pi} x$ is also finite with probability one.

Exercise 7.1.1. Let Π be a Poisson process with the mean measure (7.12). Show that $\mathbb{E}\big(\sum_{x\in\Pi} x\big)^a < \infty$ for all $0 < a < \zeta$.

Exercise 7.1.2. Let Π be a Poisson process on $(0,\infty)$ with the mean measure λdx. If we enumerate the points in Π in the increasing order, $X_1 < X_2 < \ldots$, show that the increments $X_1, X_2 - X_1, X_3 - X_2, \ldots$ are i.i.d. exponential random variables with the density $\lambda e^{-\lambda x}$ on $(0,\infty)$. *Hint:* compute the probability for $X_1,\ldots,X_n$ to be in the intervals $[x_1, x_1+\Delta x_1], [x_2, x_2+\Delta x_2], \ldots, [x_n, x_n+\Delta x_n]$ for $x_1 < x_2 < \ldots < x_n$.

Exercise 7.1.3. If Π be a Poisson process on $(0,\infty)$ with the mean measure (7.12) and $(U_x)_{x\in\Pi}$ are i.i.d. markings uniform on $[0,1]$, compute the mean measure of the process $\{xU_x : x \in \Pi\}$. Is it a Poisson process?

Exercise 7.1.4. If Π be a Poisson process on $(0,\infty)$ with the Lebesgue mean measure dx and $\zeta > 0$, show that $\{x^{-1/\zeta} : x \in \Pi\}$ is the Poisson process on $(0,\infty)$ with the mean measure (7.12).

7.2 Sums over Poisson processes

We will now consider random variables of the following type,

$$\Sigma = \sum_{x\in\Pi} f(x), \tag{7.14}$$

where Π is a Poisson process with the mean measure μ and f is a real-valued measurable function. Since the points in Π are not necessarily ordered, we understand that the series converges if it converges absolutely. If we consider disjoint subsets $S_+ = \{f > 0\}$ and $S_+ = \{f < 0\}$ then $\Pi_+ = \Pi \cap S_+$ and $\Pi_- = \Pi \cap S_-$ are independent Poisson processes and we can decompose Σ as the difference $\Sigma_+ - \Sigma_-$ of two independent random variables

$$\Sigma_+ = \sum_{x\in\Pi_+} f(x) = \sum_{x\in\Pi} f^+(x),\ \Sigma_- = -\sum_{x\in\Pi_-} f(x) = \sum_{x\in\Pi} f^-(x), \tag{7.15}$$

where $f^+ = \max(f,0)$ and $f^- = \max(-f,0)$ are positive and negative parts of f.

The distribution of a positive random variable $X \geq 0$ is determined by its *Laplace transform*

$$f(t) = \mathbb{E}e^{-tX} \text{ for } t \geq 0, \tag{7.16}$$

which can be seen as follows. Because $X \geq 0$, the function $z \to \mathbb{E}e^{zX}$ on $\mathbb{C}$ is well defined for $\Re z \leq 0$ and one can easily check that it is analytic on $\Re z < 0$. Therefore, it is uniquely determined by its values on $z \in \mathbb{R}^-$ (i.e. the Laplace transform $f(t)$) and, in particular, the characteristic function $\mathbb{E}e^{itX}$ is also uniquely determined by $f(t)$. This means that the distribution of $X \geq 0$ is determined by its Laplace transform.

Theorem 7.6 (Campbell's Theorem I). *If Π is a Poisson process with the mean measure μ and $f \geq 0$ then the Laplace transform of $\Sigma = \sum_{x\in\Pi} f(x)$ equals*

$$\mathbb{E}e^{-t\Sigma} = \exp\Bigl[-\int (1-e^{-tf(x)})\,d\mu(x)\Bigr]. \tag{7.17}$$

In particular, $\Sigma < \infty$ almost surely if and only if

$$\int \min(f(x),1)\,d\mu(x) < \infty \tag{7.18}$$

and, otherwise, $\Sigma = +\infty$ almost surely. Moreover,

$$\mathbb{E}\Sigma = \int f(x)\,d\mu(x),\ \mathrm{Var}(\Sigma) = \int f(x)^2\,d\mu(x), \tag{7.19}$$

finite or infinite. (Of course, the variance is defined only if $\mathbb{E}\Sigma < \infty$.)

Proof. First, suppose that f is a simple (step) function taking non-zero values $(f_j)_{j\leq k}$ (and possibly zero) and each level set $A_j = \{x : f(x) = f_j\}$ has finite measure, $m_j = \mu(A_j) < \infty$. Then the random variables $N_j = \mathrm{card}(\Pi \cap A_j)$ are independent Poiss(m_j) for $j \le k$, the sum $\Sigma = \sum_{x\in\Pi} f(x) = \sum_{j\le k} f_j N_j$ and, by independence,

$$\mathbb{E}e^{-t\Sigma} = \mathbb{E}e^{-t\sum_{j\le k} f_j N_j} = \prod_{j\le k} \mathbb{E}e^{-tf_jN_j}.$$

Each factor equals

$$\begin{aligned}\mathbb{E}e^{-tf_jN_j} &= \sum_{n\ge 0} e^{-tf_jn} \times \frac{m_j^n}{n!} e^{-m_j} \\ &= e^{-m_j} \sum_{n\ge 0} \frac{(m_j e^{-tf_j})^n}{n!} = \exp\Big[m_j(e^{-tf_j} - 1)\Big],\end{aligned}$$

so

$$\begin{aligned}\mathbb{E}e^{-t\Sigma} &= \exp\Big[\sum_{j\le k} m_j(e^{-tf_j} - 1)\Big] \\ &= \exp\Big[\int (e^{-tf(x)} - 1)\, d\mu(x)\Big] = \exp\Big[-\int (1 - e^{-tf(x)})\, d\mu(x)\Big].\end{aligned}$$

This proves (7.17) for simple $f \ge 0$. General $f \ge 0$ can be written as a limit of an increasing sequence of simple functions and, using monotone convergence theorem on both sides, we obtain (7.17) for general $f \ge 0$. Similarly, one can check (7.19) for simple functions directly and obtain the general case by monotone convergence theorem. Finally, it is obvious that, for $t > 0$, $\int (1 - e^{-tf(x)})\, d\mu(x) < \infty$ if and only if (7.18) holds. In particular, if this integral is infinite then $\mathbb{E}e^{-t\Sigma} = 0$, which means that $\Sigma = +\infty$ with probability one. If the integral is finite then, by the dominated convergence theorem, $\lim_{t\downarrow 0} \int (1 - e^{-tf(x)})\, d\mu(x) = 0$ and, therefore, $\mathbb{P}(\Sigma < +\infty) = \lim_{t\downarrow 0} \mathbb{E}e^{-t\Sigma} = 1$. □

The general case follow from the decomposition (7.15), although we have to replace the Laplace transform by characteristic function to make sure it is defined for both positive and negative parts.

Theorem 7.7 (Campbell's Theorem II). *If Π is a Poisson process with the mean measure μ then the series $\Sigma = \sum_{x\in\Pi} f(x)$ is absolutely convergent almost surely if and only if*

$$\int \min(|f(x)|, 1)\, d\mu(x) < \infty, \tag{7.20}$$

and in this case the characteristic function of Σ equals

$$\mathbb{E}e^{it\Sigma} = \exp\Big[\int (e^{itf(x)} - 1)\, d\mu(x)\Big]. \tag{7.21}$$

Moreover,

$$\mathbb{E}\Sigma = \int f(x)\, d\mu(x), \tag{7.22}$$

where the expectation exists if and only if the integral converges. If $\mathbb{E}\Sigma$ *converges then*

$$\mathrm{Var}(\Sigma) = \int f(x)^2\, d\mu(x), \tag{7.23}$$

finite or infinite.

Proof. All the statements follow immediately from the decomposition $\Sigma = \Sigma_+ - \Sigma_-$ in (7.15) and the independence of Σ_+ and Σ_-. For (7.21), we use that the formula (7.17) for the Laplace transform, i.e.

$$\mathbb{E}e^{z\Sigma_\pm} = \exp\Big[\int (e^{zf^\pm(x)} - 1)\, d\mu(x)\Big]$$

for $z = -t$, implies the same formula for imaginary $z = it$, since both sides are analytic on $\mathfrak{Re}\, z < 0$. □

Exercise 7.2.1. If $f_1, f_2 \in L^1(S,\mu) \cap L^2(S,\mu)$, show that

$$\mathrm{Cov}(\Sigma_1, \Sigma_2) = \int f_1(x) f_2(x)\, d\mu(x),$$

where $\Sigma_j = \sum_{x\in\Pi} f_j(x)$. *Hint:* Consider the function $f = t_1 f_1 + t_2 f_2$.

Exercise 7.2.2. Let Π be a Poisson process on $(0,\infty)$ with the mean measure $\zeta x^{-1-\zeta} dx$ for $\zeta \in (0,1)$. Show that the Laplace transform of $\sum_{x\in\Pi} x$ is of the form $\exp(-ct^\zeta)$ and find the constant c.

Exercise 7.2.3. If $\zeta \in (0,1)$, Π is a Poisson process on $(0,1)$ with the mean measure $\mu(dx) = x^{-2-\zeta} dx$, and $S_m = [1/m, 1/(m-1))$ for $m \geq 2$, show that the series

$$\sum_{m\geq 2} \Big(\sum_{x\in\Pi\cap S_m} x - \int_{S_m} x\, d\mu(x) \Big)$$

converges almost surely.

7.3 Infinitely divisible distributions: basic properties

Definition 7.1. A distribution $\mathbb{P}$ on $\mathbb{R}$ is called *infinitely divisible* if, for any $n \geq 1$, there exists a distribution $\mathbb{P}_n$ such that $\mathbb{P}$ is the n-fold convolution of $\mathbb{P}_n$, $\mathbb{P} = \mathbb{P}_n^{*n}$.

In other words, if a random variable X has distribution $\mathbb{P}$ and $X_1, \ldots, X_n$ are independent random variables with the distribution $\mathbb{P}_n$ then

$$X \stackrel{d}{=} X_1 + \ldots + X_n.$$

The motivation for this definition is to understand the so called *Lévy processes*, which are stochastic processes $X(c)$ indexed by $c \geq 0$ with $X(0) = 0$, with independent and stationary increments, and continuous in probability. This means:

(a) *(Independence of increments)* for any $0 \leq c_1 < c_2 < \cdots < c_n < \infty$, the increments $X(c_1)$, $X(c_2) - X(c_1)$, $X(c_3) - X(c_2)$, $\ldots$, $X(c_n) - X(c_{n-1})$ are independent;

(b) *(Stationary increments)* for any $c' < c$, the increment $X(c) - X(c')$ is equal in distribution to $X(c - c')$;

(c) *(Continuity in probability)* for any $\varepsilon > 0$ and $c \geq 0$, $\lim_{h \to 0} P(|X(c+h) - X(c)| > \varepsilon) = 0$.

The first two properties imply that the distribution of $X(c)$ for any given c must be infinitely divisible because $X(c)$ is equal to the sum of i.i.d. increments $X((k+1)c/n) - X(kc/n)$ for $k = 0, \ldots, n-1$.

Example 7.3.1. Most basic examples of infinitely divisible distributions which we already know include Gaussian and Poisson distributions. Another example is the Cauchy(a) distribution for $a > 0$ with the density

$$p(x) = \frac{a}{\pi(x^2 + a^2)} \quad \text{for } x \in \mathbb{R}.$$

We will leave it as an exercise below to show that the characteristic function of this distribution is $f(t) = e^{-a|t|}$ which implies that the sum of independent Cauchy(a_i) random variables is Cauchy$(\sum a_i)$. Another example is the Gamma(α, β) distribution with parameters $\alpha > 0, \beta > 0$, with the density

$$p(x) = \frac{\beta^\alpha}{\Gamma(\alpha)} x^{\alpha-1} e^{-\beta x} \text{ for } x \geq 0.$$

Again, we will leave it as an exercise below to show that the characteristic function of this distribution is $f(t) = (1 - it/\beta)^{-\alpha}$ which implies that the sum of independent Gamma(α_i, β) random variables is Gamma$(\sum \alpha_i, \beta)$. □

Below we will give a general characterization of infinitely divisible distributions, but we begin with a few basic properties. First, we will show that characteristic functions of infinitely divisible distributions do not vanish and for this we will need the following lemma.

Lemma 7.1. *If* $f(t)=\mathbb{E}e^{itX}$ *then* $1-\mathfrak{Re}\,f(2t)\le 4(1-\mathfrak{Re}\,f(t))$. *Moreover,* $|f(t)|^2$ *is also a characteristic function and, therefore,* $1-|f(2t)|^2\le 4(1-|f(t)|^2)$.

Proof. First statement follows from:

$$\begin{aligned}1-\mathfrak{Re}\,f(2t)&=\int(1-\cos 2tx)\,d\mathbb{P}(x)=2\int\sin^2 tx\,d\mathbb{P}(x)\\&=2\int(1+\cos tx)(1-\cos tx)\,d\mathbb{P}(x)\\&\le 4\int(1-\cos tx)\,d\mathbb{P}(x)=4(1-\mathfrak{Re}\,f(t)).\end{aligned}$$

The second statement follows because $\overline{f(t)}=f(-t)$ is a characteristic function and, therefore, $|f(t)|^2=\overline{f(t)}f(t)$ is also a characteristic function. □

Theorem 7.8. *A characteristic function* $f(t)=\int e^{itx}\,d\mathbb{P}(x)$ *of an infinitely divisible distribution* $\mathbb{P}$ *does not vanish, i.e.* $f(t)\ne 0$ *for any* $t\in\mathbb{R}$.

Proof. Fix any $t\in\mathbb{R}$. Given $n\ge 1$, if $\mathbb{P}=\mathbb{P}_n^{*n}$ and f_n is the characteristic function of $\mathbb{P}_n$ then $f(t)=f_n(t)^n$. Applying the above lemma k times starting with $t/2^k$ instead of t, we can write

$$1-|f_n(t)|^2\le 4^k(1-|f_n(t/2^k)|^2)=4^k(1-|f(t/2^k)|^{2/n}).$$

Since $f(0)=1$ and $f(t)$ is continuous, we can choose k large enough such that $x=|f(t/2^k)|>0$. Since $x^{2/n}\to 1$ as $n\to\infty$, choosing n large enough we can make the right hand side above smaller than $1/2$, which implies that $|f_n(t)|^2>1/2$ and, therefore, $|f(t)|=|f_n(t)|^n\ne 0$. □

If $r(t)=|f(t)|\ne 0$, since $f(0)=1$, we can represent $f(t)=r(t)e^{i\theta(t)}$ for some unique continuous $\theta(t)$ such that $\theta(0)=0$. If $f(t)=f_n(t)^n$ for some continuous $f_n(t)$ with $f_n(0)=1$, because $r_n(t):=|f_n(t)|=r(t)^{1/n}\ne 0$, we can also represent $f_n(t)=r_n(t)e^{i\theta_n(t)}$ for some unique continuous $\theta_n(t)$ such that $\theta_n(0)=0$. Clearly this means that $\theta_n(t)=\theta(t)/n$. In other words, if $f(t)$ is a characteristic function of an infinitely divisible distribution, we can take $f_n(t)=r(t)^{1/n}e^{i\theta(t)/n}$.

Next, let us state without a proof the obvious result that follows immediately from the definition.

Theorem 7.9. *Convolution of infinitely divisible distributions is infinitely divisible.*

In other words, the sum of independent infinitely divisible random variables is infinitely divisible. In particular, if $f(t)$ is a characteristic function of an infinitely divisible distribution then so is $|f(t)|^2=f(-t)f(t)$.

Theorem 7.10. *If a distribution* $\mathbb{P}$ *is a limit of infinitely divisible distributions* $\mathbb{P}_k$ *then it is also infinitely divisible.*

Proof. Let f be the c.f. of $\mathbb{P}$, let f_k be the c.f. of $\mathbb{P}_k$ and, by infinite divisibility of $\mathbb{P}_k$, let $f_{k,n}$ be the c.f. such that $f_k(t) = f_{k,n}(t)^n$. The convergence $\mathbb{P}_k \to \mathbb{P}$ implies that $f_k(t) \to f(t)$ and, therefore,

$$\lim_{k\to\infty} |f_{k,n}(t)|^2 = \lim_{k\to\infty} |f_k(t)|^{2/n} = |f(t)|^{2/n}.$$

Since $|f(t)|^{2/n}$ is continuous and is a limit of characteristic functions $|f_{k,n}(t)|^2$, it is itself a characteristic function. This means that $|f(t)|^2$ is infinitely divisible and, therefore, it does not vanish and $f(t) \neq 0$. As we mentioned above, this means that we can represent $f(t) = r(t)e^{i\theta(t)}$ for some unique continuous $\theta(t)$ with $\theta(0) = 0$ and $f_k(t) = r_k(t)e^{i\theta_k(t)}$ for some unique continuous $\theta_k(t)$ with $\theta_k(0) = 0$. The fact that $r(t) \neq 0$ and $f_k(t) \to f(t)$ implies that

$$r_k(t) \to r(t) \text{ and } e^{i\theta_k(t)} \to e^{i\theta(t)}.$$

Because the pointwise convergence of characteristic functions implies uniform convergence on compacts (see Exercise 3.2.11), we must have that $\theta_k(t) \to \theta(t)$ for all t (exercise below). This implies that

$$\lim_{k\to\infty} \sqrt[n]{f_k(t)} = \lim_{k\to\infty} r_k(t)^{1/n} e^{i\theta_k(t)/n} = \lim_{k\to\infty} r(t)^{1/n} e^{i\theta(t)/n} = \sqrt[n]{f(t)}.$$

Therefore, as a continuous limit of characteristic functions, $\sqrt[n]{f(t)}$ is itself a characteristic function of some distribution $\mathbb{Q}_n$, which proves that $\mathbb{P} = \mathbb{Q}_n^{*n}$ and $\mathbb{P}$ is infinitely divisible. □

Theorem 7.11. *If $f(t)$ is a characteristic function of an infinitely divisible distribution then so is $f(t)^c$ for any $c > 0$.*

Proof. If $c = m/n \in \mathbb{Q}$ then, for any $k \geq 1$, $f(t)^{m/n} = (\varphi_k(t))^k$, where $\varphi_k(t) = (\sqrt[nk]{f(t)})^m$. Since $f(t)$ is infinitely divisible, $\sqrt[nk]{f(t)}$ is a characteristic function and its mth power, $\varphi_k(t)$, is also a characteristic function. This proves that $f(t)^{m/n}$ is infinitely divisible. Taking a limit $m/n \to c$ and using previous theorem implies this for all $c > 0$. □

Theorem 7.12. *(i) If $(X(c))_{c\geq 0}$ is a Lévy process and $f(t)$ is the characteristic function of $X(1)$ then the characteristic function of $X(c)$ is $f(t)^c$.*

(ii) If $f(t)$ is a characteristic function of an infinitely divisible distribution then there exists a Lévy process $(X(c))_{c\geq 0}$ such that $f(t)$ is the characteristic function of $X(1)$.

Proof. (i) By the properties (a) and (b) of the increments of a Lévy process,

$$\mathbb{E}e^{itX(m/n)} = (\mathbb{E}e^{itX(1/n)})^m = f(t)^{m/n}.$$

If $m/n \to c$ then $f(t)^{m/n} \to f(t)^c$ and, by continuity in probability (c) of the Lévy process, $X(m/n) \xrightarrow{d} X(c)$. This implies that the characteristic function of $X(c)$ is $f(t)^c$.

(ii) For any

$$0 \le c_1 < c_2 < \cdots < c_n < \infty,$$

let us define the distribution of $(X(c_1),\ldots,X(c_n))$ by the conditions that $X(0)=0$ and the increments $X(c_\ell)-X(c_{\ell-1})$ for $\ell=1,\ldots,n$ are independent and have characteristic functions $f(t)^{c_\ell-c_{\ell-1}}$. These distributions are consistent because, if we remove a point c_ℓ, the increment $X(c_{\ell+1})-X(c_{\ell-1})$ must have characteristic function $f(t)^{c_{\ell+1}-c_{\ell-1}}$ which equals $f(t)^{c_{\ell+1}-c_\ell}f(t)^{c_\ell-c_{\ell-1}}$ – the characteristic function of the sum of $X(c_{\ell+1})-X(c_\ell)$ and $X(c_\ell)-X(c_{\ell-1})$. By Kolmogorov's consistency theorem, we can define a family of random variables $(X(c))_{c\ge 0}$ with these finite dimensional distributions. The properties (a) and (b) of the Lévy process hold by construction. Since $f(t)^h \to 1$ as $h \downarrow 0$, the increment $X(c+h)-X(c)$ converges in distribution to 0, which implies the continuity in probability property (c). □

We proved that Lévy processes correspond to infinitely divisible distributions by specifying the distribution of $X(1)$, so we can now focus on studying infinitely divisible distributions.

Example 7.3.2. Let us give another example which is not obviously infinitely divisible and which follows from that above properties. Given $p \in (0,1)$, let us consider a *geometric distribution* $\mathbb{P}(\{n\}) = (1-p)p^n$ for integer $n=0,1,2,\ldots$. If p is the probability of failure, this is the distribution of the number of failures before the first success occurring with probability $1-p$. To see that this distribution is infinitely divisible, let us compute the characteristic function

$$f(t) = \sum_{n\ge 0} e^{itn}(1-p)p^n = \frac{1-p}{1-pe^{it}}.$$

Using the Taylor series for $\log(1-z)$ for $|z|<1$,

$$\log f(t) = \log(1-p) - \log(1-pe^{it}) = \sum_{k\ge 1}\frac{p^k}{k}(e^{ikt}-1).$$

Now, let us recall that if X has Poisson distribution with the mean $\lambda > 0$ then its characteristic function is

$$\mathbb{E}e^{itX} = \sum_{k\ge 0} e^{itk}\frac{\lambda^k}{k!}e^{-\lambda} = e^{-\lambda}\sum_{k\ge 0}\frac{(\lambda e^{it})^k}{k!} = \exp[\lambda(e^{it}-1)].$$

This means that $\exp[\lambda(e^{ict}-1)]$ for $c\in\mathbb{R}$ is the characteristic function of cX. If we denote by $\Pi(p^k/k)$ Poisson random variables with the mean p^k/k, independent for $k\ge 1$, then it is easy to check that the series $\sum_{k\ge 1} k\Pi(p^k/k)$ converges and its characteristic function is $f(t)$ defined above. In other words, this random series gives a different representation of a geometric random variable. With this representation it is clear that the distribution is infinitely divisible because the function

$$f_n(t) = \exp\Big[\sum_{k\geq 1} \frac{p^k}{nk}(e^{ikt}-1)\Big]$$

is the characteristic function of the series $\sum_{k\geq 1} k\Pi(p^k/(nk))$. □

Definition 7.2 (Compound Poisson distributions). If N has Poiss(λ) distribution and $(X_i)_{i\geq 1}$ are i.i.d. real-valued random variables with the distribution $\mathbb{P}$ then the distribution of $\sum_{i=1}^N X_i$ is called a compound Poisson distribution.

Let us compute the characteristic function of a compound Poisson. If $f(t) = \int e^{itx}\,d\mathbb{P}(x)$ is the characteristic function of X_i's then

$$\begin{aligned}\mathbb{E}e^{it\sum_{i=1}^N X_i} &= \mathbb{E}[\mathbb{E}e^{it\sum_{i=1}^N X_i} \mid N] = \mathbb{E}f(t)^N = e^{-\lambda}\sum_{k\geq 0}\frac{(\lambda f(t))^k}{k!}\\ &= \exp[\lambda(f(t)-1)] = \exp\Big[\lambda\int(e^{itx}-1)\,d\mathbb{P}(x)\Big]. \qquad (7.24)\end{aligned}$$

We will use this to show the following fundamental fact that will be used in the next section to derive a representation for general infinitely divisible distributions.

Theorem 7.13. *Infinitely divisible distributions are limits of compound Poisson.*

Proof. If $f(t)$ is a characteristic function of some infinitely divisible distribution and $f_n(t) = \sqrt[n]{f(t)}$ then

$$\begin{aligned}n(f_n(t)-1) &= n\big(e^{\log f(t)/n}-1\big)\\ &= n\Big(1+\frac{\log f(t)}{n}+o\Big(\frac{\log f(t)}{n}\Big)-1\Big) \to \log f(t)\end{aligned}$$

as $n\to\infty$. In other words, if $\mathbb{P}_n$ is the distribution with the characteristic function $f_n(t) = \int e^{itx}\,d\mathbb{P}_n(x)$ then

$$f(t) = \lim_{n\to\infty}\exp[n(f_n(t)-1)] = \lim_{n\to\infty}\exp\Big[n\int(e^{itx}-1)\,d\mathbb{P}_n(x)\Big] \qquad (7.25)$$

is the limit of characteristic functions of compound Poisson distributions corresponding to $\lambda = n$ and the distribution $\mathbb{P}_n$. This finishes the proof. □

In the above proof, if $f(t)$ is the characteristic function of the Lévy process $X(1)$ at time $c=1$ then

$$f(t)^c = \lim_{n\to\infty}\exp\Big[cn\int(e^{itx}-1)\,d\mathbb{P}_n(x)\Big] \qquad (7.26)$$

is the characteristic function of $X(c)$ for $c\geq 0$. For a fixed n, the right hand side corresponds to the following Lévy process $X_n(c)$. If Π is a Poisson process on

$(0,\infty)$ with the mean measure $n\,dx$ and $(Y_x)_{x\in\Pi}$ are i.i.d. markings of this process from the distribution $\mathbb{P}_n$ then

$$X_n(c) = \sum_{x\in\Pi} Y_x \mathrm{I}(x \le c).$$

It is obvious that its increments are stationary, independent, and $X(c) - X(c')$ has a compound Poisson distribution with $\lambda = n(c - c')$ and $\mathbb{P} = \mathbb{P}_n$. In other words, in the sense of finite dimensional distributions, all Lévy processes are limits of such jump processes, with i.i.d. jumps Y_x occurring at the locations x of a Poisson process Π.

Exercise 7.3.1. Prove that the characteristic function of Cauchy(a) distribution is $f(t) = e^{-a|t|}$. *Hint:* compute the Fourier transform of $e^{-a|t|}$ and use the inverse Fourier transform.

Exercise 7.3.2. Prove that the characteristic function of Gamma(α,β) distribution is $f(t) = (1 - it/\beta)^{-\alpha}$. *Hint:* make the change of variables and use a contour integral to reduce back to the integral of the density.

Exercise 7.3.3. In the proof of Theorem 7.10, show that $\theta_k(t) \to \theta(t)$ for all t.

Exercise 7.3.4. If Π is a Poisson process on $\mathbb{R}$ such that $\sum_{x\in\Pi} |x| < \infty$ a.s., is the distribution of $\sum_{x\in\Pi} x$ infinitely divisible?

7.4 Canonical representations of infinitely divisible distributions

In this section we will derive a general representation for characteristic functions of infinitely divisible distributions. We will begin with the case of nonnegative infinitely divisible random variables $X \geq 0$, which is slightly simpler but which illustrates most of the ideas of the general case. In this case, we will consider the Laplace transform

$$f(t) = \mathbb{E}e^{-tX} \text{ for } t \geq 0, \tag{7.27}$$

which determines the distribution of $X \geq 0$.

Theorem 7.14. *If $X \geq 0$ is infinitely divisible then*

$$f(t) = \exp -\Big[\beta t + \int_{(0,\infty)} (1-e^{-tx})\,d\mu(x)\Big], \tag{7.28}$$

where $\beta \geq 0$ and μ is a measure on $(0,\infty)$ such that $\int_{(0,\infty)}(x\wedge 1)\,d\mu(x) < \infty$. Such representation is unique, and all such $f(t)$ correspond to some infinitely divisible distributions on $\mathbb{R}^+$.

Proof. First of all, given such $f(t)$, the existence of an infinitely divisible distribution follows from Campbell's theorem because, if Π is a Poisson process on $(0,\infty)$ with the mean measure μ, then

$$X = \beta + \sum_{x\in\Pi} x$$

has the distribution $\mathbb{P}$ with Laplace transform $f(t)$. Notice that the condition $\int_{(0,\infty)}(x\wedge 1)\,d\mu(x) < \infty$ coincides with the one in Campbell's theorem and ensures that $X < \infty$ almost surely. For any $n \geq 1$, if Π_n is a Poisson process on $(0,\infty)$ with the mean measure μ/n then the distribution $\mathbb{P}_n$ of $\beta/n + \sum_{x\in\Pi_n} x$ satisfies $\mathbb{P} = \mathbb{P}_n^{*n}$, by the superposition property of Poisson processes, or by looking at the Laplace transforms. In other words, all $f(t)$ as in (7.28) are characteristic functions of a sum of a Poisson process on $(0,\infty)$ plus a constant.

Now, suppose that $X \geq 0$ is infinitely divisible, which means that, for all $n \geq 0$, there exists a distribution $\mathbb{P}_n$ on $\mathbb{R}^+ = [0,\infty)$ such that $\mathbb{P} = \mathbb{P}_n^{*n}$, which implies that

$$f(t) = \mathbb{E}e^{-tX} = f_n(t)^n, \text{ where } f_n(t) = \int_0^\infty e^{-tx}\,d\mathbb{P}_n(x).$$

As in (7.25) above, this implies that

$$f(t) = \lim_{n\to\infty} \exp\Big[-n\int_0^\infty (1-e^{-tx})\,d\mathbb{P}_n(x)\Big].$$

First of all, if we take $t = 1$ then this implies that, for large enough n,

$$n\int_0^\infty (1-e^{-x})\,d\mathbb{P}_n(x) \leq 1 - \log f(1).$$

One can check that $1-e^{-x} \geq (1-e^{-1})(x\wedge 1)$ and, therefore, for large enough n,

$$n\int_0^\infty (x\wedge 1)\,d\mathbb{P}_n(x) \leq \frac{1-\log f(1)}{1-e^{-1}} < \infty.$$

If we consider the measures G_n on $[0,\infty)$ defined by

$$dG_n(x) := n(x\wedge 1)\,d\mathbb{P}_n(x) \tag{7.29}$$

then $G_n(\mathbb{R}^+) \leq K < \infty$ for all n, i.e. these measures are uniformly bounded.

Next, we show that these measures are also uniformly tight. For $t\leq 1$, we can write

$$\begin{aligned}G_n([1/t,\infty)) = n\mathbb{P}_n([1/t,\infty)) &\leq \frac{n}{1-e^{-1}}\int_{1/t}^\infty (1-e^{-tx})\,d\mathbb{P}_n(x)\\ &\leq \frac{n}{1-e^{-1}}\int_0^\infty (1-e^{-tx})\,d\mathbb{P}_n(x) \to -\frac{\log f(t)}{1-e^{-1}}.\end{aligned}$$

Since $\lim_{t\downarrow 0} f(t) = \lim_{t\downarrow 0}\mathbb{E}e^{-tX} = 1$, by choosing t small enough we can make the right hand side above smaller than any $\varepsilon>0$ so, for large enough n, we have $G_n([1/t,\infty))\leq\varepsilon$. This implies that (G_n) is uniformly tight.

By the Selection Theorem, we can choose a subsequence $(n_k)_{k\geq 1}$ such that $G_{n_k}\to G$ weakly for some finite measure G on $\mathbb{R}^+$. On the one hand,

$$\int_0^\infty \frac{1-e^{-tx}}{x\wedge 1}\,dG_{n_k}(x) = n_k\int_0^\infty (1-e^{-tx})\,d\mathbb{P}_{n_k}(x) \to -\log f(t)$$

as $k\to\infty$. On the other hand, if we consider the function $h(t,x) = (1-e^{-tx})/x\wedge 1$ for $x>0$ and extend it by continuity to zero, $h(t,0)=t$, then the weak convergence $G_{n_k}\to G$ implies that

$$\int_0^\infty \frac{1-e^{-tx}}{x\wedge 1}\,dG_{n_k}(x) \to \int_0^\infty h(t,x)\,dG(x)$$

as $k\to\infty$, which shows that

$$f(t) = \exp\Big[-\int_0^\infty h(t,x)\,dG(x)\Big].$$

Notice that, although $G_n(\{0\})=0$, in the limit we might have $G(\{0\})=\beta\geq 0$, so

$$f(t) = \exp-\Big[\beta t + \int_{(0,\infty)} \frac{1-e^{-tx}}{x\wedge 1}\,dG(x)\Big].$$

Finally, if we consider the measure $d\mu(x) := dG(x)/(x\wedge 1)$ on $(0,\infty)$ then $\int_{(0,\infty)}(x\wedge 1)\,d\mu(x) = G((0,\infty)) < \infty$ and the representation (7.28) holds.

To prove the uniqueness of such representation, let us write

$$-\log f(t) = \beta t + \int_{(0,\infty)} (1 - e^{-tx})\, d\mu(x),$$

from which one can easily check that, for $t \geq 1$,

$$\begin{aligned}
-\log f(t) + \frac{1}{2}\int_{t-1}^{t+1} \log f(s)\, ds &= \int_{(0,\infty)} e^{-tx}\Big(\frac{\sinh(x)}{x} - 1\Big)\, d\mu(x) \\
&= \int_{(0,\infty)} e^{-(t-1)x} p(x)\, d\mu(x)
\end{aligned}$$

where $p(x) = e^{-x}(\sinh(x)/x - 1)$. Since $0 \leq p(x) \leq 1$ for $x > 0$ and $p(x) = \mathcal{O}(x^2)$ near the origin, the assumption $\int_{(0,\infty)} (x \wedge 1)\, d\mu(x) < \infty$ implies that

$$\int_{(0,\infty)} p(x)\, d\mu(x) < \infty.$$

Therefore, the measure ν defined by $d\nu(x) := p(x)\, d\mu(x)$ is a positive bounded measure on $(0,\infty)$ and the above representation for $f(t)$ uniquely determines

$$\int_{(0,\infty)} e^{-(t-1)x}\, d\nu(x) = \int_{(0,\infty)} e^{-(t-1)x} p(x)\, d\mu(x)$$

for $t \geq 1$. In other words, we know the Laplace transform of a finite measure ν on $(0,\infty)$, which uniquely determines ν and, therefore, μ. Once we know μ, we know β, which proves that the above representation is unique. □

Next, we will consider the case of general infinitely divisible random variables and derive the canonical *Lévy-Khintchine representation* of their characteristic functions.

Theorem 7.15 (Lévy-Khintchine representation). *If X is infinitely divisible then*

$$f(t) = \mathbb{E}e^{itX} = \exp\Big[it\gamma - \frac{t^2\sigma^2}{2} + \int_{\mathbb{R}\setminus\{0\}} \big(e^{itx} - 1 - itx\mathrm{I}(|x| \leq 1)\big)\, d\mu(x)\Big], \quad (7.30)$$

where $\gamma \in \mathbb{R}$, $\sigma \geq 0$ and μ is a measure on $\mathbb{R} \setminus \{0\}$ such that $\int_{\mathbb{R}\setminus\{0\}} (x^2 \wedge 1)\, d\mu(x) < \infty$. The representation is unique.

The measure μ is called *the Lévy measure* of X. We could choose it to be a measure on $\mathbb{R}$ and set $\mu(\{0\}) = 0$ in order to still guarantee uniqueness. Notice also that in $\mathrm{I}(|x| \leq 1)$ the constant 1 can be replaced by any other positive constant, which will simply result in transferring part of the integral into the constant γ. We will prove that all such $f(t)$ as in (7.30) correspond to some infinitely divisible distributions in the next theorem, in order to emphasize the role played by various terms in the representation. For symmetric infinitely divisible random variables, we will give another concrete representation for the random variables in the exercises below.

Proof. We saw in (7.25) that, if $\mathbb{P}_n$ is the distribution with the characteristic function $f_n(t) = \sqrt[n]{f(t)}$ then

$$f(t) = \lim_{n\to\infty} \exp\Big[n \int (e^{itx} - 1)\, d\mathbb{P}_n(x)\Big]$$

and, comparing the absolute values,

$$\lim_{n\to\infty} n \int (1 - \cos tx)\, d\mathbb{P}_n(x) = -\log|f(t)|.$$

Using this with $t = 1$ we get that

$$n \int_{|x|\le 1} (1 - \cos x)\, d\mathbb{P}_n(x) \le n \int (1 - \cos x)\, d\mathbb{P}_n(x) \le 1 - \log|f(1)|$$

for large enough n and, since $1 - \cos x \ge x^2/3$ for $|x| \le 1$, we get that

$$n \int_{|x|\le 1} x^2\, d\mathbb{P}_n(x) \le K < \infty. \tag{7.31}$$

Next, let us consider the average of the above limit over $t \in [0, \delta]$. Since

$$\frac{1}{\delta}\int_0^\delta \Big[n \int (1 - \cos tx)\, d\mathbb{P}_n(x)\Big] dt = n \int \Big(1 - \frac{\sin \delta x}{\delta x}\Big)\, d\mathbb{P}_n(x),$$

we have

$$\lim_{n\to\infty} n \int \Big(1 - \frac{\sin \delta x}{\delta x}\Big)\, d\mathbb{P}_n(x) = -\frac{1}{\delta}\int_0^\delta \log|f(t)|\, dt,$$

which is bounded because $|f(t)| \ne 0$ for any infinitely divisible distribution. Since $\sin x/x \le 1$ and $1 - \sin x/x \ge 1/2$ for $|x| \ge 1$,

$$n\mathbb{P}_n(|x| \ge 1/\delta) \le 2n \int_{|x|\ge 1/\delta} \Big(1 - \frac{\sin \delta x}{\delta x}\Big)\, d\mathbb{P}_n(x) \le -\frac{2}{\delta}\int_0^\delta \log|f(t)|\, dt + \varepsilon$$

for n large enough. First of all, for $\delta = 1$ this implies that $n\mathbb{P}_n(|x| \ge 1)$ is bounded by a constant for large enough n and, combining this with (7.31) proves that

$$n \int (x^2 \wedge 1)\, d\mathbb{P}_n(x) \le K' < \infty. \tag{7.32}$$

If we consider the measure G_n on $\mathbb{R}$ defined by

$$dG_n(x) := n(x^2 \wedge 1)\, d\mathbb{P}_n(x) \tag{7.33}$$

then $G_n(\mathbb{R}) \le K'' < \infty$ for all n, i.e. total masses of these measures are uniformly bounded. Next, since $|f(t)| \approx 1$ for small t, we can choose δ above small enough so that

$$G_n(|x| \ge 1/\delta) = n\mathbb{P}_n(|x| \ge 1/\delta) \le 2\varepsilon$$

for large enough n. This means that the sequence of measures (G_n) is uniformly tight.

By the Selection Theorem, we can choose a subsequence $(n_k)_{k\geq 1}$ such that $G_{n_k} \to G$ weakly for some finite measure G on $\mathbb{R}$. On the one hand,

$$\int (e^{itx}-1)\frac{1}{x^2\wedge 1}\,dG_{n_k}(x) = n_k \int (e^{itx}-1)\,d\mathbb{P}_{n_k}(x) \to \log f(t)$$

as $k\to\infty$. On the other hand, in contrast with the case of positive infinitely divisible random variables above, the function $(e^{itx}-1)/(x^2\wedge 1)$ blows up at $x=0$ and can not be extended by continuity to zero. To fix this, let us take any $c>0$ and rewrite

$$\int (e^{itx}-1)\frac{1}{x^2\wedge 1}\,dG_{n_k}(x) = \int \big(e^{itx}-1-itx\mathrm{I}(|x|\leq c)\big)\frac{1}{x^2\wedge 1}\,dG_{n_k}(x) + it\int x\mathrm{I}(|x|\leq c)\frac{1}{x^2\wedge 1}\,dG_{n_k}(x).$$

The function $h(x) = (e^{itx}-1-itx\mathrm{I}(|x|\leq c)/(x^2\wedge 1)$ can be extended by continuity to zero, $h(0) = -t^2/2$, in which case its only points of discontinuity are $x=-c,c$. However, if we choose $c>0$ such that $-c$ and c are points of continuity of the limiting measure G then the weak convergence $G_{n_k}\to G$ implies that

$$\int h(x)\,dG_{n_k}(x) \to \int h(x)\,dG(x)$$

as $k\to\infty$. Since the sum of the two integrals above also converges (to $\log f(t)$), this forces the second integral to converge,

$$it\int x\mathrm{I}(|x|\leq c)\frac{1}{x^2\wedge 1}\,dG_{n_k}(x) \to it\gamma_c,$$

for some $\gamma_c\in\mathbb{R}$. We proved that

$$f(t) = \exp\Big[it\gamma_c + \int \big(e^{itx}-1-itx\mathrm{I}(|x|\leq c)\big)\frac{1}{x^2\wedge 1}\,dG(x)\Big].$$

Changing the indicator $\mathrm{I}(|x|\leq c)$ to $\mathrm{I}(|x|\leq 1)$ will only result in changing the term $it\gamma_c$ to $it\gamma$ for some $\gamma\in\mathbb{R}$, so

$$f(t) = \exp\Big[it\gamma + \int \big(e^{itx}-1-itx\mathrm{I}(|x|\leq 1)\big)\frac{1}{x^2\wedge 1}\,dG(x)\Big].$$

As before, although $G_n(\{0\}) = 0$, in the limit we might have $G(\{0\}) = \sigma^2\geq 0$, so

$$f(t) = \exp\Big[it\gamma - \frac{t^2\sigma^2}{2} + \int_{\mathbb{R}\setminus\{0\}} \big(e^{itx}-1-itx\mathrm{I}(|x|\leq 1)\big)\frac{1}{x^2\wedge 1}\,dG(x)\Big].$$

Finally, if we consider the measure

$$d\mu(x) := \frac{dG(x)}{x^2 \wedge 1} \tag{7.34}$$

on $\mathbb{R}\setminus\{0\}$ then $\int_{\mathbb{R}\setminus\{0\}}(x^2\wedge 1)\,d\mu(x) = G(\mathbb{R}\setminus\{0\}) < \infty$ and the representation (7.30) holds.

To prove the uniqueness of such representation, one can easily check that

$$\log f(t) - \frac{1}{2}\int_{t-1}^{t+1}\log f(s)\,ds = \int e^{itx}\Big(1-\frac{\sin x}{x}\Big)\frac{1}{x^2\wedge 1}\,dG(x) = \int e^{itx}\,dF(x),$$

where we defined the measure F on $\mathbb{R}$ by

$$dF(x) = \Big(1-\frac{\sin x}{x}\Big)\frac{1}{x^2\wedge 1}\,dG(x).$$

One can see that the prefactor in front of $dG(x)$ is strictly positive (when extended by continuity to $1/6$ at $x=0$) and bounded, so F is a positive finite measure. Since its characteristic function $\int e^{itx}\,dF(x)$ determines F, we showed that $f(t)$ determines F, and therefore G, uniquely. Once we know G, we know $\sigma^2 = G(\{0\})$ and μ, which proves that the representation is unique. □

Theorem 7.16. *All functions $f(t)$ of the form (7.30) correspond to some infinitely divisible distributions.*

Proof. We will represent the random variable X as a sum of independent components, whose characteristics functions correspond to various terms in (7.30). The term $it\gamma$, of course, corresponds to a constant γ, and the term $-t^2\sigma^2/2$ corresponds to a Gaussian $N(0,\sigma^2)$ random variable. Next, if we break the integral in (7.30) into two integrals,

$$\int_{\mathbb{R}\setminus[-1,1]}(e^{itx}-1)\,d\mu(x) \text{ and } \int_I (e^{itx}-1-itx)\,d\mu(x), \tag{7.35}$$

where $I=[-1,1]\setminus\{0\}$. By (7.24), the first integral corresponds to the compound Poisson distribution with $\lambda = \mu(\mathbb{R}\setminus[-1,1])$ and

$$d\mathbb{P}(x) = \mathrm{I}(x\in\mathbb{R}\setminus[-1,1])\frac{d\mu(x)}{\mu(\mathbb{R}\setminus[-1,1])}.$$

All of these terms so far correspond to infinitely divisible distributions, and it remains to understand why the second integral in (7.35) is a characteristic function of some infinitely divisible distribution.

Let us decompose the measure μ on I into the sum $\mu_1+\mu_2$ of a purely atomic component μ_1 and continuous component μ_2. Let x_k for $k\geq 1$ be the atoms of μ with $\mu(\{x_k\}) = \Delta_k > 0$, enumerated in an arbitrary order. Then

$$\int_I (e^{itx}-1-itx)\,d\mu_1(x) = \exp\sum_{k\geq 1}\Big[(e^{itx_k}-1)\Delta_k - itx_k\Delta_k\Big] \tag{7.36}$$

is the characteristic function of the random series

$$\Sigma_1 = \sum_{k\geq 1} \big(x_k N_k - x_k \Delta_k\big),$$

where N_k are independent $\mathrm{Poiss}(\Delta_k)$. The series converges almost surely because

$$\mathbb{E}(x_k N_k - x_k \Delta_k) = 0,\ \mathrm{Var}\big(x_k N_k\big) = x_k^2 \Delta_k,$$

and $\sum_{k\geq 1} x_k^2 \Delta_k \leq \int_I x^2 d\mu < \infty$. It is clear that the series is infinitely divisible.

Next, let $I = \cup_{m\geq 1} S_m$ be a partition of $I = [-1,1]\setminus\{0\}$ into disjoint intervals with endpoints not equal to 0. Then we can write

$$\begin{aligned} \exp \int_I (e^{itx} - 1 - itx)\, d\mu_2(x) &= \exp \sum_{m\geq 1} \int_{S_m} (e^{itx} - 1 - itx)\, d\mu_2(x) \\ = \exp \sum_{m\geq 1} \Big(\int_{S_m} (e^{itx} - 1)\, d\mu_2(x) &- it \int_{S_m} x\, d\mu_2(x) \Big), \end{aligned} \tag{7.37}$$

because on each interval S_m (separated away from zero) the measure μ_2 is finite and, thus, both integrals are well defined. Let Π be a Poisson process with the mean measure μ_2 on I, and consider the random series

$$\Sigma_2 = \sum_{m\geq 1} \Big(\sum_{x\in\Pi\cap S_m} x - \int_{S_m} x\, d\mu_2(x) \Big).$$

By Campbell's theorem,

$$\mathbb{E} \sum_{x\in\Pi\cap S_m} x = \int_{S_m} x\, d\mu_2(x),\ \mathrm{Var}\Big(\sum_{x\in\Pi\cap S_m} x \Big) = \int_{S_m} x^2\, d\mu_2(x),$$

so the sum of variances is again bounded by $\int_I x^2 d\mu < \infty$ and, therefore, the above series Σ_2 converges almost surely. Again, using Campbell's theorem, one can see that its characteristic function equals (7.37). Since the series is obviously infinitely divisible, this finishes the proof. □

Next, we will show a simple corollary of the uniqueness of the Lévy-Khintchine respresentation that will be useful in one of the exercises below.

Lemma 7.2. *If a characteristic function $f(t)$ can be represented as*

$$f(t) = \mathbb{E} e^{itX} = \exp\Big[it\gamma - \frac{t^2\sigma^2}{2} + \int_{\mathbb{R}\setminus\{0\}} \big(e^{itx} - 1 - itx\mathrm{I}(|x|\leq 1)\big)\, d\mu(x) \Big], \tag{7.38}$$

where $\gamma\in\mathbb{R}$, $\sigma\geq 0$ and μ is a signed measure on $\mathbb{R}\setminus\{0\}$ such that $\mu = \mu_1 - \mu_2$ and μ_1, μ_2 are measures on $\mathbb{R}\setminus\{0\}$ such that $\int_{\mathbb{R}\setminus\{0\}} (x^2\wedge 1)\, d\mu_j(x) < \infty$, then such representation is unique.

This means that if $f(t)$ can be represented in such a way and μ is not a positive measure (i.e. the component μ_2 is non-trivial) then $f(t)$ is not infinitely divisible.

Proof. Suppose we have two such representations with parameters $(\gamma_1, \sigma_1^2, \mu_1^1, \mu_2^1)$ and $(\gamma_2, \sigma_2^2, \mu_1^2, \mu_2^2)$. Then, rearranging the terms,

$$\begin{aligned} &it\gamma_1 - \frac{t^2\sigma_1^2}{2} + \int_{\mathbb{R}\setminus\{0\}} \left(e^{itx} - 1 - itx\mathrm{I}(|x| \leq 1)\right) d(\mu_1^1 + \mu_2^2)(x) \\ &= it\gamma_2 - \frac{t^2\sigma_2^2}{2} + \int_{\mathbb{R}\setminus\{0\}} \left(e^{itx} - 1 - itx\mathrm{I}(|x| \leq 1)\right) d(\mu_1^2 + \mu_2^1)(x). \end{aligned}$$

Both sides are of the form corresponding to a characteristic function of an infinitely divisible distribution and we proved above that we must have $\gamma_1 = \gamma_2$, $\sigma_1^2 = \sigma_2^2$ and $\mu_1^1 + \mu_2^2 = \mu_1^2 + \mu_2^1$, or $\mu_1 = \mu_1^1 - \mu_2^1 = \mu_1^2 - \mu_2^2 = \mu_2$. □

Exercise 7.4.1. Consider $a, b \in (0,1)$ and the function

$$f(t) = \frac{1-b}{1+a} \cdot \frac{1+ae^{-it}}{1-be^{it}}.$$

Find the distribution for which $f(t)$ is a characteristic function, and show that it is not infinitely divisible. *Hint:* compute $\log f(t)$ and use Lemma 7.2.

Exercise 7.4.2. In the setting of the previous exercise, suppose that $a \leq b$ and show that $|f(t)|^2$ is an infinitely divisible characteristic function. (Together with the previous exercise this implies that the sum or difference of two non-infinitely divisible random variables can be infinitely divisible.)

Exercise 7.4.3. If in Theorem 7.16 the measure μ satisfies $\int_{\mathbb{R}} (|x| \wedge 1)\, d\mu(x) < \infty$, what is the random variable corresponding to such c.f. $f(t)$?

Exercise 7.4.4. Show that if X is an infinitely divisible random variable with c.f. (7.30) then, for $p > 0$, $\mathbb{E}|X|^p < \infty$ if and only if $\int_{\{|x|>1\}} |x|^p\, d\mu(x) < \infty$. *Hint:* Consider the decomposition in the proof of Theorem 7.16 and prove that the existence of pth moment is determined by the compound Poisson component corresponding to the first integral in (7.35), because the last component (call it Y) corresponding to the second integral in (7.35) has the moment generating function

$$\mathbb{E}e^{zY} = \exp \int_{[-1,1]} (e^{zx} - 1 - zx)\, d\mu(x)$$

for all $z \in \mathbb{C}$.

Exercise 7.4.5. Suppose that X is an infinitely divisible random variable with the c.f. (7.30) with $\sigma^2 = 0$ (i.e. without the Gaussian component) and suppose that the distribution of X is symmetric. Show that, in this case,

$$f(t) = \mathbb{E}e^{itX} = \exp \int_{\mathbb{R}} (\cos(tx) - 1)\, d\mu(x). \tag{7.39}$$

Exercise 7.4.6 (Rosinski's representation). Let λ be the Lebesgue measure on $(0,\infty)$ and $\mathbb{P}$ be some probability measure on $\mathbb{R}$. Consider a measurable function $G:(0,\infty)\times\mathbb{R}\to\mathbb{R}$ and suppose that the image measure $\mu=(\lambda\times\mathbb{P})\circ G^{-1}$ satisfies $\int_{\mathbb{R}}(x^2\wedge 1)\,d\mu(x)<\infty$. Suppose that $(X_n)_{n\geq 1}$ is an *increasing enumeration* of the Poisson process with the mean measure λ on $(0,\infty)$, $(Y_n)_{n\geq 1}$ are i.i.d. random variables with the distribution $\mathbb{P}$, and $(\varepsilon_n)_{n\geq 1}$ are i.i.d. symmetric ± 1 valued random variables, and suppose all these sequences are independent of each other. Show that the series

$$S=\sum_{n\geq 1}\varepsilon_n G(X_n,Y_n)$$

converges almost surely, and S is a symmetric infinitely divisible random variable with the c.f. (7.39).

Hint: Consider a Poisson process (X_n,Y_n,ε_n) and use Campbell's theorem to compute the c.f. of S. Be careful that we do not assume absolute convergence of the above series while in Campbell's theorem we do. In other words, the condition $\int_{\mathbb{R}}(x^2\wedge 1)\,d\mu(x)<\infty$ is too weak to apply Campbell's theorem directly, so restrict the Poisson process X_n to a large interval $[0,T]$ first.

7.5 α-stable distributions and Lévy processes

Definition 7.3. An *α-stable distribution*, for $\alpha \in (0,2)$, is an infinitely divisible distribution that has a characteristic function of the form

$$f(t) = \exp\Big[it\gamma + \int_{\mathbb{R}} \big(e^{itx} - 1 - itx\mathrm{I}(|x| \le 1)\big)p(x)\,dx\Big], \tag{7.40}$$

where $\gamma \in \mathbb{R}$ and where

$$p(x) = \big(m_1\mathrm{I}(x<0) + m_2\mathrm{I}(x>0)\big)\frac{1}{|x|^{1+\alpha}}$$

for arbitrary $m_1, m_2 \ge 0$. □

Comparing with the general representation (7.30), $\sigma^2 = 0$ and $d\mu(x) = p(x)dx$. Observe that the condition $\int (x^2 \wedge 1)\,d\mu(x) < \infty$ is equivalent to $0 < \alpha < 2$ if at least one of m_1, m_2 is strictly positive. If $\gamma = 0$ and $m_1 = m_2 = m > 0$, the distribution is called *symmetric α-stable*, in which case

$$\begin{aligned}\int \big(e^{itx} - 1 - itx\mathrm{I}(|x| \le 1)\big)p(x)\,dx &= \int \big(e^{itx} - 1 - itx\mathrm{I}(|x| \le 1)\big)\frac{m}{|x|^{1+\alpha}}\,dx \\ = \int (\cos tx - 1)\frac{m}{|x|^{1+\alpha}}\,dx &= |t|^\alpha \int (\cos y - 1)\frac{m}{|y|^{1+\alpha}}\,dy = -c|t|^\alpha,\end{aligned}$$

where we made the change of variables $y = tx$ and where $c > 0$. Thus, the characteristic function of a symmetric α-stable distributions is of the form $\exp(-c|t|^\alpha)$. One can similarly explicitly express the characteristic function in the non-symmetric case as well.

The following properly expresses the stability property of α-stable distributions.

Lemma 7.3. *If $X, X_1, \ldots, X_n$ are i.i.d. α-stable random variables then*

$$X_1 + \ldots + X_n \stackrel{d}{=} n^{1/\alpha}X + K_n \tag{7.41}$$

for some constant K_n that depends on n and the parameters of the distribution. If $\gamma = 0$ and $m_1 = m_2$ then $K_n = 0$.

Proof. If $f(t)$ in (7.40) is the c.f. of X then $f(tc)$ is the c.f. of cX, which is equal to the exponential of

$$\begin{aligned}& itc\gamma + \int \big(e^{itcx} - 1 - itcx\mathrm{I}(|x| \le 1)\big)p(x)\,dx \\ &\stackrel{\{y=cx\}}{=} itc\gamma + |c|^\alpha \int \big(e^{ity} - 1 - ity\mathrm{I}(|y| \le c)\big)p(y)\,dy \\ &= itL + |c|^\alpha \int \big(e^{ity} - 1 - ity\mathrm{I}(|y| \le 1)\big)p(y)\,dy,\end{aligned}$$

where the constant

$$L = c\gamma + |c|^{\alpha} \int y\big(\mathrm{I}(|y| \le 1) - \mathrm{I}(|y| \le c)\big) p(y)\, dy.$$

If $\gamma = 0$ and $m_1 = m_2$ then $L = 0$, since we integrate an odd function. On the other hand, the c.f. of $X_1 + \ldots + X_n$ is $f(t)^n$, which is the exponential of

$$itn\gamma + n \int \big(e^{itx} - 1 - itx\mathrm{I}(|x| \le 1)\big) p(x)\, dx.$$

Therefore, if we take $c = n^{1/\alpha}$ and $K_n = n\gamma - L$, we see that the characteristic functions coincide, which finishes the proof. □

Next, we will gives a representation of symmetric α-stable random variables and corresponding Lévy processes via the so called α-stable subordinators.

Definition 7.4. If Π is a Poisson process on $(0,\infty)^2$ with the mean measure $d\mu(x)dy$ for some measure μ such that $\int_0^\infty \min(x,1)\, d\mu(x) < \infty$, then the process

$$\tau(t) = \sum_{(x,y)\in\Pi} x\mathrm{I}(y \le t) \quad \text{for } t \ge 0 \tag{7.42}$$

is called a *subordinator*. □

By Campbell's theorem, this process is well-defined because

$$\iint \min(x\mathrm{I}(y \le t), 1)\, d\mu(x)\, dy = t \int_0^\infty \min(x,1)\, d\mu(x) < \infty.$$

The process $\tau(t)$ is non-decreasing and right-continuous by our choice of a non-strict inequality $y \le t$ in the definition.

Also, notice that the increments of this process are independent, because they are defined in terms of the points in our Poisson process on disjoint subsets of $(0,\infty)^2$, and also stationary, since the Lebesgue measure dy is translation invariant, so $\tau(t)$ is a non-negative Lévy process. In fact, *any non-negative Lévy process* is of this form because any non-negative infinitely divisible random variable is a sum over a Poisson process on $(0,\infty)$.

To visualize how $\tau(t)$ looks like, let us restrict ourselves to $t \in [0,1]$. Since dy is a probability measure on $[0,1]$, we can first consider a Poisson process π on $(0,\infty)$ with the mean measure μ and then for each $x \in \pi$ take a marking y_x uniformly on $[0,1]$ independently for all x. By Marking theorem, the resulting collection of points (x, y_x) is equivalent to the Poisson process Π restricted to $(0,\infty) \times [0,1]$. The set of markings y_x is either finite, if $\mu(0,\infty) < \infty$, or countably infinite, if $\mu(0,\infty) = \infty$, in which case it is a countable dense subset of $[0,1]$ with probability one. The process $\tau(t) = \sum_{x\in\pi} x\mathrm{I}(y_x \le t)$ has jumps of size x at times $t = y_x$ in this finite or countable dense collection.

Definition 7.5. If $\zeta \in (0,1)$ then a *ζ-stable subordinator* is a subordinator corresponding to the measure $d\mu(x) = cx^{-1-\zeta}dx$ for a constant $c > 0$. □

In this case, by Campbell's theorem, the Laplace transform of $\tau(t)$ is

$$\begin{aligned}\mathbb{E}e^{-\lambda\tau(t)} &= \exp\Big[-\int_0^\infty\int_0^\infty(1-e^{-\lambda x\mathrm{I}(y\le t)})\,cx^{-1-\zeta}dy\,dx\Big]\\ &= \exp\Big[-ct\int_0^\infty(1-e^{-\lambda x})x^{-1-\zeta}\,dx\Big]\\ &= \exp\Big[-ct\lambda^\zeta\int_0^\infty(1-e^{-x})x^{-1-\zeta}\,dx\Big] = \exp(-ca(\zeta)t\lambda^\zeta),\end{aligned}$$

where $a(\zeta)=\int_0^\infty(1-e^{-x})x^{-1-\zeta}\,dx$. The fact that this Laplace transform resembles the characteristic function of a symmetric α-stable distribution allows us to represent such random variables via a time change of the Brownian motion by this subordinator. If $(W_t)_{t\ge 0}$ is a Brownian motion independent of $\tau(t)$ then $(W_{\tau(t)})_{t\ge 0}$ is a stationary process with independent symmetric α-stable increments with $\alpha=2\zeta$. Indeed,

$$\mathbb{E}e^{ixW_{\tau(t)}} = \mathbb{E}\mathbb{E}(e^{ixW_{\tau(t)}}\mid\tau(t)) = \mathbb{E}e^{-x^2\tau(t)/2} = \exp\Big(-\frac{ca(\zeta)t}{2^\zeta}|x|^{2\zeta}\Big),$$

so for any fixed t, $W_{\tau(t)}$ has an α-stable distribution with $\alpha=2\zeta$. We leave the rest to the following exercise.

Exercise 7.5.1. Show that $W_{\tau(t)}$ has independent stationary α-stable increments with $\alpha=2\zeta$.

Exercise 7.5.2. If Π is a Poisson process on $(0,\infty)^2$ with the mean measure $cx^{-1-\zeta}dxdy$ and, for $(x,y)\in\Pi$, $g_{(x,y)}\sim N(0,x)$ are independent Gaussian markings with variance x, show that the process

$$X(t)=\sum_{(x,y)\in\Pi} g_{(x,y)}\mathrm{I}(y\le t)$$

is the same Levy process as $(W_{\tau(t)})_{t\ge 0}$ above.

Exercise 7.5.3. Let $\tau(t)$ be a ζ-stable subordinator. Given a nonnegative Borel function $f\ge 0$ on $(0,\infty)$, let

$$Z(t)=\int_0^t f(s)\,d\tau(s)=\sum_{(x,y)\in\Pi} xf(y)\mathrm{I}(y\le t).$$

Prove that $Z(t)<\infty$ a.s. for all $t\ge 0$ if and only if $\int_0^t f(y)^\zeta\,dy<\infty$ for all $t\ge 0$.

Exercise 7.5.4. In the setting of the previous exercise, assuming that $\int_0^t f(y)^\zeta\,dy<\infty$ for all $t\ge 0$, compute the Laplace transform of the increments of the process $Z(t)$ and describe the relationship between $Z(t)$ and $\tau(t)$.

Chapter 8
Exchangeability

8.1 Moment problem and de Finetti's theorem for coin flips

Let us start by recalling an example in Section 5. Let $(X_i)_{i\geq 1}$ be i.i.d. with the Bernoulli distribution and the probability of success $\theta \in [0,1]$, i.e. $\mathbb{P}_\theta(X_i = 1) = \theta$ and $\mathbb{P}_\theta(X_i = 0) = 1-\theta$, and let $u\colon [0,1] \to \mathbb{R}$ be some continuous function on $[0,1]$. Then, by Theorem 2.6 in Section 5, the *Bernstein polynomials*

$$B_n(\theta) := \sum_{k=0}^{n} u\Big(\frac{k}{n}\Big)\mathbb{P}_\theta\Big(\sum_{i=1}^{n} X_i = k\Big) = \sum_{k=0}^{n} u\Big(\frac{k}{n}\Big)\binom{n}{k}\theta^k(1-\theta)^{n-k}$$

approximate $u(\theta)$ uniformly on $[0,1]$.

Moment problem. Consider a random variable X taking values in $[0,1]$ and let $\mu_k = \mathbb{E}X^k$ be its moments. Given a sequence $(c_0, c_1, c_2, \ldots)$, let us define the sequence of increments by $\Delta c_k = c_{k+1} - c_k$. Then

$$-\Delta\mu_k = \mu_k - \mu_{k+1} = \mathbb{E}(X^k - X^{k+1}) = \mathbb{E}X^k(1-X),$$

$$(-\Delta)(-\Delta\mu_k) = (-\Delta)^2\mu_k = \mathbb{E}X^k(1-X) - \mathbb{E}X^{k+1}(1-X) = \mathbb{E}X^k(1-X)^2$$

and, by induction,

$$(-\Delta)^r\mu_k = \mathbb{E}X^k(1-X)^r.$$

Clearly, $(-\Delta)^r\mu_k \geq 0$ since $X \in [0,1]$. If u is a continuous function on $[0,1]$ and B_n is the corresponding Bernstein polynomial defined above then

$$\mathbb{E}B_n(X) = \sum_{k=0}^{n} u\Big(\frac{k}{n}\Big)\binom{n}{k}\mathbb{E}X^k(1-X)^{n-k} = \sum_{k=0}^{n} u\Big(\frac{k}{n}\Big)\binom{n}{k}(-\Delta)^{n-k}\mu_k.$$

Since B_n converges uniformly to u, $\mathbb{E}B_n(X)$ converges to $\mathbb{E}u(X)$. Let us define

$$p_k^{(n)} = \binom{n}{k}(-\Delta)^{n-k}\mu_k \geq 0. \tag{8.1}$$

By taking $u \equiv 1$, it is easy to see that $\sum_{k=0}^{n} p_k^{(n)} = 1$, so we can think of $p_k^{(n)}$ as the distribution

$$\mathbb{P}\Big(X^{(n)} = \frac{k}{n}\Big) = p_k^{(n)} \tag{8.2}$$

of some random variable $X^{(n)}$. We showed that $\mathbb{E}B_n(X) = \mathbb{E}u(X^{(n)}) \to \mathbb{E}u(X)$ for any continuous function u, which means that $X^{(n)}$ converges to X in distribution. In other words, given finitely many moments of X, this construction gives an explicit approximation of the distribution of X.

If we want to express the moment μ_k for a fixed k in terms of the distribution (8.2), we can write for, $n \geq k$,

$$\begin{aligned}
\mu_k = \mathbb{E}X^k &= \mathbb{E}X^k\Big(X + 1 - X\Big)^{n-k} \\
&= \sum_{j=0}^{n-k} \binom{n-k}{j} \mathbb{E}X^{k+j}(1-X)^{n-(k+j)} \\
&= \sum_{m=k}^{n} \binom{n-k}{m-k} \mathbb{E}X^{m}(1-X)^{n-m} \\
&= \sum_{m=k}^{n} \binom{n-k}{m-k} (-\Delta)^{n-m}\mu_m = \sum_{m=k}^{n} \binom{n-k}{m-k} \binom{n}{m}^{-1} p_m^{(n)}.
\end{aligned} \tag{8.3}$$

Let us note that the part of this formula which is expressed only in terms of the sequence $(\mu_k)_{k\geq 0}$,

$$\mu_k = \sum_{m=k}^{n} \binom{n-k}{m-k} (-\Delta)^{n-m}\mu_m, \tag{8.4}$$

holds for any sequence $(\mu_k)_{k\geq 0}$ and not necessarily a sequence of moments of some random variables $X \in [0,1]$. For example, for $n = k+1$ this formula (called inversion formula) says that $\mu_k = \mu_{k+1} - \Delta\mu_k$, and it can be proved by induction for all $n \geq k$ (exercise).

Next, given a sequence (μ_k), we consider the following question: When is (μ_k) the sequence of moments of some $[0,1]$ valued random variable X? By the above, it is necessary that

$$\mu_k \geq 0, \mu_0 = 1 \text{ and } (-1)^r \Delta^r \mu_k \geq 0 \text{ for all } k, r. \tag{8.5}$$

It turns out that this is also sufficient.

Theorem 8.1 (Hausdorff). *There exists a r.v.* $X \in [0,1]$ *such that* $\mu_k = \mathbb{E}X^k$ *if and only if (8.5) holds.*

Proof. Given the sequence (μ_k), let us define $p_k^{(n)}$ by (8.1). Using the inversion formula (8.4) for $k = 0$, we get

$$\sum_{m=0}^{n} p_m^{(n)} = \mu_0 = 1$$

and, by the assumption (8.5), $p_m^{(n)} \geq 0$. Notice that, by (8.3), for any fixed k,

$$\mu_k = \sum_{m=k}^{n} \binom{n-k}{m-k} \binom{n}{m}^{-1} p_m^{(n)}$$
$$= \sum_{m=k}^{n} \frac{m(m-1)\cdots(m-k+1)}{n(n-1)\cdots(n-k+1)} p_m^{(n)} \approx \sum_{m=0}^{n} \left(\frac{m}{n}\right)^k p_m^{(n)}$$

as $n \to \infty$. If we think of $p^{(n)}$ as the probability distribution on $[0,1]$ that assigns weight $p_m^{(n)}$ to m/n then the right hand side is the kth moment of $p^{(n)}$. By the Selection Theorem, $p^{(n)}$ converges to some distribution on $[0,1]$ along some subsequence, and μ_k will obviously be the kth moment of this limiting distribution, which finishes the proof. Notice also that, because the moments determine the distribution on $[0,1]$ uniquely, the entire sequence $p^{(n)}$ converges. □

de Finetti's theorem. As a consequence of the Hausdorff theorem above, we will now prove the classical de Finetti representation for coin flips. The general case will be considered in the next section. A sequence $(X_n)_{n\geq 1}$ of $\{0,1\}$-valued random variables is called *exchangeable* if, for any $n \geq 1$ and any $x_1, \ldots, x_n \in \{0,1\}$, the probability

$$\mathbb{P}(X_1 = x_1, \ldots, X_n = x_n)$$

depends only on the 'number of successes' $x_1 + \ldots + x_n$ and does not depend on the order of 1's or 0's. Another way to say this is that, for any $n \geq 1$ and any permutation π of $\{1, \ldots, n\}$, the distribution of $(X_{\pi(1)}, \ldots, X_{\pi(n)})$ does not depend on π. Then the following holds.

Theorem 8.2 (de Finetti). *There exists a distribution F on $[0,1]$ such that, for any $n \geq 1$ and any $x_1, \ldots, x_n \in \{0,1\}$,*

$$\mathbb{P}(X_1 = x_1, \ldots, X_n = x_n) = \int_0^1 p^k (1-p)^{n-k} \, dF(p),$$

where $k = x_1 + \ldots + x_n$.

In other words, in order to generate such an exchangeable sequence of 0's and 1's, we first pick $p \in [0,1]$ from some distribution F and then generate a sequence of i.i.d Bernoulli random variables with probability of success x.

Proof. Let $\mu_0 = 1$ and, for $k \geq 1$, define

$$\mu_k = \mathbb{P}(X_1 = 1, \ldots, X_k = 1). \tag{8.6}$$

We have

$$\mathbb{P}(X_1 = 1, \ldots, X_k = 1, X_{k+1} = 0) = \mathbb{P}(X_1 = 1, \ldots, X_k = 1)$$
$$- \mathbb{P}(X_1 = 1, \ldots, X_k = 1, X_{k+1} = 1) = \mu_k - \mu_{k+1} = -\Delta\mu_k.$$

Next, using exchangeability,

$$\mathbb{P}(X_1=1,\ldots,X_k=1,X_{k+1}=0,X_{k+2}=0)=\mathbb{P}(X_1=1,\ldots,X_k=1,X_{k+1}=0)$$
$$-\mathbb{P}(X_1=1,\ldots,X_k=1,X_{k+1}=0,X_{k+2}=1)=-\Delta\mu_k-(-\Delta\mu_{k+1})=(-\Delta)^2\mu_k.$$

Similarly, by induction,

$$\mathbb{P}(X_1=1,\ldots,X_k=1,X_{k+1}=0,\ldots,X_n=0)=(-\Delta)^{n-k}\mu_k\geq 0.$$

By the Hausdorff theorem, $\mu_k=\mathbb{E}X^k$ for some r.v. $X\in[0,1]$ and, therefore,

$$\mathbb{P}(X_1=1,\ldots,X_k=1,X_{k+1}=0,\ldots,X_n=0)=(-\Delta)^{n-k}\mu_k$$
$$=\mathbb{E}X^k(1-X)^{n-k}=\int_0^1 p^k(1-p)^{n-k}\,dF(p).$$

Since, by exchangeability, changing the order of 1's and 0's does not affect the probability, this finishes the proof. □

Example 8.1.1 (Polya's urn model). Suppose we have b blue and r red balls in the urn. We pick a ball randomly and return it with c balls of the same color. Consider the random variable

$$X_i=\begin{cases}1, & \text{if the } i\text{th ball picked is blue}\\ 0, & \text{otherwise.}\end{cases}$$

X_i's are not independent but it is easy to check that they are exchangeable. For example,

$$\mathbb{P}(bbr)=\frac{b}{b+r}\times\frac{b+c}{b+r+c}\times\frac{r}{b+r+2c}$$
$$=\frac{b}{b+r}\times\frac{r}{b+r+c}\times\frac{b+c}{b+r+2c}=\mathbb{P}(brb).$$

To identify the distribution F in de Finetti's theorem, let us look at its moments μ_k in (8.6),

$$\mu_k=\mathbb{P}(\underbrace{b\ldots b}_{k\text{ times}})=\frac{b}{b+r}\times\frac{b+c}{b+r+c}\times\cdots\times\frac{b+(k-1)c}{b+r+(k-1)c}.$$

One can recognize or check that (μ_k) are the moments of Beta(α,β) distribution with the density

$$\frac{\Gamma(\alpha+\beta)}{\Gamma(\alpha)\Gamma(\beta)}x^{\alpha-1}(1-x)^{\beta-1}\mathrm{I}(0\leq x\leq 1)$$

with the parameters $\alpha=b/c,\beta=r/c$. By de Finetti's theorem, we can generate X_i's by first picking p from the distribution Beta$(b/c,r/c)$ and then generating i.i.d. Bernoulli (X_i)'s with the probability of success p. By the strong law of large numbers, applied conditionally on p, the proportion of blue balls picked in the first n trials

$$\bar{p}_n = \frac{1}{n}(X_1 + \ldots + X_n)$$

will converge to this probability of success p, i.e. in the limit it will be random with the Beta distribution. Recall that this example came up in the exercises in Section 15 on the convergence of martingales, where one showed that

$$Y_n = \frac{b + (X_1 + \ldots + X_n)c}{b + r + nc} = \frac{b + \bar{p}_n nc}{b + r + nc}$$

converges almost surely because it is a bounded martingale. Since this limits coincides with the limit of $\bar{p}_n$, this means that we identified its distribution. □

Exercise 8.1.1. Prove the formula (8.4).

Exercise 8.1.2. Compute the moments of the Beta(α, β) distribution.

Exercise 8.1.3. In the setting of de Finetti's theorem above, prove that the limit $\lim_{n\to\infty} n^{-1} \sum_{i=1}^{n} X_i$ exists almost surely and, conditionally on this limit, the sequence $(X_i)_{i\geq 1}$ is i.i.d..

8.2 General de Finetti and Aldous-Hoover representations

We will begin by proving the classical de Finetti representation for exchangeable sequences. A sequence $(s_\ell)_{\ell\geq 1}$ is called *exchangeable* if for any permutation π of finitely many indices we have equality in distribution

$$\big(s_{\pi(\ell)}\big)_{\ell\geq 1} \stackrel{d}{=} \big(s_\ell\big)_{\ell\geq 1}. \tag{8.7}$$

The following holds.

Theorem 8.3 (de Finetti). *If the sequence $(s_\ell)_{\ell\geq 1}$ is exchangeable then there exists a measurable function $g:[0,1]^2\to\mathbb{R}$ such that*

$$\big(s_\ell\big)_{\ell\geq 1} \stackrel{d}{=} \big(g(w,u_\ell)\big)_{\ell\geq 1}, \tag{8.8}$$

where w and (u_ℓ) are i.i.d. random variables with the uniform distribution on $[0,1]$.

Before we proceed to prove the above results, let us recall the following definition.

Definition 8.1. A measurable space $(\Omega,\mathscr{B})$ is called a *Borel space* if there exists a one-to-one function φ from Ω onto a Borel subset $A\subseteq[0,1]$ such that both φ and φ^{-1} are measurable. □

Perhaps, the most important example of Borel spaces are complete separable metric spaces and their Borel subsets (see e.g. Section 13.1 in R.M. Dudley "Real Analysis and Probability"). The existence of the isomorphism φ automatically implies that if we can prove the above results in the case when the elements of the sequence (s_ℓ) or array $(s_{\ell,\ell'})$ take values in $[0,1]$ then the same representation results hold when the elements take values in a Borel space. Similarly, other standard results for real-valued or $[0,1]$-valued random variables are often automatically extended to Borel spaces. For example, one can generate any real-valued random variable as a measurable function of a uniform random variable on $[0,1]$ using the quantile transform and, therefore, any random element on a Borel space can also be generated by a function of a uniform random variable on $[0,1]$.

Let us describe another typical measure theoretic argument that will be used many times below.

Lemma 8.1 (Coding Lemma). *Suppose that a random pair (X,Y) takes values in the product of a measurable space $(\Omega_1,\mathscr{B}_1)$ and a Borel space $(\Omega_2,\mathscr{B}_2)$. Then, there exists a measurable function $f:\Omega_1\times[0,1]\to\Omega_2$ such that*

$$(X,Y) \stackrel{d}{=} \big(X,f(X,u)\big), \tag{8.9}$$

where u is a uniform random variable on $[0,1]$ independent of X.

Another way to state this is to say that, conditionally on X, Y can be generated as a function $Y=f(X,u)$ of X and an independent uniform random variable u on

$[0,1]$. Rather than using the Coding Lemma itself we will often use the general ideas of conditioning and generating random variables as functions of uniform random variables on $[0,1]$ that are used in its proof, without repeating the same argument.

Proof. By the definition of a Borel space, one can easily reduce the general case to the case when $\Omega_2 = [0,1]$ equipped with the Borel σ-algebra. Since in this case the regular conditional distribution $\Pr(x,B)$ of Y given X exists, for a fixed $x \in \Omega_1$, we can define by $F(x,y) = \Pr(x,[0,y])$ the conditional distribution function of Y given x and by

$$f(x,u) = F^{-1}(x,u) = \inf\{s \in [0,1] : u \le F(x,s)\}$$

its quantile transformation. It is easy to see that f is measurable on the product space $\Omega_1 \times [0,1]$ because, for any $t \in \mathbb{R}$,

$$\{(x,u) : f(x,u) \le t\} = \{(x,u) : u \le F(x,t)\}$$

and, by the definition of the regular conditional probability, $F(x,t) = \Pr(x,[0,t])$ is measurable in x for a fixed t. If u is uniform on $[0,1]$ then, for a fixed $x \in \Omega_1$, $f(x,u)$ has the distribution $\Pr(x,\cdot)$ and this finishes the proof. □

The most important way in which the exchangeability condition (8.7) will be used is to say that for any infinite subset $I \subseteq \mathbb{N}$,

$$(s_\ell)_{\ell \in I} \stackrel{d}{=} (s_\ell)_{\ell \ge 1}. \tag{8.10}$$

Let us describe one immediate consequence of this simple observation. If

$$\mathscr{F}_I = \sigma(s_\ell : \ell \in I) \tag{8.11}$$

is the σ-algebra generated by the random variables s_ℓ for $\ell \in I$ then the following holds.

Lemma 8.2. *For any infinite subset $I \subseteq \mathbb{N}$ and $j \notin I$, the conditional expectations*

$$\mathbb{E}\big(f(s_j)\big|\mathscr{F}_I\big) = \mathbb{E}\big(f(s_j)\big|\mathscr{F}_{\mathbb{N}\setminus\{j\}}\big)$$

almost surely, for any bounded measurable function $f : \mathbb{R} \to \mathbb{R}$.

Proof. First, using the property (8.10) for $I \cup \{j\}$ instead of I implies the equality in distribution (see Exercise 1.4.3 in Section 1.4),

$$\mathbb{E}\big(f(s_j)\big|\mathscr{F}_I\big) \stackrel{d}{=} \mathbb{E}\big(f(s_j)\big|\mathscr{F}_{\mathbb{N}\setminus\{j\}}\big),$$

and, therefore, the equality of the L^2-norms,

$$\big\|\mathbb{E}\big(f(s_j)\big|\mathscr{F}_I\big)\big\|_2 = \big\|\mathbb{E}\big(f(s_j)\big|\mathscr{F}_{\mathbb{N}\setminus\{j\}}\big)\big\|_2. \tag{8.12}$$

On the other hand, $I \subseteq \mathbb{N} \setminus \{j\}$ and $\mathscr{F}_I \subseteq \mathscr{F}_{\mathbb{N}\setminus\{j\}}$, which implies that

$$\mathbb{E}\big(\mathbb{E}(f(s_j)|\mathscr{F}_I)\mathbb{E}(f(s_j)|\mathscr{F}_{\mathbb{N}\setminus\{j\}})\big) = \mathbb{E}\big(\mathbb{E}(f(s_j)|\mathscr{F}_I)^2\big) = \big\|\mathbb{E}(f(s_j)|\mathscr{F}_I)\big\|_2^2.$$

Combining this with (8.12) yields that

$$\big\|\mathbb{E}(f(s_j)|\mathscr{F}_I) - \mathbb{E}(f(s_j)|\mathscr{F}_{\mathbb{N}\setminus\{j\}})\big\|_2^2 = 0, \tag{8.13}$$

which finishes the proof. □

Proof (of Theorem 8.3). Let us take any infinite subset $I \subseteq \mathbb{N}$ such that its complement $\mathbb{N}\setminus I$ is also infinite. By (8.10), we only need to prove the representation (8.8) for $(s_\ell)_{\ell\in I}$. First, we will show that, conditionally on $(s_\ell)_{\ell\in\mathbb{N}\setminus I}$, the random variables $(s_\ell)_{\ell\in I}$ are independent. This means that, given $n \geq 1$, any distinct $\ell_1,\ldots,\ell_n \in I$, and any bounded measurable functions $f_1,\ldots,f_n : \mathbb{R}\to\mathbb{R}$,

$$\mathbb{E}\Big(\prod_{j\leq n} f_j(s_{\ell_j})\Big|\mathscr{F}_{\mathbb{N}\setminus I}\Big) = \prod_{j\leq n}\mathbb{E}\big(f_j(s_{\ell_j})\big|\mathscr{F}_{\mathbb{N}\setminus I}\big), \tag{8.14}$$

or, in other words, for any $A \in \mathscr{F}_{\mathbb{N}\setminus I}$,

$$\mathbb{E}I_A\prod_{j\leq n} f_j(s_{\ell_j}) = \mathbb{E}I_A\prod_{j\leq n}\mathbb{E}\big(f_j(s_{\ell_j})\big|\mathscr{F}_{\mathbb{N}\setminus I}\big).$$

Since I_A and $f_1(s_{\ell_1}),\ldots,f_{n-1}(s_{\ell_{n-1}})$ are $\mathscr{F}_{\mathbb{N}\setminus\{\ell_n\}}$-measurable, we can write

$$\begin{aligned}\mathbb{E}I_A\prod_{j\leq n} f_j(s_{\ell_j}) &= \mathbb{E}I_A\prod_{j\leq n-1} f_j(s_{\ell_j})\mathbb{E}\big(f_n(s_{\ell_n})\big|\mathscr{F}_{\mathbb{N}\setminus\{\ell_n\}}\big)\\ &= \mathbb{E}I_A\prod_{j\leq n-1} f_j(s_{\ell_j})\mathbb{E}\big(f_n(s_{\ell_n})\big|\mathscr{F}_{\mathbb{N}\setminus I}\big),\end{aligned}$$

where the second equality follows from Lemma 8.2. This implies that

$$\mathbb{E}\Big(\prod_{j\leq n} f_j(s_{\ell_j})\Big|\mathscr{F}_{\mathbb{N}\setminus I}\Big) = \mathbb{E}\Big(\prod_{j\leq n-1} f_j(s_{\ell_j})\Big|\mathscr{F}_{\mathbb{N}\setminus I}\Big)\mathbb{E}\big(f_n(s_{\ell_n})\big|\mathscr{F}_{\mathbb{N}\setminus I}\big)$$

and (8.14) follows by induction on n. Let us also observe that, because of (8.10), the distribution of the array $(s_\ell,(s_j)_{j\in\mathbb{N}\setminus I})$ does not depend on $\ell\in I$. This implies that, conditionally on $(s_j)_{j\in\mathbb{N}\setminus I}$, the random variables $(s_\ell)_{\ell\in I}$ are identically distributed in addition to being independent. The product space $[0,1]^\infty$ with the Borel σ-algebra generated by the product topology is a Borel space (recall that, equipped with the usual metric, it becomes a complete separable metric space). Therefore, in order to conclude the proof, it remains to generate $X = (s_\ell)_{\ell\in\mathbb{N}\setminus I}$ as a function of a uniform random variable w on $[0,1]$, $X = X(w)$, and then use the argument in the Coding Lemma 8.1 to generate the sequence $(s_\ell)_{\ell\in I}$ as $(f(X(w),u_\ell))_{\ell\in I}$, where $(u_\ell)_{\ell\in I}$ are i.i.d. random variables uniform on $[0,1]$. This finishes the proof. □

Comments about de Finetti's theorem. Consider an exchangeable sequence $(s_\ell)_{\ell\geq 1}$ taking values in some complete separable space Ω. It is equal in distribution to $(g(w,u_\ell))_{\ell\geq 1}$. For a fixed w, this is an i.i.d. sequence from the distribution

$$\mu_w = \lambda \circ g(w,\cdot)^{-1},$$

where λ is the Lebesgue measure on $[0,1]$. By the strong law of large numbers for empirical measures (Varadarajan's Theorem 4.5 in Section 4.2),

$$\mu_w = \lim_{n\to\infty} \frac{1}{n}\sum_{\ell=1}^{n} \delta_{g(w,u_\ell)}.$$

The limit on the right hand side is taken on the complete separable metric space $\mathscr{P}(\Omega,\mathscr{B})$ of probability measures on Ω equipped, for example, with the bounded Lipschitz metric β or Levy-Prokhorov metric ρ, and this limit exists almost surely over $(u_\ell)_{\ell\geq 1}$. This implies that

(i) the limit $\mu = \lim_{n\to\infty}\frac{1}{n}\sum_{\ell=1}^{n}\delta_{s_\ell}$ exists almost surely;
(ii) μ is a (random) probability measure on Ω, i.e. a random element in $\mathscr{P}(\Omega,\mathscr{B})$;
(iii) given μ, the sequence $(s_\ell)_{\ell\geq 1}$ is i.i.d. from with the distribution μ.

The measure μ is called the *empirical measure* of the sequence $(s_\ell)_{\ell\geq 1}$. One can now interpret de Finetti's representation as follows. First, we generate μ as a function of a uniform random variable w on [0,1] and then generate s_ℓ as a function of μ and i.i.d. random variables u_ℓ uniform on $[0,1]$, using the Coding Lemma. Combining two steps, we generate s_ℓ as a function of w and u_ℓ.

Remark 8.1. There are two equivalent definitions of a random probability measure on a complete separable metric space Ω with the Borel σ-algebra $\mathscr{B}$. On the one hand, as above, these are just random elements taking values in the space $\mathscr{P}(\Omega,\mathscr{B})$ of probability measures on $(\Omega,\mathscr{B})$ equipped with the topology of weak convergence or a metric that metrizes weak convergence. On the other hand, we can think of a random measure η as a probability kernel, i.e. as a function $\eta = \eta(x,A)$ of a generic point $x\in X$ for some probability space $(X,\mathscr{F},\Pr)$ and a measurable set $A\in\mathscr{B}$ such that for a fixed x, $\eta(x,\cdot)$ is a probability measure and for a fixed A, $\eta(\cdot,A)$ is a measurable function on $(X,\mathscr{F})$. It is well known that these two definitions coincide (see, e.g. Lemma 1.37 and Theorem A2.3 in Kallenberg's "Foundations of Modern Probability").

Example 8.2.1 (de Finetti's representation for $\{0,1\}$-valued sequences). Coming back to the example considered in the previous section, consider an exchangeable sequence $(s_\ell)_{\ell\geq 1}$ of random variables that take values $\{0,1\}$. Then the empirical measure is a random probability measure on $\{0,1\}$. It is encoded by one (random) parameter $p = \mu(\{1\})\in[0,1]$. To fix μ means to fix p and, given p, the sequence $(s_\ell)_{\ell\geq 1}$ is i.i.d. Bernoulli with the probability of success equal to p. If η is the distribution of p then, for any $n\geq 1$ and any $\varepsilon_1,\ldots,\varepsilon_n\in\{0,1\}$,

$$\mathbb{P}(s_1 = \varepsilon_1, \ldots, s_n = \varepsilon_n) = \int_{[0,1]} p^k (1-p)^{n-k} \, d\eta(p),$$

where $k = \varepsilon_1 + \ldots + \varepsilon_n$. □

Another piece of information that can be deduced from de Finetti's theorem and is often useful is the following. Suppose that in addition to the sequence $(s_\ell)_{\ell \geq 1}$ we are given another random element Z taking values in a complete separable metric space, and their joint distribution is not affected by permutations of the sequence,

$$\big(Z, (s_\ell)_{\ell \geq 1}\big) \stackrel{d}{=} \big(Z, \big(s_{\pi(\ell)}\big)_{\ell \geq 1}\big). \tag{8.15}$$

Suppose that μ is the empirical measure of $(s_\ell)_{\ell \geq 1}$.

Theorem 8.4. *Conditionally on μ, the sequence $(s_\ell)_{\ell \geq 1}$ is i.i.d. with the distribution μ and independent of Z.*

For example, one can deduce Theorem 8.7 in the exercise below from this statement rather directly.

Proof (of Theorem 8.4). If we define $t_\ell = (Z, s_\ell)$ then (8.15) implies that $(t_\ell)_{\ell \geq 1}$ is exchangeable. Let

$$\nu = \lim_{n \to \infty} \frac{1}{n} \sum_{\ell=1}^{n} \delta_{t_\ell} = \lim_{n \to \infty} \frac{1}{n} \sum_{\ell=1}^{n} \delta_{(Z, s_\ell)}$$

be the empirical measure of this sequence. Obviously, $\nu = \delta_Z \times \mu$. Since, given ν, the sequence (Z, s_ℓ) is i.i.d. from ν, this means that, given ν, the sequence (s_ℓ) is i.i.d. from μ. In other words, given Z and μ, the sequence (s_ℓ) is i.i.d. from μ. The statement then follows from the following simple lemma, which appears as one of the exercises in Section 1.4. □

Lemma 8.3. *Suppose that three random elements X, Y and Z take values on some complete separable metric spaces. The conditional distribution of X given the pair (Y, Z) depends only on Y ($\sigma(Y)$-measurable) if and only if X and Z are independent conditionally on Y.*

The Aldous-Hoover representation. We will now prove two analogues of de Finetti's theorem for two-dimensional arrays. Let us consider an infinite random array $s = (s_{\ell,\ell'})_{\ell,\ell' \geq 1}$. The array s is called an *exchangeable array* if for any permutations π and ρ of finitely many indices we have equality in distribution,

$$\big(s_{\pi(\ell),\rho(\ell')}\big)_{\ell,\ell' \geq 1} \stackrel{d}{=} \big(s_{\ell,\ell'}\big)_{\ell,\ell' \geq 1}, \tag{8.16}$$

in the sense that their finite dimensional distributions are equal. Here is one natural example of an exchangeable array. Given a measurable function $\sigma : [0,1]^4 \to \mathbb{R}$ and sequences of i.i.d. random variables w, (u_ℓ), $(v_{\ell'})$, $(x_{\ell,\ell'})$ that have the uniform distribution on $[0,1]$, the array

$$\big(\sigma(w, u_\ell, v_{\ell'}, x_{\ell,\ell'})\big)_{\ell,\ell' \geq 1} \tag{8.17}$$

is, obviously, exchangeable. It turns out that all exchangeable arrays are of this form.

Theorem 8.5 (Aldous-Hoover). *Any infinite exchangeable array $(s_{\ell,\ell'})_{\ell,\ell'\geq 1}$ is equal in distribution to (8.17) for some function σ.*

Another version of the Aldous-Hoover representation holds in the symmetric case. A symmetric array s is called *weakly exchangeable* if for any permutation π of finitely many indices we have equality in distribution

$$\big(s_{\pi(\ell),\pi(\ell')}\big)_{\ell,\ell'\geq 1} \stackrel{d}{=} \big(s_{\ell,\ell'}\big)_{\ell,\ell'\geq 1}. \tag{8.18}$$

Because of the symmetry, we can consider only half of the array indexed by (ℓ,ℓ') such that $\ell \leq \ell'$ or, equivalently, consider the array indexed by sets $\{\ell,\ell'\}$ and rewrite (8.18) as

$$\big(s_{\{\pi(\ell),\pi(\ell')\}}\big)_{\ell,\ell'\geq 1} \stackrel{d}{=} \big(s_{\{\ell,\ell'\}}\big)_{\ell,\ell'\geq 1}. \tag{8.19}$$

Notice that, compared to (8.16), the diagonal elements now play somewhat different roles from the rest of the array. One natural example of a weakly exchangeable array is given by

$$s_{\{\ell,\ell\}} = g(w,u_\ell) \text{ and } s_{\{\ell,\ell'\}} = f\big(w,u_\ell,u_{\ell'},x_{\{\ell,\ell'\}}\big) \text{ for } \ell \neq \ell', \tag{8.20}$$

for any measurable functions $g : [0,1]^2 \to \mathbb{R}$ and $f : [0,1]^4 \to \mathbb{R}$, which is symmetric in its middle two coordinates $u_\ell, u_{\ell'}$, and i.i.d. random variables w, (u_ℓ), $(x_{\{\ell,\ell'\}})$ with the uniform distribution on $[0,1]$. Again, it turns out that such examples cover all possible weakly exchangeable arrays.

Theorem 8.6 (Aldous-Hoover). *Any infinite weakly exchangeable array is equal in distribution to the array (8.20) for some functions g and f, which is symmetric in its two middle coordinates.*

After we give a proof of Theorem 8.6, we will leave a similar proof of Theorem 8.5 as an exercise (with some hints). We will then give another, quite different, proof of Theorem 8.5. The proof of the Dovbysh-Sudakov representation in the next section will be based on Theorem 8.5.

The most important way in which the exchangeability condition (8.18) will be used is to say that, for any infinite subset $I \subseteq \mathbb{N}$,

$$\big(s_{\{\ell,\ell'\}}\big)_{\ell,\ell'\in I} \stackrel{d}{=} \big(s_{\{\ell,\ell'\}}\big)_{\ell,\ell'\geq 1}. \tag{8.21}$$

Again, one important consequence of this observation will be the following. Given $j, j' \in I$ such that $j \neq j'$, let us now define the σ-algebra

$$\mathscr{F}_I(j,j') = \sigma\big(s_{\{\ell,\ell'\}} : \ell,\ell' \in I, \{\ell,\ell'\} \neq \{j,j'\}\big). \tag{8.22}$$

In other words, this σ-algebra is generated by all elements $s_{\{\ell,\ell'\}}$ with both indices ℓ and ℓ' in I, excluding $s_{\{j,j'\}}$. The following analogue of Lemma 8.2 holds.

Lemma 8.4. *For any infinite subset $I \subseteq \mathbb{N}$ and any $j, j' \in I$ such that $j \neq j'$, the conditional expectations*

$$\mathbb{E}\big(f(s_{\{j,j'\}})\big|\mathscr{F}_I(j,j')\big) = \mathbb{E}\big(f(s_{\{j,j'\}})\big|\mathscr{F}_{\mathbb{N}}(j,j')\big)$$

almost surely, for any bounded measurable function $f:\mathbb{R}\to\mathbb{R}$.

Proof. The proof is almost identical to the proof of Lemma 8.2. The property (8.21) implies the equality in distribution,

$$\mathbb{E}\big(f(s_{\{j,j'\}})\big|\mathscr{F}_I(j,j')\big) \stackrel{d}{=} \mathbb{E}\big(f(s_{\{j,j'\}})\big|\mathscr{F}_{\mathbb{N}}(j,j')\big),$$

and, therefore, the equality of the L^2-norms,

$$\big\|\mathbb{E}\big(f(s_{\{j,j'\}})\big|\mathscr{F}_I(j,j')\big)\big\|_2 = \big\|\mathbb{E}\big(f(s_{\{j,j'\}})\big|\mathscr{F}_{\mathbb{N}}(j,j')\big)\big\|_2.$$

On the other hand, since $\mathscr{F}_I(j,j') \subseteq \mathscr{F}_{\mathbb{N}}(j,j')$, as in (8.13), this implies that

$$\big\|\mathbb{E}\big(f(s_{\{j,j'\}})\big|\mathscr{F}_I(j,j')\big) - \mathbb{E}\big(f(s_{\{j,j'\}})\big|\mathscr{F}_{\mathbb{N}}(j,j')\big)\big\|_2^2 = 0,$$

which finishes the proof. □

Proof (of Theorem 8.6). Let us take an infinite subset $I \subseteq \mathbb{N}$ such that its complement $\mathbb{N}\setminus I$ is also infinite. By (8.21), we only need to prove the representation (8.20) for $(s_{\{\ell,\ell'\}})_{\ell,\ell'\in I}$. For each $j \in I$, let us consider the array

$$S_j = \big(s_{\{\ell,\ell'\}}\big)_{\ell,\ell'\in(\mathbb{N}\setminus I)\cup\{j\}} = \Big(s_{\{j,j\}}, \big(s_{\{j,\ell\}}\big)_{\ell\in\mathbb{N}\setminus I}, \big(s_{\{\ell,\ell'\}}\big)_{\ell,\ell'\in\mathbb{N}\setminus I}\Big). \tag{8.23}$$

It is obvious that the weak exchangeability (8.19) implies that the sequence $(S_j)_{j\in I}$ is exchangeable, since any permutation π of finitely many indices from I in the array (8.19) results in the corresponding permutation of the sequence $(S_j)_{j\in I}$. We can view each array S_j as an element of the Borel space $[0,1]^\infty$ and de Finetti's Theorem 8.3 implies that

$$\big(S_j\big)_{j\in I} \stackrel{d}{=} \big(X(w,u_j)\big)_{j\in I} \tag{8.24}$$

for some measurable function X on $[0,1]^2$ taking values in the space of such arrays. Next, we will show that, conditionally on the sequence $(S_j)_{j\in I}$, the off-diagonal elements $s_{\{j,j'\}}$ for $j,j'\in I, j\neq j'$, are independent and, moreover, the conditional distribution of $s_{\{j,j'\}}$ depends only on S_j and $S_{j'}$. This means that if we consider the σ-algebras

$$\mathscr{F} = \sigma\big((S_j)_{j\in I}\big),\ \mathscr{F}_{j,j'} = \sigma(S_j,S_{j'}) \text{ for } j,j'\in I, j\neq j', \tag{8.25}$$

then we would like to show that, for any finite set $\mathscr{C}$ of indices $\{\ell,\ell'\}$ for $\ell,\ell'\in I$ such that $\ell\neq\ell'$ and any bounded measurable functions $f_{\ell,\ell'}$ corresponding to the indices $\{\ell,\ell'\}\in\mathscr{C}$, we have

$$\mathbb{E}\Big(\prod_{\{\ell,\ell'\}\in\mathscr{C}} f_{\ell,\ell'}(s_{\{\ell,\ell'\}})\Big|\mathscr{F}\Big)=\prod_{\{\ell,\ell'\}\in\mathscr{C}}\mathbb{E}\big(f_{\ell,\ell'}(s_{\{\ell,\ell'\}})\big|\mathscr{F}_{\ell,\ell'}\big). \tag{8.26}$$

Notice that the definitions (8.22) and (8.25) imply that $\mathscr{F}_{j,j'}=\mathscr{F}_{(\mathbb{N}\setminus I)\cup\{j,j'\}}(j,j')$, since all the elements $s_{\{\ell,\ell'\}}$ with both indices in $(\mathbb{N}\setminus I)\cup\{j,j'\}$, except for $s_{\{j,j'\}}$, appear as one of the coordinates in the arrays S_j or $S_{j'}$. Therefore, by Lemma 8.4,

$$\mathbb{E}\big(f(s_{\{j,j'\}})\big|\mathscr{F}_{j,j'}\big)=\mathbb{E}\big(f(s_{\{j,j'\}})\big|\mathscr{F}_{\mathbb{N}}(j,j')\big). \tag{8.27}$$

Let us fix any $\{j,j'\}\in\mathscr{C}$, let $\mathscr{C}'=\mathscr{C}\setminus\{\{j,j'\}\}$ and consider an arbitrary set $A\in\mathscr{F}$. Since I_A and $f_{\ell,\ell'}(s_{\{\ell,\ell'\}})$ for $\{\ell,\ell'\}\in\mathscr{C}'$ are $\mathscr{F}_{\mathbb{N}}(j,j')$-measurable,

$$\begin{aligned}
&\mathbb{E}I_A\prod_{\{\ell,\ell'\}\in\mathscr{C}'} f_{\ell,\ell'}(s_{\{\ell,\ell'\}})f_{j,j'}(s_{\{j,j'\}})\\
&=\mathbb{E}\Big(I_A\prod_{\{\ell,\ell'\}\in\mathscr{C}'} f_{\ell,\ell'}(s_{\{\ell,\ell'\}})\mathbb{E}\big(f_{j,j'}(s_{\{j,j'\}})\big|\mathscr{F}_{\mathbb{N}}(j,j')\big)\Big)\\
&=\mathbb{E}\Big(I_A\prod_{\{\ell,\ell'\}\in\mathscr{C}'} f_{\ell,\ell'}(s_{\{\ell,\ell'\}})\mathbb{E}\big(f_{j,j'}(s_{\{j,j'\}})\big|\mathscr{F}_{j,j'}\big)\Big),
\end{aligned}$$

where the second equality follows from (8.27). Since $\mathscr{F}_{j,j'}\subseteq\mathscr{F}$, this implies that

$$\mathbb{E}\Big(\prod_{\{\ell,\ell'\}\in\mathscr{C}} f_{\ell,\ell'}(s_{\{\ell,\ell'\}})\Big|\mathscr{F}\Big)=\mathbb{E}\Big(\prod_{\{\ell,\ell'\}\in\mathscr{C}'} f_{\ell,\ell'}(s_{\{\ell,\ell'\}})\Big|\mathscr{F}\Big)\mathbb{E}\big(f_{j,j'}(s_{\{j,j'\}})\big|\mathscr{F}_{j,j'}\big)$$

and (8.26) follows by induction on the cardinality of $\mathscr{C}$. By the argument in the Coding Lemma 8.1, (8.26) implies that, conditionally on $\mathscr{F}$, we can generate

$$\big(s_{\{j,j'\}}\big)_{j\neq j'\in I}\stackrel{d}{=}\big(h(S_j,S_{j'},x_{\{j,j'\}})\big)_{j\neq j'\in I}, \tag{8.28}$$

for some measurable function h and i.i.d. uniform random variables $x_{\{j,j'\}}$ on $[0,1]$. The reason why the function h can be chosen to be the same for all $\{j,j'\}$ is because, by symmetry, the distribution of $(S_j,S_{j'},s_{\{j,j'\}})$ does not depend on $\{j,j'\}$ and the conditional distribution of $s_{\{j,j'\}}$ given S_j and $S_{j'}$ does not depend on $\{j,j'\}$. Also, the arrays $(S_j,S_{j'},s_{\{j,j'\}})$ and $(S_{j'},S_j,s_{\{j,j'\}})$ are equal in distribution and, therefore, the function h is symmetric in the coordinates S_j and $S_{j'}$. Finally, let us recall (8.24) and define the function

$$f\big(w,u_j,u_{j'},x_{\{j,j'\}}\big)=h\big(X(w,u_j),X(w,u_{j'}),x_{\{j,j'\}}\big),$$

which is, obviously, symmetric in u_j and $u_{j'}$. Then, the equations (8.24) and (8.28) imply that

$$\Big((S_j)_{j\in I}, (s_{\{j,j'\}})_{j\neq j'\in I}\Big) \stackrel{d}{=} \Big(\big(X(w,u_j)\big)_{j\in I}, \big(f(w,u_j,u_{j'},x_{\{j,j'\}})\big)_{j\neq j'\in I}\Big).$$

In particular, if we denote by g the first coordinate of the map X corresponding to the element $s_{\{j,j\}}$ in the array S_j in (8.23), this proves that

$$\Big((s_{\{j,j\}})_{j\in I}, (s_{\{j,j'\}})_{j\neq j'\in I}\Big) \stackrel{d}{=} \Big(\big(g(w,u_j)\big)_{j\in I}, \big(f(w,u_j,u_{j'},x_{\{j,j'\}})\big)_{j\neq j'\in I}\Big).$$

Recalling (8.21) finishes the proof of the representation (8.20). □

Proof of Theorem 8.5. One proof of Theorem 8.5 similar to the proof of Theorem 8.6 is sketched in an exercise below. We will now give a different proof, and the main part of the proof will be based on the following observation. Suppose that we have an exchangeable sequence of pairs $(t_\ell, s_\ell)_{\ell\geq 1}$ with coordinates in complete separable metric spaces. Given the sequence $(t_\ell)_{\ell\geq 1}$, how can we generate the sequence of second coordinates $(s_\ell)_{\ell\geq 1}$? Consider the empirical measures

$$\mu = \lim_{n\to\infty} \frac{1}{n}\sum_{\ell=1}^n \delta_{(t_\ell,s_\ell)},\ \mu_1 = \lim_{n\to\infty} \frac{1}{n}\sum_{\ell=1}^n \delta_{t_\ell}.$$

Obviously, μ_1 is the marginal of μ on the first coordinate and, moreover, μ_1 is a measurable function of the sequence $(t_\ell)_{\ell\geq 1}$, so if we are given this sequence we automatically know μ_1. The following holds.

Lemma 8.5. *Given $(t_\ell)_{\ell\geq 1}$, we can generate the sequence $(s_\ell)_{\ell\geq 1}$ in distribution as*

$$(s_\ell)_{\ell\geq 1} \stackrel{d}{=} \big(f(\mu_1, t_\ell, v, x_\ell)\big)_{\ell\geq 1}$$

for some measurable function f and i.i.d. uniform random variables v and $(x_\ell)_{\ell\geq 1}$ on $[0,1]$.

Proof. First, let us note how to generate the empirical measure μ given the sequence $t = (t_\ell)_{\ell\geq 1}$. Given μ, the sequence $(t_\ell, s_\ell)_{\ell\geq 1}$ is i.i.d. from μ and, since μ_1 is the first marginal of μ, $(t_\ell)_{\ell\geq 1}$ are i.i.d. from μ_1. This means that, if we consider the triple (μ, μ_1, t) then the conditional distribution of t given (μ, μ_1) depends only on μ_1,

$$\mathbb{P}(t\in\cdot \mid \mu,\mu_1) = \mathbb{P}(t\in\cdot \mid \mu_1).$$

By Lemma 8.3, t and μ are independent given μ_1. Therefore, again by Lemma 8.3,

$$\mathbb{P}(\mu\in\cdot \mid t,\mu_1) = \mathbb{P}(\mu\in\cdot \mid \mu_1).$$

On the other hand, μ_1 is a function of t, so $\mathbb{P}(\mu\in\cdot \mid t,\mu_1) = \mathbb{P}(\mu\in\cdot \mid t)$ and, thus,

$$\mathbb{P}(\mu\in\cdot \mid t) = \mathbb{P}(\mu\in\cdot \mid \mu_1).$$

In other words, to generate μ given t, we can simply compute μ_1 and generate μ given μ_1. By the Coding Lemma, we can generate $\mu = g(\mu_1, v)$ as a function of μ_1 and independent uniform random variable v on $[0,1]$.

Now, recall that, given μ, the sequence $(t_\ell, s_\ell)_{\ell\geq 1}$ is i.i.d. with the distribution μ so, given t_ℓ and μ, we can simply generate s_ℓ from the conditional distribution $\mu(s_\ell \in \cdot \mid t_\ell)$ independently from each other. Again, using the Coding Lemma, we can generate $s_\ell = h(\mu, t_\ell, x_\ell)$ as a function of i.i.d. x_ℓ uniform random variables on $[0,1]$. Finally, recalling that $\mu = g(\mu_1, v)$, we can write

$$s_\ell = h(g(\mu_1, v), t_\ell, x_\ell) = f(\mu_1, t_\ell, v, x_\ell).$$

This finishes the proof. □

Proof (of Theorem 8.5). Let us for convenience index the array $s_{\ell,\ell'}$ by $\ell \geq 1$ and $\ell' \in \mathbb{Z}$ instead of $\ell' \geq 1$. Let us denote by

$$X_{\ell'} = \big(s_{\ell,\ell'}\big)_{\ell\geq 1} \text{ and } X = \big(X_{\ell'}\big)_{\ell'\leq 0}$$

the ℓ'-th column and 'left half' of this array. Since the sequence of columns $(X_{\ell'})_{\ell'\in\mathbb{Z}}$ is exchangeable, we showed in the proof of de Finetti's theorem that, conditionally on X, the columns $(X_{\ell'})_{\ell'\geq 1}$ in the 'right half' of the array are i.i.d. If we describe the distribution of one column X_1 given X then we can generate all columns $(X_{\ell'})_{\ell'\geq 1}$ independently from this distribution. Therefore, our strategy will be to describe the distribution of X_1 given X, and then combine it with the structure of the distribution of X. Both steps will use exchangeability with respect to permutations of rows, because so far we have only used exchangeability with respect to permutations of columns. Let

$$Y_\ell = \big(s_{\ell,\ell'}\big)_{\ell'\leq 0}$$

be the elements in the ℓ-th row of the 'left half' X of the array. We want to describe the distribution of $X_1 = (s_{\ell,1})_{\ell\geq 1}$ given $X = (Y_\ell)_{\ell\geq 1}$ and we will use the fact that the sequence $(Y_\ell, s_{\ell,1})_{\ell\geq 1}$ is exchangeable. By Lemma 8.5, conditionally on $X = (Y_\ell)_{\ell\geq 1}$, $(s_{\ell,1})_{\ell\geq 1}$ can be generated as

$$s_{\ell,1} = f(\mu_1, Y_\ell, v_1, x_{\ell,1}),$$

where μ_1 is the empirical measure of (Y_ℓ) and where instead of v and (x_ℓ) we wrote v_1 and $(x_{\ell,1})$ to emphasize the first column index 1. Since, conditionally on X, the columns $(X_{\ell'})_{\ell'\geq 1}$ in the 'right half' of the array are i.i.d., we can generate

$$s_{\ell,\ell'} = f(\mu_1, Y_\ell, v_{\ell'}, x_{\ell,\ell'}),$$

where $v_{\ell'}$ and $x_{\ell,\ell'}$ are i.i.d. uniform random variables on $[0,1]$. Finally, since $(Y_\ell)_{\ell\geq 1}$ are i.i.d. given the empirical distribution μ_1, we can generate $\mu_1 = h(w)$ as a function of a uniform random variable w on $[0,1]$ and then, using the Coding Lemma, generate $Y_\ell = Y(\mu_1, u_\ell) = Y(h(w), u_\ell)$ as a function of μ_1 and i.i.d. uniform random variables u_ℓ on $[0,1]$. Plugging these into f above gives $s_{\ell,\ell'} = \sigma(w, u_\ell, v_{\ell'}, x_{\ell,\ell'})$ for some function σ, which finishes the proof. □

Exercise 8.2.1. Prove Lemma 8.3.

Exercise 8.2.2. Give a similar proof of Theorem 8.5, using only global symmetry considerations. *Hints:* Since the row and column indices play a different role in this case, the first step will be slightly different (the second step will be essentially the same). One has to consider two sequences indexed by $j \in I$,

$$S_j^1 = \Big((s_{\ell,j})_{\ell \in \mathbb{N}\setminus I}, (s_{\ell,\ell'})_{\ell,\ell' \in \mathbb{N}\setminus I} \Big) \text{ and } S_j^2 = \Big((s_{j,\ell})_{\ell \in \mathbb{N}\setminus I}, (s_{\ell,\ell'})_{\ell,\ell' \in \mathbb{N}\setminus I} \Big).$$

These two sequences are *separately exchangeable*, which means that

$$\Big((S^1_{\pi(j)})_{j\in I}, (S^2_{\rho(j)})_{j\in I} \Big) \stackrel{d}{=} \Big((S^1_j)_{j\in I}, (S^2_j)_{j\in I} \Big)$$

for any permutations π and ρ of finitely many indices. Notice that these sequences are not independent. In this case one needs to prove (as a part of the exercise) the following modification of de Finetti's representation.

Theorem 8.7. *If the sequences $(s_\ell^1)_{\ell\geq 1}$ and $(s_\ell^2)_{\ell\geq 1}$ are separately exchangeable then there exist measurable functions $g_1, g_2 : [0,1]^2 \to \mathbb{R}$ such that*

$$\Big((s_\ell^1)_{\ell\geq 1}, (s_\ell^2)_{\ell\geq 1} \Big) \stackrel{d}{=} \Big((g_1(w,u_\ell))_{\ell\geq 1}, (g_2(w,v_\ell))_{\ell\geq 1} \Big) \tag{8.29}$$

where w, (u_ℓ) and (v_ℓ) are i.i.d. random variables uniform on $[0,1]$.

Exercise 8.2.3. Consider a pair of random arrays

$$(s^1_{\ell,\ell'})_{\ell,\ell'\geq 1} \text{ and } (s^2_{\ell,\ell'})_{\ell,\ell'\geq 1}$$

that are separately exchangeable in the first coordinate and jointly exchangeable in the second coordinate, that is,

$$\Big((s^1_{\pi_1(\ell),\rho(\ell')})_{\ell,\ell'\geq 1}, (s^2_{\pi_2(\ell),\rho(\ell')})_{\ell,\ell'\geq 1} \Big) \stackrel{d}{=} \Big((s^1_{\ell,\ell'})_{\ell,\ell'\geq 1}, (s^2_{\ell,\ell'})_{\ell,\ell'\geq 1} \Big)$$

for any permutations π_1, π_2, ρ of finitely many coordinates. Show that there exist two functions σ_1, σ_2 such that these arrays can be generated in distribution by

$$s^1_{\ell,\ell'} = \sigma_1(w, u^1_\ell, v_{\ell'}, x^1_{\ell,\ell'}),\ s^2_{\ell,\ell'} = \sigma_2(w, u^2_\ell, v_{\ell'}, x^2_{\ell,\ell'}),$$

where all the arguments are i.i.d. uniform random variables on $[0,1]$.

8.3 The Dovbysh-Sudakov representation for Gram-de Finetti arrays

Let us consider an infinite symmetric random array $R = (R_{\ell,\ell'})_{\ell,\ell'\geq 1}$, which is weakly exchangeable in the sense defined in (8.18), i.e. for any permutation π of finitely many indices we have equality in distribution

$$\big(R_{\pi(\ell),\pi(\ell')}\big)_{\ell,\ell'\geq 1} \stackrel{d}{=} \big(R_{\ell,\ell'}\big)_{\ell,\ell'\geq 1}. \tag{8.30}$$

In addition, suppose that R is positive definite with probability one, where by positive definite we will always mean non-negative definite. Such weakly exchangeable positive definite arrays are called *Gram-de Finetti arrays*. It turns out that all such arrays are generated essentially as the covariance matrix of an i.i.d. sample from a random measure on a Hilbert space. Let H be the Hilbert space $L^2([0,1],dv)$, where dv denotes the Lebesgue measure on $[0,1]$.

Theorem 8.8 (Dovbysh-Sudakov). *There exists a random probability measure η on $H\times\mathbb{R}^+$ such that the array $R = (R_{\ell,\ell'})_{\ell,\ell'\geq 1}$ is equal in distribution to*

$$\big(h_\ell\cdot h_{\ell'} + a_\ell\,\delta_{\ell,\ell'}\big)_{\ell,\ell'\geq 1}, \tag{8.31}$$

where, conditionally on η, $(h_\ell,a_\ell)_{\ell\geq 1}$ is a sequence of i.i.d. random variables with the distribution η and $h\cdot h'$ denotes the scalar product on H.

In particular, the marginal G of η on H can be used to generate the sequence (h_ℓ) and the off-diagonal elements of the array R.

Proof (of Theorem 8.8.). Since the array R is positive definite, conditionally on R, we can generate a Gaussian vector g in $\mathbb{R}^{\mathbb{N}}$ with the covariance equal to R. Now, also conditionally on R, let $(g_i)_{i\geq 1}$ be independent copies of g. If, for each $i\geq 1$, we denote the coordinates of g_i by $g_{\ell,i}$ for $\ell\geq 1$ then, since the array $R = (R_{\ell,\ell'})_{\ell,\ell'\geq 1}$ is weakly exchangeable, it should be obvious that the array $(g_{\ell,i})_{\ell,i\geq 1}$ is exchangeable in the sense of (8.16). By Theorem 8.5, there exists a measurable function $\sigma : [0,1]^4 \to \mathbb{R}$ such that

$$\big(g_{\ell,i}\big)_{\ell,i\geq 1} \stackrel{d}{=} \big(\sigma(w,u_\ell,v_i,x_{\ell,i})\big)_{\ell,i\geq 1}, \tag{8.32}$$

where w, (u_ℓ), (v_i), $(x_{\ell,i})$ are i.i.d. random variables with the uniform distribution on $[0,1]$. By the strong law of large numbers (applied conditionally on R), for any $\ell,\ell'\geq 1$,

$$\frac{1}{n}\sum_{i=1}^n g_{\ell,i}g_{\ell',i} \to R_{\ell,\ell'}$$

almost surely as $n\to\infty$. Similarly, by the strong law of large numbers (now applied conditionally on w and $(u_\ell)_{\ell\geq 1}$),

$$\frac{1}{n}\sum_{i=1}^n \sigma(w,u_\ell,v_i,x_{\ell,i})\sigma(w,u_{\ell'},v_i,x_{\ell',i}) \to \mathbb{E}'\sigma(w,u_\ell,v_1,x_{\ell,1})\sigma(w,u_{\ell'},v_1,x_{\ell',1})$$

almost surely, where $\mathbb{E}'$ denotes the expectation with respect to the random variables v_1 and $(x_{\ell,1})_{\ell\geq 1}$. Therefore, (8.32) implies that

$$\big(R_{\ell,\ell'}\big)_{\ell,\ell'\geq 1} \stackrel{d}{=} \big(\mathbb{E}'\sigma(w,u_\ell,v_1,x_{\ell,1})\sigma(w,u_{\ell'},v_1,x_{\ell',1})\big)_{\ell,\ell'\geq 1}. \tag{8.33}$$

If we denote

$$\sigma^{(1)}(w,u,v) = \int \sigma(w,u,v,x)\,dx,\ \sigma^{(2)}(w,u,v) = \int \sigma(w,u,v,x)^2\,dx,$$

then the off-diagonal and diagonal elements on the right hand side of (8.33) are given by

$$\int \sigma^{(1)}(w,u_\ell,v)\sigma^{(1)}(w,u_{\ell'},v)\,dv \ \text{ and } \int \sigma^{(2)}(w,u_\ell,v)\,dv$$

correspondingly. Notice that, for almost all w and u, the function $v \to \sigma^{(1)}(w,u,v)$ is in $H = L^2([0,1],dv)$, since

$$\int \sigma^{(1)}(w,u,v)^2\,dv \leq \int \sigma^{(2)}(w,u,v)\,dv$$

and, by (8.33), the right hand side is equal in distribution to $R_{1,1}$. Therefore, if we denote

$$h_\ell = \sigma^{(1)}(w,u_\ell,\cdot),\ a_\ell = \int \sigma^{(2)}(w,u_\ell,v)\,dv - h_\ell\cdot h_\ell,$$

then (8.33) becomes

$$\big(R_{\ell,\ell'}\big)_{\ell,\ell'\geq 1} \stackrel{d}{=} \big(h_\ell\cdot h_{\ell'} + a_\ell\delta_{\ell,\ell'}\big)_{\ell,\ell'\geq 1}. \tag{8.34}$$

It remains to observe that $(h_\ell,a_\ell)_{\ell\geq 1}$ is an i.i.d. sequence from the random measure η on $H\times\mathbb{R}^+$ given by the image of the Lebesgue measure du on $[0,1]$ by the map

$$u \to \Big(\sigma^{(1)}(w,u,\cdot), \int \sigma^{(2)}(w,u,v)\,dv - \int \sigma^{(1)}(w,u,v)\sigma^{(1)}(w,u,v)\,dv\Big).$$

This finishes the proof. □

Let us now suppose that there is equality instead of equality in distribution in (8.31), i.e. the array R is generated by an i.i.d. sample (h_ℓ,a_ℓ) from the random measure η. To complete the picture, let us show that one can reconstruct η, up to an orthogonal transformation of its marginal on H, as a measurable function of the array R itself, with values in the set $\mathscr{P}(H\times\mathbb{R}^+)$ of all probability measures on $H\times\mathbb{R}^+$ equipped with the topology of weak convergence.

Lemma 8.6. *There exists a measurable function $\eta' = \eta'(R)$ of the array $(R_{\ell,\ell'})_{\ell,\ell'\geq 1}$ with values in $\mathscr{P}(H\times\mathbb{R}^+)$ such that $\eta' = \eta\circ(U,\mathrm{id})^{-1}$ almost surely for some orthogonal operator U on H that depends on the sequence $(h_\ell)_{\ell\geq 1}$.*

Proof. Let us begin by showing that the norms $\|h_\ell\|$ can be reconstructed almost surely from the array R. Consider a sequence (g_ℓ) on H such that $g_\ell \cdot g_{\ell'} = R_{\ell,\ell'}$ for all $\ell, \ell' \geq 1$. In other words, $\|g_\ell\|^2 = \|h_\ell\|^2 + a_\ell$ and $g_\ell \cdot g_{\ell'} = h_\ell \cdot h_{\ell'}$ for all $\ell \neq \ell'$. Without loss of generality, let us assume that $g_\ell = h_\ell + \sqrt{a_\ell} e_\ell$, where $(e_\ell)_{\ell \geq 1}$ is an orthonormal sequence orthogonal to the closed span of (h_ℓ). If necessary, we identify H with $H \oplus H$ to choose such a sequence (e_ℓ). Since (h_ℓ) is an i.i.d. sequence from the marginal G of the measure η on H, with probability one, there are elements in the sequence $(h_\ell)_{\ell \geq 2}$ arbitrarily close to h_1 and, therefore, the length of the orthogonal projection of h_1 onto the closed span of $(h_\ell)_{\ell \geq 2}$ is equal to $\|h_1\|$. As a result, the length of the orthogonal projection of g_1 onto the closed span of $(g_\ell)_{\ell \geq 2}$ is also equal to $\|h_1\|$, and it is obvious that this length is a measurable function of the array $g_\ell \cdot g_{\ell'} = R_{\ell,\ell'}$. Similarly, we can reconstruct all the norms $\|h_\ell\|$ as measurable functions of the array R and, thus, all $a_\ell = R_{\ell,\ell} - \|h_\ell\|^2$. Therefore,

$$\big(h_\ell \cdot h_{\ell'}\big)_{\ell,\ell' \geq 1} \text{ and } (a_\ell)_{\ell \geq 1} \tag{8.35}$$

are both measurable functions of the array R. Given the matrix $(h_\ell \cdot h_{\ell'})$, we can find a sequence (x_ℓ) in H isometric to (h_ℓ), for example, by choosing x_ℓ to be in the span of the first ℓ elements of some fixed orthonormal basis. This means that all x_ℓ are measurable functions of R and that there exists an orthogonal operator $U = U((h_\ell)_{\ell \geq 1})$ on H such that

$$x_\ell = U h_\ell \text{ for } \ell \geq 1. \tag{8.36}$$

Since $(h_\ell, a_\ell)_{\ell \geq 1}$ is an i.i.d. sequence from distribution η, by the strong law of large numbers for empirical measures (Varadarajan's theorem in Section 17),

$$\frac{1}{n} \sum_{1 \leq \ell \leq n} \delta_{(h_\ell, a_\ell)} \to \eta \text{ weakly}$$

almost surely and, therefore, (8.36) implies that

$$\frac{1}{n} \sum_{1 \leq \ell \leq n} \delta_{(x_\ell, a_\ell)} \to \eta \circ (U, \mathrm{id})^{-1} \text{ weakly} \tag{8.37}$$

almost surely. The left hand side is, obviously, a measurable function of the array R in the space of all probability measures on $H \times \mathbb{R}^+$ equipped with the topology of weak convergence and, therefore, as a limit, $\eta' = \eta \circ (U, \mathrm{id})^{-1}$ is also a measurable function of R. This finishes the proof. □

Exercise 8.3.1. Consider a sequence of pairs $(G_N^1, G_N^2)_{N \geq 1}$ of random probability measures on $\{-1,+1\}^N$ that are not necessarily independent. Let $(\sigma^\ell)_{\ell \leq 0}$ be an i.i.d. sample of replicas from G_N^1 and $(\sigma^\ell)_{\ell \geq 1}$ be an i.i.d. sample of replicas from G_N^2 (for convenience of notation, we denote both by σ^ℓ but use difference sets of indices for ℓ). Let $R^N = (R^N_{\ell,\ell'})_{\ell,\ell' \in \mathbb{Z}}$ be the array of the so-called overlaps

$$R^N_{\ell,\ell'} = \frac{1}{N}\sum_{i=1}^N \sigma_i^\ell \sigma_i^{\ell'}.$$

Suppose that R^N converges in distributions to an array $R = (R_{\ell,\ell'})_{\ell,\ell'\in\mathbb{Z}}$. Notice that this array is weakly exchangeable,

$$\big(R_{\pi(\ell),\pi(\ell')}\big)_{\ell,\ell'\in\mathbb{Z}} \stackrel{d}{=} \big(R_{\ell,\ell'}\big)_{\ell,\ell'\in\mathbb{Z}},$$

but not under all permutations of integers $\mathbb{Z}$, but only those permutations that map positive integers into positive and non-positive into non-positive. Prove that there exists a pair of random measures G^1 and G^2 on a separable Hilbert space H (not necessarily independent) such that

$$\big(R_{\ell,\ell'}\big)_{\ell\neq\ell'\in\mathbb{Z}} \stackrel{d}{=} \big(h_\ell\cdot h_{\ell'}\big)_{\ell\neq\ell'\in\mathbb{Z}},$$

where $(h_\ell)_{\ell\leq 0}$ is an i.i.d. sample from G^1 and $(h_\ell)_{\ell\geq 1}$ is an i.i.d. sample from G^2.

Chapter 9
Finite State Markov Chains

9.1 Definitions and basic properties

In this chapter, we will study finite state Markov chains, namely, sequences $(X_i)_{i\geq 1}$ of random variables taking values in a finite set

$$S = \{s_1, \ldots, s_m\} \tag{9.1}$$

(elements of this set will be called *states*), satisfying

$$\mathbb{P}\big(X_{k+1} = x_{k+1} \mid X_1 = x_1, \ldots, X_k = x_k\big) = \mathbb{P}\big(X_{k+1} = x_{k+1} \mid X_k = x_k\big). \tag{9.2}$$

This means that the conditional distribution of the next outcome X_{k+1} given the preceding outcomes $X_1 = x_1, \ldots, X_k = x_k$ depends only on the most recent outcome $X_k = x_k$. This property is called a *Markov property* of the sequence, also called a *memoryless property*, in the sense that we do not need to remember the entire past and only need to know the most recent outcome to know the chances of the next outcome. The conditional distribution in (9.2) is also called *transition probability*. Markov chain is called *homogeneous* if transition probabilities $\mathbb{P}(X_{k+1} = a \mid X_k = b)$ do not depend on the index k. For any $n \geq 1$, the joint distribution of $X_1, \ldots, X_n$ can be constructed sequentially

$$\mathbb{P}\big(X_1 = x_1, \ldots, X_n = x_n\big) = \prod_{k=0}^{n-1} \mathbb{P}\big(X_{k+1} = x_{k+1} \mid X_k = x_k\big), \tag{9.3}$$

where the first factor for $k = 0$ is just $\mathbb{P}(X_1 = x_1)$.

The *transition matrix* P is the matrix

$$P := \big[p_{ij}\big]_{1\leq i,j\leq n} = \big[p(s_i, s_j)\big]_{1\leq i,j\leq n}, \tag{9.4}$$

whose entries are the transition probabilities

$$p_{ij} = p(s_i, s_j) = \mathbb{P}\big(X_{k+1} = s_j \mid X_k = s_i\big). \tag{9.5}$$

It is customary (and convenient in terms of notation) to start the Markov chain at 'time zero', i.e. $k = 0$. The distribution of X_0, which we will denote by μ,

$$\mu_i = \mu(s_i) := \mathbb{P}(X_0 = s_i) \text{ for } s_i \in S, \tag{9.6}$$

is called the *initial distribution* of the Markov chain. With this notation, the definition of the (homogeneous finite state) Markov chain can be rewritten as

$$\mathbb{P}\big(X_0 = x_0, X_1 = x_1, \ldots, X_n = x_n\big) = \mu(x_0)p(x_0,x_1)p(x_1,x_2)\cdots p(x_{n-1},x_n). \tag{9.7}$$

The initial distribution μ will vary depending on how we want to start the chain, so we can also express this definition entirely in terms of the transition matrix,

$$\mathbb{P}\big(X_1 = x_1, \ldots, X_n = x_n \mid X_0 = x_0\big) = p(x_0,x_1)p(x_1,x_2)\cdots p(x_{n-1},x_n). \tag{9.8}$$

If we start the chain in the state x_0, then the probability that the chain will visit states $x_1, \ldots, x_n$ in the next n steps is the product of transition probabilities along the path. If we multiply by $\mu(x_0)$, we recover the previous equation.

It is also convenient to visualize transition probabilities by a *transition graph*. For example, the transition matrix

$$P = \begin{bmatrix} 0.25 & 0.25 & 0 & 0 & 0.5 & 0 & 0 \\ 0.4 & 0 & 0 & 0 & 0.6 & 0 & 0 \\ 0 & 0.3 & 0 & 0.7 & 0 & 0 & 0 \\ 0 & 0 & 0.25 & 0 & 0.75 & 0 & 0 \\ 0 & 0.9 & 0 & 0 & 0.1 & 0 & 0 \\ 0 & 0 & 0 & 0 & 0 & 0 & 0.5 \\ 0 & 0 & 0 & 0 & 0 & 0.5 & 0 \end{bmatrix} \tag{9.9}$$

can be depicted as in Figure 9.1, with dots representing states, and arrows representing non-zero transition probabilities. Notice that, for example, $p_{12} \neq p_{21}$.

This graphical representation of transition probabilities suggests that we can think of Markov chains as random walks on the set of states, where at each step we pick one of the neighbours of the current state s_i with probabilities p_{ij} and move there. If $p_{ii} \neq 0$, we can end up staying at s_i.

A state s_i is called *inessential* if we can reach from it another state s_j with positive probability (not necessarily in one step), from which we cannot come back to s_i. For example, in Figure 9.1, the states s_3 and s_4 are inessential. Such states are also called *transient*, although this terminology is better suited for infinite state Markov chains. For finite state chains, once we reach s_j, we can never come back to s_i, so for the long-term behaviour of the Markov chain these states do not matter.

Two states s_i and s_j are called *communicating*, denoted $s_i \leftrightarrow s_j$, if we can go from each of them to the other with positive probability. If

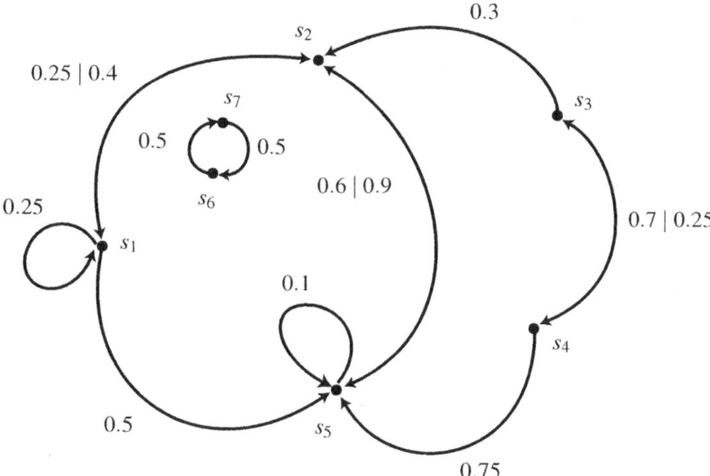

Fig. 9.1 Transition graph of a Markov chain with the transition matrix (9.9). States s_3, s_4 are inessential, and essential states are divided into two clusters, $\{s_1, s_2, s_5\}$ and $\{s_6, s_7\}$.

$$S_i = \{s \in S : s_i \leftrightarrow s\}$$

is the set of states communicating with s_i, and if s_i communicates with s_j then, obviously, $S_i = S_j$, because any state that communicates with s_i communicates with s_j and vice versa. This means that communicating states can be divided into disjoint *clusters*. For example, in Figure 9.1, there are two clusters of essential states, $\{s_1, s_2, s_5\}$ and $\{s_6, s_7\}$, and one cluster of inessential states, $\{s_3, s_4\}$. A state s_i is called *absorbing state* if $p_{ii} = 1$, because, once we reach this state, we can never leave. Transition probabilities on a given cluster of essential states define a Markov chain on that cluster.

Markov chain is called *irreducible* if all its states communicate with each other, which means that there are no inessential states and there is only one cluster of essential states. Below we will mostly study irreducible Markov chains.

The transition matrix P is much more useful than just as a representation of the transition probabilities. If we denote

$$p_{ij}(n) = \left(P^n\right)_{ij} \tag{9.10}$$

the (i, j)th element of P^n then the following holds.

Lemma 9.1 (n-step transition probabilities). *For all $n \geq 1$,*

$$\mathbb{P}(X_n = s_j \mid X_0 = s_i) = p_{ij}(n) \tag{9.11}$$

and, in particular,

$$\mathbb{P}(X_n = s_j) = \sum_{i=1}^{n} \mu_i p_{ij}(n), \tag{9.12}$$

where μ is the initial distribution defined in (9.6).

If we write the last equation in the vector form,

$$\big(\mathbb{P}(X_n = s_1), \ldots, \mathbb{P}(X_n = s_m)\big) = \mu P^n, \tag{9.13}$$

where $\mu = (\mu_1, \ldots, \mu_m)$, we can compute the distribution of the chain at time n by multiplying the initial distribution μ by P^n. It is important that we multiply the row vector μ by P^n on the right and, for example, $P^n \mu^T$ does not have the same meaning. The equation (9.11) also shows that

$$\sum_{j=1}^{m} p_{ij}(n) = 1,$$

so P^n can be viewed as an n-step transition matrix.

Proof. We start with $n = 2$. In order to compute

$$\mathbb{P}(X_2 = s_j \mid X_0 = s_i),$$

we can sum over possible outcomes of $X_1 = s_k$,

$$\begin{aligned} \mathbb{P}(X_2 = s_j \mid X_0 = s_i) &= \sum_{k=1}^{m} \mathbb{P}(X_1 = s_k, X_2 = s_j \mid X_0 = s_i) \\ &= \sum_{k=1}^{m} p(s_i, s_k) p(s_k, s_j) = \sum_{k=1}^{m} p_{ik} p_{kj}, \end{aligned}$$

which is the (i, j)th entry of P^2. The general case follows by induction, using a similar calculation. Assuming that (9.11) holds,

$$\begin{aligned} \mathbb{P}(X_{n+1} = s_j \mid X_0 = s_i) &= \sum_{k=1}^{m} \mathbb{P}(X_n = s_k, X_{n+1} = s_j \mid X_0 = s_i) \\ &= \sum_{k=1}^{m} \mathbb{P}(X_{n+1} = s_j \mid X_n = s_k) \mathbb{P}(X_n = s_k \mid X_0 = s_i) \\ &= \sum_{k=1}^{m} p_{ik}(n) p_{kj} = p_{ij}(n+1). \end{aligned}$$

If we multiply this by $\mu_i = \mathbb{P}(X_0 = s_i)$ and sum over $i \le m$, we get (9.12) for X_{n+1}, so the proof of the induction step is complete. □

The *period d_i of a state s_i* is defined by

$$d_i = \gcd\{n \ge 1 \,:\, p_{ii}(n) > 0\}, \tag{9.14}$$

the greatest common divisor all the times n when a walk starting at s_i can return to s_i. If $d_i = 1$ then the state is called *aperiodic*. We call a Markov chain *aperiodic* if all periods $d_i = 1$. It turns out that for irreducible chains this is the same as requiring only one period d_i to be equal to 1.

Lemma 9.2. *If a Markov chain is irreducible then all periods d_i are equal.*

Proof. Let us suppose that s_j can be reached from s_i in N steps and s_i can be reached from s_j in M steps with positive probabilities, $p_{ij}(N) > 0$ and $p_{ji}(M) > 0$. If $p_{jj}(n) > 0$ then

$$p_{ii}(N+M+n) \geq p_{ij}(N)p_{jj}(n)p_{ji}(M) > 0,$$

because we can reach s_j in N steps, then come back to s_j in n steps, and then reach s_i in M steps, so the probability to return to s_i in $N+M+n$ steps is positive. This is also true for $n = 0$, because we can go to s_j and come back to s_i in $N+M$ steps. Since d_i is the period of s_i, by definition, d_i divides all numbers $N+M+n$ as above. In particular, it divides $N+M$, which implies that it divides all n such that $p_{jj}(n) > 0$. This means that d_i divides d_j, because d_j is the greatest common divisor of all such n. Similarly, d_j divides d_i, so they must be equal. □

The following property of irreducible aperiodic Markov chains will be very important in Section 9.3.

Lemma 9.3. *If a Markov chain is irreducible and aperiodic then there exists $N \geq 1$ such that, for all $n \geq N$,*

$$p_{ij}(n) > 0 \text{ for all } i, j. \tag{9.15}$$

In other words, for large n, all the entries of P^n are strictly positive.

Proof. Let $T(s_1) = \{n \geq 1 : p_{11}(n) > 0\}$ be the set of all times the chain starting at s_1 can come back to s_1. Let $d \geq 1$ be the smallest positive integer that can be written as

$$d = a_1 n_1 + \ldots + a_k n_k,$$

for $k \geq 1$, $n_i \in T(s_1)$ and $a_i \in \mathbb{Z}$. Then d must divide all $n \in T(s_1)$, because if it does not divide some $n \in T(s_1)$ then

$$n = ad + r,$$

where the remainder $r \geq 1$ is strictly less than d. However,

$$r = n - ad = n - a_1 n_1 - \ldots - a_k n_k$$

is also a linear combination with integer coefficients of the times in $T(s_1)$, which contradicts that d was the smallest such number. This proves that d divides the period of s_1 and, since the chain is aperiodic, $d = 1$. We showed that

$$1 = a_1 n_1 + \ldots + a_k n_k, \tag{9.16}$$

for some $k \geq 1$, $n_i \in T(s_1)$ and $a_i \in \mathbb{Z}$. Some a_i of course can be negative.

Let us take the largest $|a_i|$ and, for certainty, suppose that $|a_1|$ is the largest. If

$$N_1 = |a_1|n_1(n_1 + \ldots + n_k),$$

we will now show that all $n \geq N_1$ can be written as a linear combinations

$$n = c_1 n_1 + \ldots + c_k n_k,$$

with non-negative integer coefficients c_i. This will prove that $p_{11}(n) > 0$ for such n, because the chain starting from the state s_1 can come back to s_1 in n_i number of steps, so we just need to repeat these loops c_i times for all $i \leq k$ to see that the chain can come back in n steps with positive probability.

Let us first consider any n between N_1 and $N_1 + n_1$, which means that $n = N_1 + \ell$ for some $\ell \leq n_1$. By (9.16), we can write

$$n = N_1 + \ell = |a_1|n_1(n_1 + \ldots + n_k) + \ell(a_1 n_1 + \ldots + a_k n_k) = c_1 n_1 + \ldots + c_k n_k,$$

where

$$c_i = |a_1|n_1 + \ell a_i \geq |a_1|\ell + \ell a_i = (|a_1| + a_i)\ell \geq 0$$

as we wished, because we assumed that $|a_1|$ is greater or equal to $|a_i|$. Since

$$N_1 + n_1 = (|a_1|n_1 + 1)n_1 + |a_1|n_1(n_2 + \ldots + n_k),$$

we can repeat the same argument for n in between $N_1 + n_1$ and $N_1 + 2n_1$, and so on.

Similarly, for each state s_i, we can find N_i such that $p_{ii}(n) > 0$ for $n \geq N_i$. If we take $N' = \max(N_1, \ldots, N_m)$ then $p_{ii}(n) > 0$ for all i and all $n \geq N'$. Since the chain is irreducible, we can always go from one state s_i to another state s_j in under m steps with positive probability (see exercise below), which implies that $p_{ij}(n) > 0$ for all $n \geq N' + m$. □

Exercise 9.1.1. Let $U_1, U_2, \ldots$ be i.i.d. random variables with the uniform distribution on $\{1, \ldots, m\}$ and let

$$X_n = \max_{i \leq n} U_i.$$

Show that $X_1, X_2, \ldots$ is a Markov chain and find its transition probabilities. What are the essential states of this chain?

Exercise 9.1.2. Suppose that m while balls and m black balls are mixed together and divided evenly between two baskets. At each step two balls are chosen at random, one from each basket, and switched. We say that the system is in the state s_i if there are i white balls in the first basket. Find the transition probabilities of this chain.

Exercise 9.1.3. If a Markov chain on m states is irreducible, show that, for any $i, j \leq m$, $p_{ij}(k) > 0$ for some $k \leq m$.

Exercise 9.1.4. If a Markov chain is irreducible and $p_{11} > 0$, what is the period d_2 of the state s_2?

Exercise 9.1.5. If a Markov chain $(X_n)_{n\geq 0}$ is irreducible and has period d, show that $Y_n = X_{dn}$ for $n \geq 0$ is a Markov chain and that it is aperiodic. Does it have to be irreducible?

Exercise 9.1.6. Let $N \sim \text{Poiss}(\lambda)$ be a Poisson random variable. Let us run N independent Markov chains starting from the state s_1, and let $N_n(i)$ be the number of these chains in the state s_i at time n. Show that $N_n(i) \sim \text{Poiss}(\lambda p_{1i}(n))$.

9.2 Stationary distributions

In Lemma 9.1 we showed that if the vector $\mu = (\mu_1, \ldots, \mu_m)$ describes the distribution of the chain X_0 at time zero, then μP^n is the distribution of X_n at time n. The distribution μ on the set of states is called *stationary* if

$$\mu = \mu P, \tag{9.17}$$

i.e. the distribution of X_1 is the same as that of X_0. Of course, this implies that $\mu = \mu P^n$, so the distribution of all X_n is the same. We will show that a stationary distribution always exists, and for an irreducible Markov chain it is unique. For irreducible chains, we will also derive a representation of the stationary distribution in terms of expected return times. To find a stationary distribution in practice, one can just solve a system of linear equations $\mu = \mu P$.

Lemma 9.4 (Existence). *For any finite state Markov chain, there exists at least one stationary distribution.*

Proof. Let us consider a sequence of matrices

$$A_n = \frac{1}{n}(1 + P + \ldots + P^n).$$

Since each matrix P^k is a transition matrix and its rows sum up to 1, the rows of A_n also sum up to one, so its entries are all between 0 and 1. By the Bolzano–Weierstrass theorem, there exists a convergent subsequence $A_{n_k} \to A$. Notice that the limit A is also a transition matrix because the entries must be nonnegative and rows add up to 1. If we consider

$$A_n P = \frac{1}{n}(P + \ldots + P^n + P^{n+1}),$$

we see that the difference goes to a zero matrix,

$$A_n P - A_n = \frac{1}{n}(-1 + P^{n+1}) \to 0.$$

This implies that

$$A = \lim_{k\to\infty} A_{n_k} = \lim_{k\to\infty} A_{n_k} P = AP,$$

which shows that $A = AP$. This means that each row a of this matrix satisfies $a = aP$. Since each row is a probability distribution on states (entries are nonnegative and add up to one), each row is a stationary distribution. □

Notice that if the stationary distribution is unique, all the rows of A in the above proof must be equal.

Next, we will show the following.

Lemma 9.5 (Uniqueness). *If a Markov chain is irreducible then the stationary distribution is unique.*

Proof. Previous lemma shows that $P-I$ is not invertible, since there exists a non-zero solution of $\mu(P-I)=0$. Let us first show that, when P is the transition matrix of an irreducible Markov chain, the rank of $P-I$ is $m-1$. Consider the linear system $(P-I)v^T=0$ for $v=(v_1,\ldots,v_n)$. Let us show that irreducibility implies that

$$v_1=v_2=\ldots=v_n.$$

Consider the largest coordinate of v, let us say, v_1. The first equation in the system reads

$$\sum_{j=1}^{m} p_{1j}v_j=v_1.$$

On the other hand, since all $v_j\leq v_1$ and $p_{1j}v_j\leq p_{1j}v_1$,

$$v_1=\sum_{j=1}^{m} p_{1j}v_j\leq\sum_{j=1}^{m} p_{1j}v_1=v_1.$$

So the equality is actually an equality, and all $p_{1j}v_j=p_{1j}v_1$. This means that $v_j=v_1$ for all j such that $p_{1j}\neq 0$. In other words, all the states s_j that can be reached from the state s_1 in one step must have $v_j=v_1$. Then, repeating the same calculation for these immediate neighbours we can see that all their neighbours also much have $v_j=v_1$. Since the chain is irreducible, we can reach all states in finitely many steps, which proves that all v_j's are equal.

This proves that the kernel of $P-I$ is one dimensional,

$$\ker(P-I)=\{c(1,\ldots,1)\,:\,c\in\mathbb{R}\},\tag{9.18}$$

spanned by the vector $(1,\ldots,1)$, which means that the rank of $P-I$ equals to $m-1$. On the other hand, if we had more than one stationary distribution, the system $\mu=\mu P$ would have at least two linearly independent solutions, which would mean that the rank of $P-I$ would be not greater than $m-2$. This proves that the stationary distribution is unique. □

Next, we will derive a more descriptive formula for the stationary distribution of irreducible Markov chains. First, let us give a heuristic non-rigorous explanation and then check carefully that our intuitive guess is correct. In the proof of Lemma 9.4 we considered the sequence of matrices

$$A_n=\frac{1}{n}(1+P+\ldots+P^n)$$

and showed that any limit A over some subsequence consists of rows which are stationary distributions. For irreducible chains, we just showed that the stationary distribution μ is unique, so we must have that

$$A = \begin{bmatrix} \mu \\ \vdots \\ \mu \end{bmatrix} = \begin{bmatrix} \mu_1 & \cdots & \mu_m \\ \vdots & \ddots & \vdots \\ \mu_1 & \cdots & \mu_m \end{bmatrix}.$$

In particular, this means that the limits over all convergent subsequences are equal to A and, therefore, the limit exists over the entire sequence,

$$\lim_{n\to\infty} \frac{1}{n}(1+P+\ldots+P^n) = \begin{bmatrix} \mu_1 & \cdots & \mu_m \\ \vdots & \ddots & \vdots \\ \mu_1 & \cdots & \mu_m \end{bmatrix}.$$

Since the (i,j)th entry of P^k is

$$(P^k)_{ij} = p_{ij}(k) = \mathbb{P}(X_k = s_j \mid X_0 = s_i),$$

this can be written as

$$\lim_{n\to\infty} \frac{1}{n} \sum_{k=1}^{n} \mathbb{P}(X_k = s_j \mid X_0 = s_i) = \mu_j. \tag{9.19}$$

If we multiply both sides by $\mathbb{P}(X_0 = s_i)$, sum over $i \le m$ and use that

$$\sum_{i=1}^{m} \mathbb{P}(X_k = s_j \mid X_0 = s_i)\mathbb{P}(X_0 = s_i) = \mathbb{P}(X_k = s_j),$$

we get

$$\lim_{n\to\infty} \frac{1}{n} \sum_{k=1}^{n} \mathbb{P}(X_k = s_j) = \mu_j. \tag{9.20}$$

In other words, this equation holds no matter how we start the chain, i.e. no matter what the distribution of X_0 is. Since

$$\mathbb{P}(X_k = s_j) = \mathbb{E}\,\mathrm{I}(X_k = s_j),$$

by the linearity of expectation,

$$\lim_{n\to\infty} \mathbb{E}\frac{1}{n} \sum_{k=1}^{n} \mathrm{I}(X_k = s_j) = \mu_j. \tag{9.21}$$

Clearly, this equation can be interpreted by saying that μ_j is the *expected proportion of time the chain visits the state* s_j, if we looks at these proportions over long periods of time.

There is another interpretation of the last equation. Since it does not matter how we start the chain, let us start it in the state s_1. Then, let us divide the time between 1 and n into intervals between consecutive visits of the state s_1. Let us say that M visits of s_1 occurred between 1 and n, with intervals $T_1, T_2, \ldots, T_M$, so that

$$T_1 + \ldots + T_M \approx n$$

in the sense that their ratio is close to 1. Let $N_1(j), \ldots, N_M(j)$ be the number of times we visited s_j in between these consecutive returns to s_1, so that

$$N_1(j) + \ldots + N_M(j) \approx \sum_{k=1}^{n} \mathrm{I}(X_k = s_j).$$

Then

$$\frac{1}{n}\sum_{k=1}^{n} \mathrm{I}(X_k = s_j) \approx \frac{N_1(j) + \ldots + N_M(j)}{T_1 + \ldots + T_M}.$$

If we know that the Markov chain is visiting s_1, the future does not depend on the past because of the memoryless property. This suggests that what happens in between consecutive visits is independent of each other and the pairs

$$(N_1(j), T_1), \ldots, (N_M(j), T_M)$$

are actually independent and identically distributed. To make this statement rigorous, once needs to prove the so called *strong Markov property*, which we are not going to go into here. However, if we use it as a guiding intuition and divide both numerator and denominator above by M, the law of large numbers tells us that

$$\frac{N_1(j) + \ldots + N_M(j)}{T_1 + \ldots + T_M} \approx \frac{\mathbb{E}N_1(j)}{\mathbb{E}T_1}.$$

This suggests another representation,

$$\mu_j = \frac{\mathbb{E}N_1(j)}{\mathbb{E}T_1}, \tag{9.22}$$

where we assume that the chain starts in the state s_1,

$$T_1 = \min\{n \geq 1 : X_n = s_1\} \tag{9.23}$$

is the *first return time* to s_1, and

$$N_1(j) = \sum_{k=0}^{T_1 - 1} \mathrm{I}(X_k = s_j) \tag{9.24}$$

is the number of times the chain visits s_j before returning to s_1.

The first return time T_1 is an example of a *stopping time*. In general, an integer valued random variable $T \geq 0$ is called a stopping time if the event $\{T = n\}$ depends only on $X_0, \ldots, X_n$. In other words, we decide whether to stop at time n based on what was observed up to time n. The return time T_1 to s_1 is a stopping time in this sense, because

$$\{T_1 = n\} = \{X_1 \neq s_1, \ldots, X_{n-1} \neq s_1, X_n = s_1\}.$$

Notice also that $N_1(1) = 1$, because the chain is in the state s_1 only at time zero before the return, so (9.22) implies that

$$\mu_1 = \frac{1}{\mathbb{E}T_1}. \tag{9.25}$$

Since it did not matter how we start the chain, if we start it at s_j and define

$$T_j = \min\{n \geq 1 : X_n = s_j\} \tag{9.26}$$

then the same logic suggests that

$$\mu_j = \frac{1}{\mathbb{E}T_j}. \tag{9.27}$$

In order to prove the representation (9.22), we need to check two things. First, we will need to check that $\mathbb{E}T_1 < \infty$. Because the number of visits to different states before time T_1 adds up to T_1,

$$\sum_{j=1}^{m} N_1(j) = T_1,$$

and therefore $N_1(j) \leq T_1$, proving $\mathbb{E}T_1 < \infty$ would also imply that $\mathbb{E}N_1(j) < \infty$. This would show that the numbers in (9.22) are well defined, and they define a distribution because

$$\sum_{j=1}^{m} \mu_j = \sum_{j=1}^{m} \frac{\mathbb{E}N_1(j)}{\mathbb{E}T_1} = \frac{\mathbb{E}T_1}{\mathbb{E}T_1} = 1.$$

After that we will need to check that $\mu P = \mu$, so it is the stationary distribution.

Lemma 9.6. *If a Markov chain is irreducible then $\mathbb{E}T_1 < \infty$, where T_1 defined in (9.23) is the first return time to the state s_1 for the chain that starts at s_1.*

In the proof we will use the Markov property (9.2) in the following way. Let us consider two vectors

$$X^1 = (X_1, \ldots, X_{t_1}),\ X^2 = (X_{t_1+1}, \ldots, X_{t_2})$$

consisting of the Markov chain up to time t_1 and in between t_1 and t_2. Let us consider some subsets

$$A \subseteq \{s_1, \ldots, s_m\}^{t_1},\ B \subseteq \{s_1, \ldots, s_m\}^{t_2 - t_1}$$

of possible outcomes for these two blocks of the Markov chain, and consider the conditional probability

$$\mathbb{P}(X^2 \in B \mid X^1 \in A).$$

We can not use the Markov property directly, because the condition is a set, but we can rewrite this as

$$\begin{aligned}
\mathbb{P}(X^2 \in B \mid X^1 \in A) &= \frac{\mathbb{P}(X^2 \in B, X^1 \in A)}{\mathbb{P}(X^1 \in A)} \\
&= \sum_{a \in A} \frac{\mathbb{P}(X^2 \in B, X^1 = a)}{\mathbb{P}(X^1 \in A)} \\
&= \sum_{a \in A} \frac{\mathbb{P}(X^2 \in B \mid X^1 = a)\mathbb{P}(X^1 = a)}{\mathbb{P}(X^1 \in A)} \\
&= \sum_{a \in A} \frac{\mathbb{P}(X^2 \in B \mid X_{t_1} = a_{t_1})\mathbb{P}(X^1 = a)}{\mathbb{P}(X^1 \in A)},
\end{aligned} \tag{9.28}$$

where in the last step we used the Markov property to write the condition in terms of the last outcome $X_{t_1} = a_{t_1}$. This is very useful because, for example, if we can control the probability

$$\mathbb{P}(X^2 \in B \mid X_{t_1} = a_{t_1}) \leq p$$

for all possible outcomes $X_{t_1} = a_{t_1}$ then the above equation implies that

$$\mathbb{P}(X^2 \in B \mid X^1 \in A) \leq p \sum_{a \in A} \frac{\mathbb{P}(X^1 = a)}{\mathbb{P}(X^1 \in A)} = p. \tag{9.29}$$

In other words, Markov property can be used in this way even if the condition is a set.

Proof (of Lemma 9.6). Since the chain is irreducible, if the chain is in the state s_i at time k, $X_k = s_i$, we can reach the state s_1 in under m steps, which means that

$$p_{i1}(\ell) = \mathbb{P}(X_{k+\ell} = s_1 \mid X_k = s_i) > 0$$

for some $\ell \leq m$. This implies that the probability that we do not reach the state s_1 in the next m steps is strictly smaller than 1,

$$\begin{aligned}
&\mathbb{P}(X_{k+1} \neq s_1, \ldots, X_{k+m} \neq s_1 \mid X_k = s_i) \\
&\leq \mathbb{P}(X_{k+\ell} \neq s_1 \mid X_k = s_i) = 1 - \mathbb{P}(X_{k+\ell} = s_1 \mid X_k = s_i) < 1.
\end{aligned}$$

This is true for any s_i, so

$$\mathbb{P}(X_{k+1} \neq s_1, \ldots, X_{k+m} \neq s_1 \mid X_k = s_i) \leq 1 - \varepsilon, \tag{9.30}$$

for some small enough $\varepsilon > 0$, for all $i \leq m$.

Using Markov property, this will imply that the probability that the chain does not come back to s_1 in Nm steps (i.e. N cycles of m steps) will be smaller than $(1-\varepsilon)^N$,

$$\mathbb{P}(T_1 > Nm) \le (1-\varepsilon)^N. \tag{9.31}$$

To see this, let us write

$$\mathbb{P}(T_1 > Nm) = \mathbb{P}(X_1 \ne s_1, \ldots, X_{(N-1)m} \ne s_1, \ldots, X_{Nm} \ne s_1)$$

and let us decompose this sequence into two blocks

$$X^1 = (X_1, \ldots, X_{(N-1)m}),\ X^2 = (X_{(N-1)m+1}, \ldots, X_{Nm}).$$

If we take

$$A = \{s_2, \ldots, s_m\}^{(N-1)m},\ B = \{s_2, \ldots, s_m\}^m$$

to be the sets consisting of all the paths that avoid the state s_1 and have lengths $(N-1)m$ and m then

$$\mathbb{P}(T_1 > Nm) = \mathbb{P}(X^2 \in B, X^1 \in A) = \mathbb{P}(X^2 \in B \mid X^1 \in A)\mathbb{P}(X^1 \in A).$$

The equation (9.30) implies that, no matter what the value of $X_{(N-1)m}$ is, the conditional probability that all coordinates of X^2 avoid the state s_1 is smaller than $1-\varepsilon$. By the discussion before the proof and (9.29), the Markov property implies

$$\mathbb{P}(X^2 \in B \mid X^1 \in A) \le 1-\varepsilon,$$

and, therefore,

$$\mathbb{P}(T_1 > Nm) \le (1-\varepsilon)\mathbb{P}(X^1 \in A) = (1-\varepsilon)\mathbb{P}(T_1 > (N-1)m).$$

By induction on N, the equation (9.31) follows.

Next, let us denote $\delta = (1-\varepsilon)^{1/m} < 1$ and consider any $k \ge 1$. Take $N \ge 0$ such that $Nm < k \le (N+1)m$. Then

$$\mathbb{P}(T_1 \ge k) \le \mathbb{P}(T_1 > Nm) \le (1-\varepsilon)^N = \delta^{Nm} \le \delta^{k-m}.$$

This implies that

$$\mathbb{E}T_1 = \sum_{k=1}^{\infty} \mathbb{P}(T_1 \ge k) \le \sum_{k=1}^{\infty} \delta^{k-m} = \frac{\delta^{1-m}}{1-\delta} < \infty,$$

which finishes the proof. □

Recall the equations 9.22), (9.23), (9.24),

$$\mu_j = \frac{\mathbb{E}N_1(j)}{\mathbb{E}T_1}, \tag{9.32}$$

$$T_1 = \min\Big\{n \ge 1 : X_n = s_1\Big\}, \tag{9.33}$$

$$N_1(j) = \sum_{k=0}^{T_1-1} \mathrm{I}(X_k = s_j), \tag{9.34}$$

where we assumed that the chain starts in the state s_1. To check that our heuristic discussion above gives the correct answer, it remains to prove the following.

Theorem 9.1 (Representation of stationary distribution). *If a Markov chain is irreducible then μ defined in (9.32) is the stationary distribution.*

Proof. Showing that $\mu = \mu P$ is equivalent to showing

$$\mathbb{E}N_1(j) = \sum_{i=1}^{m} p_{ij}\mathbb{E}N_1(i), \tag{9.35}$$

because the two equations only differ by a factor of $1/\mathbb{E}T_1$. To prove (9.35), first let us consider $j \neq 1$. Because the chain starts at s_1, we have $X_0 = s_1 \neq s_j$ and, in the definition of $N_1(j)$, we can start the summation from $k = 1$,

$$N_1(j) = \sum_{k=0}^{T_1-1} \mathrm{I}(X_k = s_j) = \sum_{k=1}^{T_1-1} \mathrm{I}(X_k = s_j).$$

We can also write this as

$$N_1(j) = \sum_{k=1}^{\infty} \mathrm{I}(X_k = s_j, k < T_1) = \sum_{k=1}^{\infty} \mathrm{I}(X_k = s_j, k-1 < T_1), \tag{9.36}$$

where we used that

$$\{X_k = s_j, k < T_1\} = \{X_k = s_j, k-1 < T_1\},$$

because $X_k = s_j \neq s_1$, so it is impossible that $T_1 = k$. When $j = 1$, the equation (9.36) also holds, for a different reason. In this case, $N_1(1) = 1$ by definition, while the right hand side of (9.36) equals to 1 because $k - 1 < T_1$ is the same as $k \leq T_1$ and, between times 1 and T_1, the chain visits s_1 once at $k = T_1$.

If we take expectations of both sides of (9.36),

$$\begin{aligned} \mathbb{E}N_1(j) &= \sum_{k=1}^{\infty} \mathbb{P}(X_k = s_j, k-1 < T_1) \\ &= \sum_{k=1}^{\infty}\sum_{i=1}^{m} \mathbb{P}(X_{k-1} = s_i, X_k = s_j, k-1 < T_1), \end{aligned} \tag{9.37}$$

where we also partitioned the event into different outcomes $X_{k-1} = s_i$. The key observation we need to make is that the event

$$\{k-1 < T_1\} = \{X_1 \neq s_1, \ldots, X_{k-1} \neq s_1\}$$

depends only on the random variables $X_1, \ldots, X_{k-1}$ up to time $k-1$. Let us write one term in the above sum by conditioning on the first $k-1$ random variables,

$$\begin{aligned}
&\mathbb{P}(X_{k-1}=s_i, X_k=s_j, k-1<T_1)\\
&=\mathbb{P}(X_k=s_j \mid X_{k-1}=s_i, k-1<T_1)\mathbb{P}(X_{k-1}=s_i, k-1<T_1).
\end{aligned}$$

If we use the Markov property in the form (9.28), we can drop the condition $k-1<T_1$ and rewrite this as

$$\begin{aligned}
&\mathbb{P}(X_{k-1}=s_i, X_k=s_j, k-1<T_1)\\
&=\mathbb{P}(X_k=s_j \mid X_{k-1}=s_i)\mathbb{P}(X_{k-1}=s_i, k-1<T_1)\\
&=p_{ij}\mathbb{P}(X_{k-1}=s_i, k-1<T_1).
\end{aligned}$$

Plugging this back into (9.37) and rearranging terms,

$$\begin{aligned}
\mathbb{E}N_1(j) &= \sum_{k=1}^{\infty}\sum_{i=1}^{m} p_{ij}\mathbb{P}(X_{k-1}=s_i, k-1<T_1)\\
&= \sum_{i=1}^{m} p_{ij}\sum_{k=1}^{\infty}\mathbb{P}(X_{k-1}=s_i, k-1<T_1)\\
\{\ell=k-1\} &= \sum_{i=1}^{m} p_{ij}\sum_{\ell=0}^{\infty}\mathbb{P}(X_\ell=s_i, \ell<T_1)\\
&= \sum_{i=1}^{m} p_{ij}\mathbb{E}\sum_{\ell=0}^{T_1-1}\mathrm{I}(X_\ell=s_i) = \sum_{i=1}^{m} p_{ij}\mathbb{E}N_1(i).
\end{aligned}$$

This proves (9.35) and finishes the proof. □

Exercise 9.2.1. Show that if a Markov chain has two different stationary distributions then there exist infinitely many stationary distributions.

Exercise 9.2.2. Find the stationary distribution of Markov chain with the transition matrix

$$P=\begin{bmatrix} 0 & 0.5 & 0.5\\ 0 & 0 & 1\\ 0.5 & 0.5 & 0\end{bmatrix}.$$

If the chain starts at s_2, what is the expectation of the first return time to s_2?

Exercise 9.2.3. If $\mu=(\mu_1,\ldots,\mu_m)$ is the stationary distribution of an irreducible Markov chain, show that $\mu_i \neq 0$ for all $1\le i\le m$.

Exercise 9.2.4 (Another proof of uniqueness). Consider an irreducible Markov chain and suppose that there exist two stationary distributions μ^1 and μ^2. Let j be any minimizer

$$\frac{\mu_j^1}{\mu_j^2}=\min_{i\le m}\frac{\mu_i^1}{\mu_i^2}.$$

Show that

$$\frac{\mu_j^1}{\mu_j^2} = \frac{\mu_i^1}{\mu_i^2}$$

for any i such that $p_{ij} > 0$. From this, derive that $\mu^1 = \mu^2$.

Exercise 9.2.5. If $p_{1j} = p_{2j}$ for all j and

$$Y_n = \begin{cases} s_j, & \text{if } X_n = s_j \text{ for } j \geq 3, \\ s^*, & \text{if } X_n = s_j \text{ for } j \leq 2, \end{cases}$$

show that $Y_1, Y_2, \ldots$ is a Markov chain on the new state space $S = \{s^*, s_3, \ldots, s_m\}$.

Exercise 9.2.6. If a Markov chain is irreducible, prove that all moments $\mathbb{E}T_1^p < \infty$ for $p \geq 1$, where T_1 is defined in (9.23).

Exercise 9.2.7. In Toronto, it rains (or snows) on about 38% of days per year. Let us consider a Markov chain model of weather with two states $s_1 =$ 'rain', $s_2 =$ 'no rain', and the transition matrix

$$P = \begin{bmatrix} p & 1-p \\ 1-q & q \end{bmatrix},$$

for some $p, q \in (0,1)$, so that if it rains today then it will rain tomorrow with probability p, and if it does not rain today then it will not rain tomorrow with probability q. Find all p and q for which the expected proportion of rainy days over long periods of time is 0.38.

Exercise 9.2.8. Suppose that the state s_1 is inessential and let S_1 be the cluster of states that communicate with s_1. Suppose that the chain starts at s_1, and let T be the first time we exit the cluster S_1,

$$T = \min\{n \geq 1 : X_n \notin S\}.$$

Prove that $\mathbb{E}T < \infty$.

Exercise 9.2.9. Let us consider an irreducible Markov chain $X_0, X_1, \ldots,$ and let $A \subseteq S$ be a non-empty subset of its states. If the chain starts at s_i then let

$$T_A(s_i) = \min\{n \geq 0 : X_n \in A\}$$

be the first time the chain visits one of the states in A and let $t_i = \mathbb{E}T_A(s_i)$ be its expectation. Prove that

(a) If $s_i \in A$ then $t_i = 0$.
(b) If $s_i \notin A$ then $t_i = 1 + \sum_{j=1}^m p_{ij} t_j$.
(c) The equations in parts (a) and (b) determine $t = (t_1, \ldots, t_m)$ uniquely. *Hint:* consider equation (9.18).

9.3 Convergence theorem

In Lemma 9.4 in the last section we considered the sequence of matrices

$$A_n = \frac{1}{n}(1+P+\ldots+P^n)$$

and showed that any subsequential limit A consists of rows which are stationary distributions. For irreducible chains, we also showed that the stationary distribution μ is unique and, therefore,

$$\lim_{n\to\infty}\frac{1}{n}(1+P+\ldots+P^n) = \begin{bmatrix} \mu_1 & \cdots & \mu_m \\ \vdots & \ddots & \vdots \\ \mu_1 & \cdots & \mu_m \end{bmatrix}.$$

Can we expect a stronger statement that

$$\lim_{n\to\infty} P^n = \begin{bmatrix} \mu_1 & \cdots & \mu_m \\ \vdots & \ddots & \vdots \\ \mu_1 & \cdots & \mu_m \end{bmatrix}?$$

First, we need to check that this is, indeed, a stronger statement.

Exercise 9.3.1. If the limit $\lim_{n\to\infty} P^n$ exists then the limit $\lim_{n\to\infty}\frac{1}{n}(1+P+\ldots+P^n)$ exists and is the same.

It is easy to see that this can not be true in general. For example, if a Markov chain has two states s_1, s_2 and it has the transition probabilities $p_{12} = p_{21} = 1$ then,

$$P^{2n} = \begin{bmatrix} 0 & 1 \\ 1 & 0 \end{bmatrix},\ P^{2n+1} = \begin{bmatrix} 1 & 0 \\ 0 & 1 \end{bmatrix}.$$

Another way to put it is that, if we start the chain in the state s_i, it can be back in s_i only at even times. The problem with this example is that this chain has period 2, which results in this periodic behaviour. However, if the chain is aperiodic then the convergence holds.

Theorem 9.2 (Convergence Theorem). *If a Markov chain is irreducible and aperiodic then*

$$\lim_{n\to\infty} P^n = \begin{bmatrix} \mu_1 & \cdots & \mu_m \\ \vdots & \ddots & \vdots \\ \mu_1 & \cdots & \mu_m \end{bmatrix}, \tag{9.38}$$

where $\mu = (\mu_1, \ldots, \mu_m)$ is its stationary distribution.

Notice that this is equivalent to the following statement: if $\nu = (\nu_1, \ldots, \nu_m)$ is any distribution on the set of states then

$$\lim_{n\to\infty} \nu P^n = \mu. \tag{9.39}$$

In one direction, multiplying the right hand side of (9.38) by ν on the left, obviously, gives μ. In the other direction, if we take $\nu = (0,\dots,1,\dots,0)$ where the ith coordinate is 1, then

$$\lim_{n\to\infty} \nu P^n = \mu$$

is the ith row of the limit $\lim_{n\to\infty} P^n$, so all the rows of this limit are equal to μ.

Recall that if ν is the initial distribution of the chain, i.e. the distribution of X_0, then νP^n is the distribution of X_n. Hence, the convergence theorem in the form (9.39) states that the distribution of X_n converges to the stationary distribution μ no matter what the initial distribution is.

Proof (Theorem 9.2). We will be proving the formula (9.39). Let us consider the L^1-distance between vectors on $\mathbb{R}^m$,

$$\|x-y\|_1 = \sum_{i=1}^{m} |x_i - y_i|.$$

Let us notice that multiplying by any transition matrix P on the right does not increase this distance, because

$$\begin{aligned}\|xP-yP\|_1 &= \sum_{j=1}^{m}\Big|\sum_{i=1}^{m} p_{ij}(x_i-y_i)\Big| \le \sum_{j=1}^{m}\sum_{i=1}^{m} p_{ij}|x_i-y_i| \\ &= \sum_{i=1}^{m}\sum_{j=1}^{m} p_{ij}|x_i-y_i| \\ &= \sum_{i=1}^{m} |x_j-y_j| = \|x-y\|_1.\end{aligned}$$

If we apply this to distributions ν and μ and use that $\mu = \mu P$, we get that

$$\|\nu P-\mu\|_1 \le \|\nu-\mu\|_1.$$

This suggests that the distributions get closer to each other after one step but, in order to prove convergence, we would like to have a strict inequality and then iterate it.

This is where the assumption that our chain is aperiodic comes into play. By Lemma 9.3 in Section 9.1, for large enough N, all entries $p_{ij}(N)$ of P^N are strictly positive. Let $\varepsilon > 0$ be the smallest among its entries. If we apply the above calculation to the transition matrix P^N and distributions ν and μ, the improvement comes from the fact that

$$\sum_{i=1}^{m} \varepsilon(\nu_i - \mu_i) = \varepsilon(1-1) = 0$$

and, therefore,

$$\sum_{i=1}^{m} p_{ij}(N)(\nu_i - \mu_i) = \sum_{i=1}^{m} (p_{ij}(N) - \varepsilon)(\nu_i - \mu_i).$$

Notice that all $p_{ij}(N) - \varepsilon \geq 0$, because ε was the smallest among all entries, so

$$\begin{aligned}
\|\nu P^N - \mu P^N\|_1 &= \sum_{j=1}^{m} \Big| \sum_{i=1}^{m} p_{ij}(N)(\nu_i - \mu_i) \Big| \\
&= \sum_{j=1}^{m} \Big| \sum_{i=1}^{m} (p_{ij}(N) - \varepsilon)(\nu_i - \mu_i) \Big| \\
&\leq \sum_{j=1}^{m} \sum_{i=1}^{m} (p_{ij}(N) - \varepsilon)|\nu_i - \mu_i| \\
&= \sum_{i=1}^{m} \sum_{j=1}^{m} (p_{ij}(N) - \varepsilon)|\nu_i - \mu_i| \\
&= (1 - m\varepsilon) \sum_{i=1}^{m} |\nu_j - \mu_j|.
\end{aligned}$$

Thus, $\|\nu P^N - \mu P^N\|_1 \leq (1 - m\varepsilon)\|\nu - \mu\|_1$ and, by induction,

$$\|\nu P^{Nk} - \mu P^{Nk}\|_1 \leq (1 - m\varepsilon)^k \|\nu - \mu\|_1.$$

For powers in between the multiples of N, of the form $Nk + \ell$ for $\ell \leq N$, we use

$$\|\nu P^{Nk+\ell} - \mu P^{Nk+\ell}\|_1 \leq (1 - m\varepsilon)^k \|\nu P^\ell - \mu P^\ell\|_1 \leq (1 - m\varepsilon)^k \|\nu - \mu\|_1.$$

This proves that $\|\nu P^n - \mu P^n\|_1 \to 0$ for any two distributions ν and μ. When μ is the stationary distribution, $\mu P^n = \mu$, so $\|\nu P^n - \mu\|_1 \to 0$. This finishes the proof.□

Exercise 9.3.2. Given $p \in (0,1)$, consider a Markov chain with the transition probabilities

$$p_{i,i-1} = 1 - p, p_{i,i+1} = p \text{ for } i = 2, \ldots, m-1$$

and $p_{1,1} = 1 - p$, $p_{1,2} = p$, $p_{m,m-1} = 1 - p$, $p_{m,m} = p$. What is the limit $\lim_{n\to\infty} P^n$. *Hint:* solve the system $\mu = \mu P$ explicitly.

Exercise 9.3.3. Show that if a Markov chain is irreducible and aperiodic then, for any function $f: S \to \mathbb{R}$, the expected value $\mathbb{E} f(X_n)$ converges.

Exercise 9.3.4. Suppose that m people sit at the round table and all have plates with rice in front of them. At the same time, each person divides his of her rice in half and evens out one half with the half of the person on the right (i.e. they split the total of their halves evenly), and another half with the person on the left. If they keep repeating this, what will happen in the long run?

Exercise 9.3.5. Let us consider the L^∞ norm on $\mathbb{R}^m$,

$$\|x-y\|_\infty = \max_{i\le m} |x_i - y_i|.$$

If P is a Markov transition matrix P, show that

$$\|Px - Py\|_\infty \le \|x-y\|_\infty.$$

9.4 Reversible Markov chains

Let us consider a Markov chain with the transition matrix P. A probability distribution $\mu = (\mu_1, \dots, \mu_m)$ on the set of states is called *reversible for the chain* if

$$\mu_i p_{ij} = \mu_j p_{ji} \tag{9.40}$$

for all i and j. A Markov chain is called *reversible* if it has a reversible distribution. The equations (9.40) are called the *detailed balance equations*. If we start the chain with the distribution μ then (9.40) states that the first two outcomes (s_i, s_j) and (s_j, s_i) are equally likely. Moreover,

$$\begin{aligned}\mathbb{P}(X_0 = s_i, X_1 = s_j, X_2 = s_k) &= \mu_i p_{ij} p_{jk} \\ = p_{ji}\mu_j p_{jk} = p_{ji} p_{kj} \mu_k &= \mathbb{P}(X_0 = s_k, X_1 = s_j, X_2 = s_i),\end{aligned}$$

which means that the first three outcomes (s_i, s_j, s_k) and (s_k, s_j, s_i) are equally likely. The same calculation show that any sequence of outcomes and the same sequence in reverse order are equally likely, hence the name reversible. It is easy to check that a reversible distribution is stationary.

Lemma 9.7. *A reversible distribution μ is stationary.*

Proof. Summing (9.40) over i, we get

$$\sum_{i=1}^{m} \mu_i p_{ij} = \mu_j \sum_{i=1}^{m} p_{ji} = \mu_j,$$

which finishes the proof. □

There are a number of examples, where the equations (9.40) are easy to guess or easy to check, so this is one good way to find a stationary distribution. Let us consider some classical examples of reversible Markov chains.

Example 9.4.1 (Birth-and-death processes). Let us consider a Markov chain that in one step moves from a state s_i only to s_{i+1} or s_{i-1}. Of course, since our state space is finite, this means that from s_1 it only moves up to s_2 and from s_m it only moves down to s_{m-1}. The chain can also stay in each state s_i. Another way to say this is that

$$p_{ij} > 0 \text{ if } |i-j| = 1, \text{ and } p_{ij} = 0 \text{ if } |i-j| > 1.$$

The probabilities p_{ii} may be positive or zero. Such Markov chains are called *birth-and-death* chains, and they are all reversible. To find a reversible distribution, let us set $\mu_1 = x$ as an unknown variable. If (9.40) holds then

$$\mu_2 = \frac{\mu_1 p_{12}}{p_{21}} = \frac{p_{12}}{p_{21}} x.$$

Given μ_2, we can find μ_3,

$$\mu_3 = \frac{\mu_2 p_{23}}{p_{32}} = \frac{p_{12}}{p_{21}} \frac{p_{23}}{p_{32}} x,$$

and we can continue, by induction,

$$\mu_i = \frac{p_{12}}{p_{21}} \cdots \frac{p_{i-1,i}}{p_{i,i-1}} x.$$

Since all μ_i are of the form $c_i x$ for some positive $c_i > 0$, to find x we can use that the probabilities add up to 1, so

$$x = \frac{1}{c_1 + \ldots + c_m}.$$

Of course, the same calculation works for any chain with the transition graph that looks like a single path between two 'endpoint' vertices, and we can relabel the states of such chain to turn it into a formal birth-and-death chain. □

Example 9.4.2 (The Ehrenfest model of diffusion). One example of a birth-and-death Markov chain is Ehrenfest's model of diffusion. Suppose there are m particles inside two connected containers. The particles can move between the two containers, and suppose that the next particle that moves is chosen uniformly at random. This process can be described by a Markov chain with $m+1$ states s_i for $i = 0, \ldots, m$, where the system is in the state i if there are i particles in the first container and $m - i$ in the second. The transition probabilities are given by

$$p_{i,i-1} = \frac{i}{m}, \; p_{i,i+1} = \frac{m-i}{m}. \tag{9.41}$$

Since this is a birth-and-death chain, it is reversible and it turns out that its reversible distribution is the Binomial $B(m, \frac{1}{2})$ distribution,

$$\mu_i = \binom{m}{i} \frac{1}{2^m}.$$

We will leave it as an exercise to check that the detailed balance equations (9.40) are satisfied. □

Example 9.4.3 (Random walks on graphs). Let us consider a connected graph G on the set of vertices $\{s_1, \ldots, s_m\}$. This means that all vertices are connected by a path on edges of the graph. Let N_i be the number of neighbours of s_i. Let us consider a Markov chain with the transition probabilities

$$p_{ij} = \begin{cases} \frac{1}{N_i}, & \text{if } s_j \text{ is a neighbour of } s_i, \\ 0, & \text{otherwise.} \end{cases} \tag{9.42}$$

In other words, in each state s_i the chain picks a neighbour uniformly at random and moves there. A distribution μ is reversible for this chain if

$$\mu_i \frac{1}{N_i} = \mu_j \frac{1}{N_j}.$$

Therefore, all the ratios $\frac{\mu_i}{N_i}$ must be equal to the same constant c, so $\mu_i = cN_i$. Since the probabilities must add up to 1, the constant $c = \frac{1}{N}$, where $N = N_1 + \ldots + N_m$, and

$$\mu_i = \frac{N_i}{N}. \tag{9.43}$$

Since the graph is connected, the Markov chain is irreducible and μ is the unique stationary distribution. □

The most important examples of reversible Markov chains appear in the context of *Markov Chain Monte Carlo (MCMC)* algorithms. Let us describe one such algorithm.

Example 9.4.4 (Metropolis algorithm). As in the previous example, let us consider a connected graph G on the set of vertices $S = \{s_1, \ldots, s_m\}$. Suppose that we are interested in some probability distribution μ on S, but μ is defined by a complicated model that makes the computation of its coordinates μ_i impractical. On the other hand, for any neighbours s_i and s_j on the graph, μ_i and μ_j are related in some simple way, so that we know the ratio

$$r_{ij} = \frac{\mu_i}{\mu_j}$$

without knowing μ_i and μ_j. Of course, given r_{ij}, the distribution μ can in principle be reconstructed by setting $\mu_1 = x$ as unknown parameter, then finding $\mu_i = r_{i1}x$ in terms of x for all its neighbours, and propagating this through the graph to express all $\mu_i = xc_i$ in terms of x. Then we find

$$x = \frac{1}{c_1 + \ldots + c_m},$$

because the probabilities must add up to one. However, in many applications the graph is so large (say $m = 2^{20}$) that it is not even feasible to add up m numbers.

If we only know the above ratios r_{ij}, how can we compute, for example, the expectation of some function $f\colon S \to \mathbb{R}$ on S with respect to the distribution μ,

$$\mathbb{E}_\mu f := \sum_{i=1}^{m} f(s_i)\mu_i,$$

without computing μ? The idea of the Markov Chain Monte Carlo method is to combine the law of large numbers with the convergence theorem for Markov chains. Suppose that we can construct a Markov chain with the stationary distribution μ, whose transition probabilities depends only on the ratios r_{ij}. Then we can pick the initial state at random and run the chain for a large number of steps, n, to obtain a random variable X_n with the distribution close to μ. If we repeat this procedure N times, we will get N i.i.d. random variables $X_n^1,\ldots,X_n^N$ with the distribution close to μ. By the law of large numbers,

$$\frac{1}{N}\sum_{k=1}^{N} f(X_n^k) \approx \mathbb{E}f(X_n^1) \approx \sum_{i=1}^{m} f(s_i)\mu_i.$$

In other words, this procedure allows us to approximate the expectation of f with respect to the distribution μ, when it is not feasible to compute μ directly.

Here is a classical construction of a Markov chain, called the *Metropolis* chain, with the stationary distribution μ and the transition probabilities that depend only on the ratios r_{ij}. Let us write $i \sim j$ to denote that s_i and s_j are neighbours. For each vertex s_i, we define

$$p_{ij} = \frac{1}{N_i}\min\Big(\frac{\mu_j N_i}{\mu_i N_j}, 1\Big) \tag{9.44}$$

if $j \sim i$ and define

$$p_{ii} = 1 - \sum_{j\sim i} p_{ij}. \tag{9.45}$$

All other $p_{ij} = 0$. Notice that $p_{ii} \geq 0$, because

$$\sum_{j\sim i} p_{ij} \leq \sum_{j\sim i}\frac{1}{N_i} = 1,$$

so this definition makes sense. Notice also that the transition probabilities depend only on the ratios r_{ij}.

To show that μ is the stationary distribution of this Markov chain, we will check that μ satisfies the detailed balance equations (9.40). For $i = j$ there is nothing to check, and if $i \not\sim j$ then both sides are zero, so we only need to check this for $i \sim j$. If $\mu_j N_i \leq \mu_i N_j$ then, by (9.44),

$$p_{ij} = \frac{1}{N_i} \text{ and } p_{ji} = \frac{1}{N_j}\frac{\mu_i N_j}{\mu_j N_i} = \frac{\mu_i}{\mu_j N_i},$$

which immediately implies (9.40). The case $\mu_j N_i \geq \mu_i N_j$ is exactly the same, so the Metropolis chain has the stationary distribution μ.

If at least one $p_{ii} > 0$ then the chain is aperiodic. If the chain is periodic, a slight modification will make it periodic without affecting the stationary distribution μ (see exercise below). Then the convergence theorem applies and we can use this chain to produce an i.i.d. sample with the distribution close to μ. □

Exercise 9.4.1. Show that if μ is a reversible distribution for P, it is also a reversible distribution for $\lambda I + (1-\lambda)P$ for any $\lambda \in [0,1]$.

Exercise 9.4.2. Show that if μ is a reversible distribution for P, it is also a reversible distribution for P^n.

Exercise 9.4.3. Show that a birth-and-death Markov chain is aperiodic if and only if at least one $p_{ii} > 0$.

Exercise 9.4.4. Show that a Markov chain with the following transition matrix is not reversible:

$$P = \begin{bmatrix} 0 & 0.75 & 0.25 \\ 0.25 & 0 & 0.75 \\ 0.75 & 0.25 & 0 \end{bmatrix}.$$

Exercise 9.4.5. Check that the Binomial distribution $B(m, \frac{1}{2})$ satisfies the detailed balance equations (9.40) for the Ehrenfest chain (9.41).

Exercise 9.4.6. Suppose that two containers contain a total of m while balls and m black balls. At each step a ball is chosen at random from all $2m$ balls and put into the other container. We say that a system is in the state s_i if there are i white balls in the first container. Find the transition probabilities of this chain and compute $\lim_{n\to\infty} P^n$.

Exercise 9.4.7. Consider the Ehrenfest chain (9.41) and let D_n be the difference of particles in the two containers at time n. This means that if the chain is in the state s_i at time n then $D_n = 2i - m$. Prove that

$$\mathbb{E}D_{n+1} = \frac{m-2}{m}\mathbb{E}D_n = \left(\frac{m-2}{m}\right)^n \mathbb{E}D_0.$$

Exercise 9.4.8. Consider a random walk on the following graph:

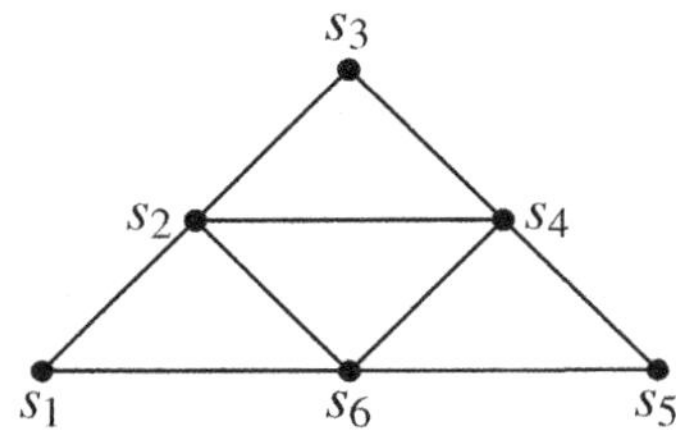

If we start the chain at s_1, what is the expected time of the first return to s_1? What is the expected number of visits to s_4 before the first return to s_1? *Hint:* recall the results in Section 9.2.

Chapter 10
Gaussian Distributions

10.1 Gaussian integration by parts, interpolation, and concentration

In this section, we will describe several Gaussian techniques, such as the Gaussian integration by parts, interpolation and concentration. We will begin with the Gaussian integration by parts. Let g be a centered Gaussian random variable with variance v^2 and let us denote the density function of its distribution by

$$\varphi_v(x) = \frac{1}{\sqrt{2\pi}v}\exp\Big(-\frac{x^2}{2v^2}\Big). \tag{10.1}$$

Since $x\varphi_v(x) = -v^2\varphi_v'(x)$, given a differentiable function $F\colon \mathbb{R}\to\mathbb{R}$, we can formally integrate by parts,

$$\begin{aligned}\mathbb{E}gF(g) = \int xF(x)\varphi_v(x)\,dx &= -v^2F(x)\varphi_v(x)\Big|_{-\infty}^{+\infty} + v^2\int F'(x)\varphi_v(x)\,dx\\ &= v^2\int F'(x)\varphi_v(x)\,dx = v^2\mathbb{E}F'(g),\end{aligned}$$

if the limits $\lim_{x\to\pm\infty}F(x)\varphi_v(x) = 0$ and the expectations on both sides are finite. In fact, this formula holds under the assumption that $\mathbb{E}|F'(g)| < \infty$, i.e. the right hand side is well defined.

Lemma 10.1 (Gaussian integration by parts). *If $g\sim N(0,v^2)$ then*

$$\mathbb{E}gF(g) = \mathbb{E}g^2\,\mathbb{E}F'(g) = v^2\,\mathbb{E}F'(g), \tag{10.2}$$

whenever $\mathbb{E}|F'(g)| < \infty$.

Proof. By making the change of variables $Z = g/v$ and considering the function $f(x) = F(vx)$, it is enough to prove that $Z\sim N(0,1)$ satisfies

$$\mathbb{E}Zf(Z) = \mathbb{E}f'(Z), \tag{10.3}$$

whenever $\mathbb{E}|f'(Z)| < \infty$. Let us start by writing

$$\begin{aligned}\mathbb{E}f'(Z) &= \frac{1}{\sqrt{2\pi}}\int_{-\infty}^{+\infty} f'(z)e^{-\frac{z^2}{2}}\,dz \\ &= \frac{1}{\sqrt{2\pi}}\int_{-\infty}^{0} f'(z)e^{-\frac{z^2}{2}}\,dz + \frac{1}{\sqrt{2\pi}}\int_{0}^{+\infty} f'(z)e^{-\frac{z^2}{2}}\,dz.\end{aligned}$$

If we represent $e^{-\frac{z^2}{2}} = -\int_{-\infty}^{z} xe^{-\frac{x^2}{2}}\,dx$ for $z \le 0$ then

$$\begin{aligned}\frac{1}{\sqrt{2\pi}}\int_{-\infty}^{0} f'(z)e^{-\frac{z^2}{2}}\,dz &= -\frac{1}{\sqrt{2\pi}}\int_{-\infty}^{0}\int_{-\infty}^{z} f'(z)xe^{-\frac{x^2}{2}}\,dxdz \\ &= -\frac{1}{\sqrt{2\pi}}\int_{-\infty}^{0}\int_{x}^{0} f'(z)xe^{-\frac{x^2}{2}}\,dzdx \\ &= \frac{1}{\sqrt{2\pi}}\int_{-\infty}^{0} \big(f(x)-f(0)\big)xe^{-\frac{x^2}{2}}\,dx,\end{aligned}$$

where we could switch the order of integration, by Fubini's theorem, because the integrand is absolutely integrable over the region $x \le z \le 0$ by the assumption that $\mathbb{E}|f'(Z)| < \infty$. Similarly, representing

$$e^{-\frac{z^2}{2}} = \int_{z}^{\infty} xe^{-\frac{x^2}{2}}\,dx \ \text{ for } z \ge 0,$$

we can write the second integral as

$$\frac{1}{\sqrt{2\pi}}\int_{0}^{+\infty} f'(z)e^{-\frac{z^2}{2}}\,dz = \frac{1}{\sqrt{2\pi}}\int_{0}^{\infty} \big(f(x)-f(0)\big)xe^{-\frac{x^2}{2}}\,dx.$$

Adding up the two integrals, we get $\mathbb{E}f'(Z) = \mathbb{E}Z\big(f(Z)-f(0)\big) = \mathbb{E}Zf(Z)$, which finishes the proof. □

This computation can be generalized to Gaussian vectors. Let $g = (g_\ell)_{1\le\ell\le n}$ be a Gaussian vector with the covariance C, $g \sim N(0,C)$, and consider a differentiable function $F = F\big((x_\ell)_{1\le\ell\le n}\big)\colon \mathbb{R}^n \to \mathbb{R}$. Let us denote

$$F_\ell = \frac{\partial F}{\partial x_\ell}. \tag{10.4}$$

Then the following holds.

Lemma 10.2 (Gaussian integration by parts). *If $g \sim N(0,C)$ is a centered Gaussian vector on $\mathbb{R}^n$ then*

$$\mathbb{E}g_1F(g) = \sum_{\ell\le n} \mathbb{E}(g_1g_\ell)\,\mathbb{E}F_\ell(g), \tag{10.5}$$

if, for all $\ell \le n$, either $\mathbb{E}|F_\ell(g)| < \infty$ or $\mathbb{E}g_1g_\ell = 0$.

Proof. If $v^2 = \mathbb{E}g_1^2$ then the Gaussian vector $g' = (g'_\ell)_{1\le \ell\le n}$ defined by

$$g'_\ell = g_\ell - \lambda_\ell g_1, \text{ where } \lambda_\ell = v^{-2}\mathbb{E}g_1 g_\ell, \tag{10.6}$$

is independent of g_1, since $\mathbb{E}g_1 g'_\ell = \mathbb{E}g_1 g_\ell - \lambda_\ell v^2 = 0$. If we denote $\lambda = (\lambda_\ell)_{1\le\ell\le n}$ then we can write $g = g' + g_1\lambda$. If $\mathbb{E}_1$ denotes the expectation in g_1 only then using (10.2) conditionally on g' implies that

$$\mathbb{E}_1 g_1 F(g) = \mathbb{E}_1 g_1 F(g' + g_1\lambda) = v^2\,\mathbb{E}_1 \frac{\partial F}{\partial x}(g' + x\lambda)\Big|_{x=g_1} \tag{10.7}$$

if the right hand side is well defined. Since

$$\frac{\partial F}{\partial x}(g' + x\lambda)\Big|_{x=g_1} = \sum_{\ell\le n} \lambda_\ell \frac{\partial F}{\partial x_\ell}(g' + x\lambda)\Big|_{x=g_1} = \sum_{\ell\le n}\lambda_\ell F_\ell(g), \tag{10.8}$$

our assumption that, for all $\ell \le n$, either $\mathbb{E}|F_\ell(g)| < \infty$ or $\lambda_\ell = 0$ implies (10.7). Moreover, by this assumption, (10.7) is absolutely integrable and integrating in g' finishes the proof. □

We will encounter two types of examples, when the function $F = F((x_\ell)_{1\le\ell\le n})$ and all its partial derivatives are bounded, or, when they grow at most exponentially fast, that is, for some constants $c_1, c_2 > 0$, for $\ell \le n$,

$$\Big|\frac{\partial F}{\partial x_\ell}(x)\Big| \le c_1 e^{c_2|x|}. \tag{10.9}$$

Most commonly, Gaussian integration by parts will be used in various calculations involving *Gaussian interpolation*. Consider two independent Gaussian random vectors $X = (X_i)_{i\le n}$ and $Y = (Y_i)_{i\le n}$ with the covariances

$$a_{i,j} = \mathbb{E}X_iX_j \text{ and } b_{i,j} = \mathbb{E}Y_iY_j. \tag{10.10}$$

Since X and Y are independent, for $0 \le t \le 1$,

$$Z(t) = \sqrt{t}X + \sqrt{1-t}Y \tag{10.11}$$

is a Gaussian random vector $Z(t) = (Z_i(t))_{i\le n}$ with the covariance

$$\mathbb{E}Z_i(t)Z_j(t) = ta_{i,j} + (1-t)b_{i,j},$$

which corresponds to a linear interpolation between the covariances of X and Y. Let us consider a function

$$f(t) = \mathbb{E}F(Z(t)) = \mathbb{E}F(\sqrt{t}X + \sqrt{1-t}Y). \tag{10.12}$$

The end points of this interpolation are $f(0) = \mathbb{E}F(Y)$ and $f(1) = \mathbb{E}F(X)$, and the derivative along the interpolation can be computed as follows.

Lemma 10.3 (Gaussian interpolation). *If for some* $c_1, c_2 > 0$ *and all* $1 \le i, j \le n$,

$$\Big|\frac{\partial F}{\partial x_i}(x)\Big| \le c_1 e^{c_2|x|} \text{ and, either } \Big|\frac{\partial^2 F}{\partial x_i \partial x_j}(x)\Big| \le c_1 e^{c_2|x|} \text{ or } a_{i,j} = b_{i,j}, \tag{10.13}$$

then

$$f'(t) = \frac{1}{2}\sum_{i,j\le n}(a_{i,j} - b_{i,j})\mathbb{E}\frac{\partial^2 F}{\partial x_i \partial x_j}(Z(t)). \tag{10.14}$$

Proof. When the partial derivatives have at most exponential growth, it is easy to check that one can interchange the derivative and integral to write

$$f'(t) = \mathbb{E}\frac{d}{dt}F(Z(t)) = \mathbb{E}\sum_{i\le n}\frac{\partial F}{\partial x_i}(Z(t))Z_i'(t) = \sum_{i\le n}\mathbb{E}\frac{\partial F}{\partial x_i}(Z(t))Z_i'(t).$$

Let us apply Gaussian integration by parts to each of the terms on the right hand side. Since the covariance

$$\mathbb{E}Z_i'(t)Z_j(t) = \mathbb{E}\Big(\frac{1}{2\sqrt{t}}X_i - \frac{1}{2\sqrt{1-t}}Y_i\Big)\big(\sqrt{t}X_j + \sqrt{1-t}Y_j\big) = \frac{1}{2}(a_{i,j} - b_{i,j}),$$

the Gaussian integration by parts formula (10.5) implies that

$$\mathbb{E}\frac{\partial F}{\partial x_i}(Z(t))Z_i'(t) = \frac{1}{2}\sum_{j\le n}(a_{i,j} - b_{i,j})\mathbb{E}\frac{\partial^2 F}{\partial x_i \partial x_j}(Z(t)),$$

because of the assumption (10.13). Adding up over $i \le n$ finishes the proof. □

Using the above interpolation, one can prove the following classical Gaussian concentration inequality for Lipschitz functions. Let us consider a function $F = F\big((x_\ell)_{1\le\ell\le n}\big)\colon \mathbb{R}^n \to \mathbb{R}$ such that, for some $L > 0$,

$$|F(x) - F(y)| \le L|x - y| \text{ for all } x, y \in \mathbb{R}^n. \tag{10.15}$$

The smallest such L is called the Lipschitz seminorm $\|F\|_{\mathrm{Lip}}$ of F.

Theorem 10.1 (Gaussian concentration). *If* $g = (g_i)_{i\le n}$ *is standard Gaussian on* $\mathbb{R}^n$ *then, for any* $t \ge 0$,

$$\mathbb{P}\Big(\big|F(g) - \mathbb{E}F(g)\big| \ge t\Big) \le 2\exp\Big(-\frac{t^2}{4\|F\|_{\mathrm{Lip}}^2}\Big). \tag{10.16}$$

Proof. First, let us suppose that F is differentiable and its gradient is bounded by L, $|\nabla F| \le L$. Take any $\lambda \ge 0$. The Gaussian interpolation we would like to consider is of the form

$$f(t) = \mathbb{E}\exp\lambda\Big(F(\sqrt{t}g^1 + \sqrt{1-t}g) - F(\sqrt{t}g^2 + \sqrt{1-t}g)\Big), \tag{10.17}$$

where $g = (g_i)_{i\le n}$, $g^1 = (g_i^1)_{i\le n}$ and $g^2 = (g_i^2)_{i\le n}$ are three independent standard Gaussian vectors on $\mathbb{R}^n$. If we consider a function $G\colon \mathbb{R}^{2n} \to \mathbb{R}$ given by the formula

$$G(x_1,\ldots,x_n,x_{n+1},\ldots,x_{2n}) = \exp\lambda\Big(F(x_1,\ldots,x_n) - F(x_{n+1},\ldots,x_{2n})\Big),$$

and define Gaussian random vectors X and Y on $\mathbb{R}^{2n}$ by $X = (g^1, g^2)$ and $Y = (g,g)$, then the above interpolation can be rewritten in the form (10.12) as

$$f(t) = \mathbb{E}G(\sqrt{t}X + \sqrt{1-t}Y).$$

Notice that the covariance matrices of X and Y are equal to

$$A = \mathrm{Cov}(X) = \begin{bmatrix} I_n & 0 \\ 0 & I_n \end{bmatrix},\ B = \mathrm{Cov}(Y) = \begin{bmatrix} I_n & I_n \\ I_n & I_n \end{bmatrix},$$

where I_n is $n \times n$ identity matrix. Therefore,

$$A - B = \begin{bmatrix} 0 & -I_n \\ -I_n & 0 \end{bmatrix},$$

and the difference $a_{i,j} - b_{i,j}$ is non-zero only when $1 \le i \le n$ and $j = i+n$, or $1 \le j \le n$ and $i = j+n$. In both cases, $a_{i,j} - b_{i,j} = -1$. If we introduce the notation $F_i = \frac{\partial F}{\partial x_i}$ then it is easy to see that

$$\frac{\partial^2 G}{\partial x_i \partial x_{i+n}} = \lambda^2 G F_i(x_1,\ldots,x_n) F_i(x_{n+1},\ldots,x_{2n}).$$

By our assumption, $|\nabla F| \le L$, so the partial derivatives F_i are all bounded, and F grows at most linearly, $|F(x)| \le c_1 + c_2|x|$. Hence, the growth conditions in (10.13) (for G) are satisfied, and the formula (10.14) gives

$$f'(t) = \lambda^2 \mathbb{E}G(\sqrt{t}X + \sqrt{1-t}Y) \sum_{i=1}^n F_i(\sqrt{t}g^1 + \sqrt{1-t}g) F_i(\sqrt{t}g^2 + \sqrt{1-t}g).$$

By the Cauchy-Schwarz inequality,

$$\begin{aligned}&\sum_{i=1}^n F_i(\sqrt{t}g^1 + \sqrt{1-t}g) F_i(\sqrt{t}g^2 + \sqrt{1-t}g) \\ &\le |\nabla F(\sqrt{t}g^1 + \sqrt{1-t}g)||\nabla F(\sqrt{t}g^2 + \sqrt{1-t}g)| \le L^2,\end{aligned}$$

which implies that (since G is positive)

$$f'(t) \le \lambda^2 L^2 \mathbb{E}G(\sqrt{t}X + \sqrt{1-t}Y) = \lambda^2 L^2 f(t).$$

Therefore, the derivative

$$\big(f(t)e^{-\lambda^2L^2t}\big)' = e^{-\lambda^2L^2t}\big(f'(t) - \lambda^2L^2 f(t)\big) \le 0,$$

so $f(t)e^{-\lambda^2L^2t}$ is decreasing and $f(1) \le e^{\lambda^2L^2}f(0)$. Recalling the definition (10.17), we see that $f(0) = 1$ and, therefore, we proved that

$$f(1) = \mathbb{E}\exp\lambda\big(F(g^1) - F(g^2)\big) \le e^{\lambda^2L^2}.$$

The Gaussian interpolation and the assumption on the gradient, $|\nabla F| \le L$, have played their roles.

Since g^1 and g^2 are independent, integrating in g^2 first (let us denote this integral $\mathbb{E}_2$) and using Jensen's inequality,

$$\exp\lambda\big(F(g^1) - \mathbb{E}_2F(g^2)\big) \le \mathbb{E}_2\exp\lambda\big(F(g^1) - F(g^2)\big).$$

Integrating this in g^1, and using that g^1, g^2 have the same distribution as g, we get

$$\mathbb{E}\exp\lambda\big(F(g) - \mathbb{E}F(g)\big) \le \mathbb{E}\exp\lambda\big(F(g^1) - F(g^2)\big) \le e^{\lambda^2L^2}.$$

Using Markov's inequality,

$$\mathbb{P}\big(F(g) - \mathbb{E}F(g) \ge t\big) \le \mathbb{E}e^{-\lambda t}\exp\lambda\big(F(g) - \mathbb{E}F(g)\big) \le e^{-\lambda t + \lambda^2L^2}.$$

This inequality holds for any $\lambda \ge 0$, and minimizing the right hand side over λ (in other words, setting $\lambda = t/2L^2$) proves that

$$\mathbb{P}\Big(F(g) - \mathbb{E}F(g) \ge t\Big) \le \exp\Big(-\frac{t^2}{4\|F\|_{\mathrm{Lip}}^2}\Big). \tag{10.18}$$

Applying this to $-F$ and using the union bound proves (10.16).

Finally, in the general case when we do not assume differentiability and only assume (10.15), one can use the standard smoothing technique. Namely, for $\varepsilon > 0$, we define

$$F_\varepsilon(x) = \mathbb{E}F(x + \varepsilon g), \tag{10.19}$$

where g is a standard Gaussian vector on $\mathbb{R}^n$. This function is also Lipschitz,

$$|F_\varepsilon(x) - F_\varepsilon(y)| \le \mathbb{E}|F(x+\varepsilon g) - F(y+\varepsilon g)| \le L|x-y|,$$

but it is also differentiable (even smooth), because

$$F_\varepsilon(x) = \frac{1}{(\sqrt{2\pi})^n}\int_{\mathbb{R}^n} F(x+\varepsilon y)e^{-|y|^2/2}\,dy = \frac{1}{(\varepsilon\sqrt{2\pi})^n}\int_{\mathbb{R}^n} F(y)e^{-|y-x|^2/2\varepsilon^2}\,dy.$$

The case proved above shows that

$$\mathbb{P}\Big(\big|F_\varepsilon(g) - \mathbb{E}F_\varepsilon(g)\big| \geq t\Big) \leq 2\exp\Big(-\frac{t^2}{4\|F\|_{\mathrm{Lip}}^2}\Big). \tag{10.20}$$

Since F_ε approximates F uniformly,

$$|F(x) - F_\varepsilon(x)| \leq \mathbb{E}|F(x) - F(x+\varepsilon g)| \leq L\varepsilon\mathbb{E}\|g\|,$$

letting $\varepsilon \downarrow 0$ finishes the proof of the general case. □

10.2 Examples of concentration inequalities

Example 10.2.1. Our first example will be a supremum of linear functionals in $\mathbb{R}^n$. Let us consider a bounded set A in $\mathbb{R}^n$, and consider the function

$$F(x) = \sup_{a\in A}(a,x) = \sup_{a\in A}\big(a_1x_1+\ldots+a_nx_n\big).$$

Since

$$\begin{aligned}|F(x)-F(y)| &= \Big|\sup_{a\in A}\big(a_1x_1+\ldots+a_nx_n\big)-\sup_{a\in A}\big(a_1y_1+\ldots+a_ny_n\big)\Big| \\ &\leq \sup_{a\in A}\Big|\big(a_1(x_1-y_1)+\ldots+a_n(x_n-y_n)\big)\Big| \leq \sup_{a\in A}\|a\|\,\|x-y\|,\end{aligned}$$

the Lipschitz constant of F is bounded by $\|F\|_{Lip}\leq \sup_{a\in A}\|a\|$. Therefore, if g is a standard Gaussian vector in $\mathbb{R}^n$ then

$$\mathbb{P}\Big(\big|\sup_{a\in A}(a,g)-\mathbb{E}\sup_{a\in A}(a,g)\big|\geq t\Big)\leq 2\exp\Big(-\frac{t^2}{4\sup_{a\in A}\|a\|^2}\Big). \tag{10.21}$$

Let us give a couple of special cases of this inequality.

Example 10.2.2. Let X by a Gaussian random vector in $\mathbb{R}^n$ with arbitrary covariance matrix. We know that X is equal in distribution to a linear transformation Cg of the standard Gaussian random vector g on $\mathbb{R}^n$. If $C_i=(c_{i,j})_{j\leq n}$ is the ith row of C then $X_i=C_ig$ and

$$\mathbb{E}X_i^2=\mathbb{E}(C_ig)^2=\sum_{j=1}^n c_{i,j}^2=\|C_i\|^2.$$

If we apply (10.21) with A given by the collection of rows $C_1,\ldots,C_n$, we get

$$\mathbb{P}\Big(\big|\max_{i\leq n}X_i-\mathbb{E}\max_{i\leq n}X_i\big|\geq t\Big)\leq 2\exp\Big(-\frac{t^2}{4\max_{i\leq n}\mathbb{E}X_i^2}\Big). \tag{10.22}$$

Example 10.2.3 (Concentration of finite dimensional norms of Gaussian vectors). Let us consider n-dimensional normed vector space E with the norm $\|\cdot\|_E$ and let $b_1,\ldots,b_n$ be any basis of E. If g is a standard Gaussian vector in $\mathbb{R}^n$, then

$$X=g_1b_1+\ldots+g_nb_n \tag{10.23}$$

is one way to produce a random vector in E. The norm of a vector $x\in E$ can be written as the supremum over the unit ball of the dual space E^* of linear functionals

$$\|x\|_E=\sup\Big\{\zeta(x)\,:\,\zeta\in E^*,\|\zeta\|_{E^*}\leq 1\Big\}, \tag{10.24}$$

and, in particular, the norm of the random vector X in (10.23) can be written as

$$\|X\|_E = \sup\Big\{g_1\zeta(b_1)+\ldots+g_n\zeta(b_n) : \zeta\in E^*, \|\zeta\|_{E^*}\le 1\Big\}. \tag{10.25}$$

This functional is of the same form as in (10.21), with the set A in $\mathbb{R}^n$ given by

$$A = \Big\{(\zeta(b_1),\ldots,\zeta(b_n) : \zeta\in E^*, \|\zeta\|_{E^*}\le 1\Big\}.$$

The Lipschitz norm of this function is bounded by $\sup_{a\in A}\|a\|$, denoted by

$$\sigma(X) := \sup\Big\{\big(\zeta(b_1)^2+\ldots+\zeta(b_n)^2\big)^{1/2} : \zeta\in E^*, \|\zeta\|_{E^*}\le 1\Big\}. \tag{10.26}$$

This shows that the norm of a random vector X with Gaussian coordinates satisfies the following concentration inequality,

$$\mathbb{P}\Big(\big|\|X\|_E - \mathbb{E}\|X\|_E\big| \ge t\Big) \le 2\exp\Big(-\frac{t^2}{4\sigma(X)^2}\Big). \tag{10.27}$$

Another common way to write this, replacing t by $t\mathbb{E}\|X\|_E$,

$$\mathbb{P}\Big(\big|\|X\|_E - \mathbb{E}\|X\|_E\big| \ge t\mathbb{E}\|X\|_E\Big) \le 2\exp\Big(-\frac{t^2}{4}\Big(\frac{\mathbb{E}\|X\|_E}{\sigma(X)}\Big)^2\Big). \tag{10.28}$$

The quantity $d(X) := \big(\frac{\mathbb{E}\|X\|_E}{\sigma(X)}\big)^2$ is called the *concentration dimension* of X.

Example 10.2.4 (Moment comparison for norms of Gaussian vectors). One useful consequence of the concentration inequality in previous example is that the L^p-norms for $p\ge 1$ of the norm $\|X\|_E$ of the Gaussian vector X in (10.23) are comparable to each other,

$$\big(\mathbb{E}\|X\|_E\big)^p \le \mathbb{E}\|X\|_E^p \le c_p\big(\mathbb{E}\|X\|_E\big)^p, \tag{10.29}$$

where c_p is some constant that depends only on p. The first inequality is just Jensen's inequality, so we only need to show the second one.

If we denote $x^+ = \max(0,x)$ the positive part of x, then we can bound

$$\|X\|_E \le \mathbb{E}\|X\|_E + (\|X\|_E - \mathbb{E}\|X\|_E)^+.$$

Let us denote $Z = (\|X\|_E - \mathbb{E}\|X\|_E)^+$. Using the inequality $(a+b)^p \le 2^{p-1}(a^p + b^p)$, we get

$$\mathbb{E}\|X\|_E^p \le 2^{p-1}(\mathbb{E}\|X\|_E)^p + 2^{p-1}\mathbb{E}Z^p.$$

For simplicity of notation, let us write σ instead of $\sigma(X)$. Recall that, if $p>0$ and a random variable $Z\ge 0$ is nonnegative then we can write the pth moment of Z as

$$\mathbb{E}Z^p = p\int_0^\infty x^{p-1}\mathbb{P}(Z\ge x)\,dx. \tag{10.30}$$

By (10.27), we know that, $\mathbb{P}(Z \geq x) \leq e^{-x^2/4\sigma^2}$ for $x \geq 0$ and, using (10.30),

$$\mathbb{E}Z^p \leq p\int_0^\infty x^{p-1}e^{-x^2/4\sigma^2}\,dx = \sigma^p p\int_0^\infty t^{p-1}e^{-t^2/4}\,dt = a_p\sigma^p,$$

where we made the change of variables $t = \sigma x$ and then denoted by a_p the constant $p\int_0^\infty t^{p-1}e^{-t^2/4}\,dt$. Using the concentration inequality in the previous example, we proved that

$$\mathbb{E}\|X\|_E^p \leq 2^{p-1}(\mathbb{E}\|X\|_E)^p + 2^{p-1}a_p\sigma(X)^p. \tag{10.31}$$

It remains to understand how to bound $\sigma(X)$ in (10.26).

For $\zeta \in E^*$, the random variable

$$\zeta(X) = g_1\zeta(b_1) + \ldots + g_n\zeta(b_n)$$

is Gaussian with the variance $\mathbb{E}\zeta(X)^2 = \zeta(b_1)^2 + \ldots + \zeta(b_n)^2$. If Y is an arbitrary (centred) Gaussian random variable $\sim N(0,v^2)$ then

$$\mathbb{E}|Y| = \frac{1}{\sqrt{2\pi}v}\int_{-\infty}^\infty |x|e^{-x^2/2v^2}\,dx = \frac{v}{\sqrt{2\pi}}\int_{-\infty}^\infty |y|e^{-y^2/2}\,dy = v\sqrt{2/\pi},$$

which implies that $v^2 = \mathbb{E}Y^2 = \frac{\pi}{2}(\mathbb{E}|Y|)^2$. Using this for $Y = \zeta(X)$, we get

$$\zeta(b_1)^2 + \ldots + \zeta(b_n)^2 = \mathbb{E}\zeta(X)^2 = \frac{\pi}{2}(\mathbb{E}|\zeta(X)|)^2.$$

Taking supremum over $\zeta \in E^*$ such that $\|\zeta\|_{E^*} \leq 1$, we get that

$$\sigma(X)^2 = \frac{\pi}{2}\big(\sup_\zeta \mathbb{E}|\zeta(X)|\big)^2 \leq \frac{\pi}{2}\big(\mathbb{E}\sup_\zeta|\zeta(X)|\big)^2 = \frac{\pi}{2}\big(\mathbb{E}\|X\|_E\big)^2.$$

This means that $\sigma(X) \leq \sqrt{\pi/2}\mathbb{E}\|X\|_E$, and plugging this into (10.31), finally proves (10.29) with the constant $c_p = 2^{p-1}(1 + a_p(\pi/2)^{p/2})$.

Exercise 10.2.1. Let $X \sim N(0,C)$ is a Gaussian random vector in $\mathbb{R}^n$, show that

$$\mathbb{P}\Big(\Big|\log\sum_{i=1}^n e^{X_i} - \mathbb{E}\log\sum_{i=1}^n e^{X_i}\Big| \geq t\Big) \leq 2\exp\Big(-\frac{t^2}{4\max_{i\leq n}\mathbb{E}X_i^2}\Big). \tag{10.32}$$

Exercise 10.2.2. If $g \sim N(0,I)$ is standard Gaussian on $\mathbb{R}^n$ and $\|g\|$ is its Euclidean norm, show that

$$\mathbb{P}\Big(\big|\|g\| - \mathbb{E}\|g\|\big| \geq t\Big) \leq 2\exp\Big(-\frac{t^2}{4}\Big).$$

10.3 Gaussian comparison inequalities

First, we will prove the so-called *Slepian-Fernique inequalities.*

Theorem 10.2 (Slepian-Fernique inequality I). *Let $X = (X_i)_{i\le n}$ and $Y = (Y_i)_{i\le n}$ be two Gaussian vectors on $\mathbb{R}^n$ such that*

1. $\mathbb{E}X_i^2 = \mathbb{E}Y_i^2$ *for all* $i \le n$,
2. $\mathbb{E}X_iX_j \le \mathbb{E}Y_iY_j$ *for all* $i, j \le n$.

Then, for any choice of parameters $(\lambda_i)_{i\le n} \in \mathbb{R}^n$,

$$\mathbb{P}\Big(\bigcap_{i=1}^n \{X_i \le \lambda_i\}\Big) \le \mathbb{P}\Big(\bigcap_{i=1}^n \{Y_i \le \lambda_i\}\Big) \tag{10.33}$$

and

$$\mathbb{E}\max_i X_i \ge \mathbb{E}\max_i Y_i. \tag{10.34}$$

What this means is that, if the coordinates of X have the same variance, but are less correlated, then they are less likely to stay below given thresholds (λ_i) and, therefore, their maximum is bigger on average. After we prove this result, we will give another proof of the second statement (10.34) under less restrictive assumptions.

Proof. Let us rewrite the indicator

$$\mathrm{I}\Big(\bigcap_{i=1}^n \{x_i \le \lambda_i\}\Big) = \prod_{i=1}^n \mathrm{I}(x_i \le \lambda_i).$$

Let us approximate each indicator $\mathrm{I}(x_i \le \lambda_i)$ by a smooth nonnegative decreasing function $\varphi_i(x_i)$. Define

$$\varphi(x) = \prod_{i=1}^n \varphi_i(x_i).$$

If we consider the interpolation $f(t) = \mathbb{E}\varphi(\sqrt{t}X + \sqrt{1-t}Y)$ and use the Gaussian interpolation formula (10.14), we will now check that the assumptions of the theorem about the covariances imply that $f'(t) \le 0$. Indeed, for $j \ne i$,

$$\frac{\partial^2}{\partial x_i \partial x_j} \prod_{\ell=1}^n \varphi_\ell(x_\ell) \ge 0,$$

because the derivatives applied to the factors i and j in the product will both be negative, since all functions φ_ℓ are decreasing. On the other hand, by assumption, the difference of the covariances $\mathbb{E}X_iX_j - \mathbb{E}Y_iY_j \le 0$ is negative in this case, so the corresponding term in (10.14) will be negative. The derivatives $\partial^2/\partial x_i^2$ are not important because the variances are equal and $\mathbb{E}X_i^2 - \mathbb{E}Y_i^2 = 0$. This proves that $f'(t) \le 0$ and, therefore,

$$f(1) = \mathbb{E}\varphi(X) \le f(0) = \mathbb{E}\varphi(Y).$$

Now, letting $\varphi_{i,j}$'s converge to the corresponding indicators, proves that

$$\mathbb{E}\mathrm{I}\Big(\bigcap_{i=1}^{n}\{X_i \le \lambda_i\}\Big) \le \mathbb{E}\mathrm{I}\Big(\bigcap_{i=1}^{n}\{Y_i \le \lambda_i\}\Big),$$

which is the same as (10.33). Let us show how this implies (10.34).

Notice that, if we take all $\lambda_i = \lambda$ then (10.33) can be rewritten as

$$\mathbb{P}\Big(\max_{i=1} X_i \le \lambda\Big) \le \mathbb{P}\Big(\max_{i=1} Y_i \le \lambda\Big). \tag{10.35}$$

Then the inequality (10.34) for the averages is an immediate consequence of the representation $\mathbb{E}Z = \int_0^\infty \mathbb{P}(Z \ge x)\,dx$ for $Z \ge 0$. Let us define $X = \max_{i=1} X_i$ and $Y = \max_{i=1} Y_i$ and let us decompose $X = X^+ - X^-$ and $Y = Y^+ - Y^-$ into positive and negative parts. If we take $\lambda \ge 0$, and apply the inequality (10.35) with λ and $-\lambda$, we get

$$\mathbb{P}(X^+ \ge \lambda) \ge \mathbb{P}(Y^+ \ge \lambda),\ \mathbb{P}(X^- \ge \lambda) \le \mathbb{P}(Y^- \ge \lambda).$$

This implies that $\mathbb{E}X^+ \ge \mathbb{E}Y^+$, $\mathbb{E}X^- \le \mathbb{E}Y^-$, and subtracting two inequalities we get (10.34). □

Next, we will prove (10.34) under less restrictive assumptions. If we write

$$\mathbb{E}(X_i - X_j)^2 = \mathbb{E}X_i^2 + \mathbb{E}X_j^2 - 2\mathbb{E}X_iX_j,\ \mathbb{E}(Y_i - Y_j)^2 = \mathbb{E}Y_i^2 + \mathbb{E}Y_j^2 - 2\mathbb{E}Y_iY_j$$

then, under the first assumption that all variances are equal, $\mathbb{E}X_i^2 = \mathbb{E}Y_i^2$, the second assumption that $\mathbb{E}X_iX_j \le \mathbb{E}Y_iY_j$ for all $i,j \le n$ is equivalent to

$$\mathbb{E}(X_i - X_j)^2 \ge \mathbb{E}(Y_i - Y_j)^2 \text{ for all } i,j \le n.$$

We will now show that with this reformulation of the second assumption, we can drop the first assumption, which is often convenient in applications.

Theorem 10.3 (Slepian-Fernique inequality II). *Let $X = (X_i)_{i\le n}$ and $Y = (Y_i)_{i\le n}$ be two Gaussian vectors on $\mathbb{R}^n$ such that*

$$\mathbb{E}(X_i - X_j)^2 \ge \mathbb{E}(Y_i - Y_j)^2 \textit{ for all } i,j \le n.$$

Then, the inequality (10.34) holds, i.e.

$$\mathbb{E}\max_i X_i \ge \mathbb{E}\max_i Y_i. \tag{10.36}$$

Proof. We will apply the same interpolation as in the above proof, but to a special smooth approximation of the maximum function $\max_{i\le n} x_i$ given by

$$F(x) = \frac{1}{\beta} \log \sum_{i \le n} e^{\beta x_i},$$

for positive parameter $\beta > 0$. The reason we can view this as a smooth approximation of the maximum is because

$$\max_{i \le n} x_i \le F(x) = \frac{1}{\beta} \log \sum_{i \le n} e^{\beta x_i} \le \max_{i \le n} x_i + \frac{\log n}{\beta}.$$

Indeed, if we keep only the largest term in the sum $\sum_{i \le n} e^{\beta x_i}$ we get the lower bound, and if we replace each term in this sum by the largest one, we get the upper bound. Now, we can make the second term $\log n / \beta$ becomes as small as we like by taking β large. If we prove that $\mathbb{E}F(X) \ge \mathbb{E}F(Y)$, we recover (10.36) by letting β go to infinity. To use the Gaussian interpolation (10.14), let us compute the derivatives of F. First of all,

$$p_i(x) := \frac{\partial F}{\partial x_i} = \frac{e^{\beta x_i}}{\sum_{j \le n} e^{\beta x_j}}.$$

Differentiating this, we see that

$$\frac{\partial^2 F}{\partial x_i^2} = \beta(p_i(x) - p_i(x)^2), \quad \frac{\partial^2 F}{\partial x_i \partial x_j} = -\beta p_i(x) p_j(x) \text{ if } i \ne j.$$

Therefore, the Gaussian interpolation formula (10.14) gives (to simplify notation, we write p_i instead of $p_i(Z(t))$)

$$\begin{aligned} f'(t) = \frac{d}{dt}\mathbb{E}F(Z(t)) &= \frac{1}{2} \sum_{i,j \le n} (\mathbb{E}X_i X_j - \mathbb{E}Y_i Y_j) \mathbb{E} \frac{\partial^2 F}{\partial x_i \partial x_j}(Z(t)) \\ &= \frac{\beta}{2} \sum_{i \le n} (\mathbb{E}X_i^2 - \mathbb{E}Y_i^2) \mathbb{E}(p_i - p_i^2) - \frac{\beta}{2} \sum_{i \ne j} (\mathbb{E}X_i X_j - \mathbb{E}Y_i Y_j) \mathbb{E} p_i p_j. \end{aligned}$$

However, we chose F in such a way that,

$$\sum_{i \le n} p_i(x) = \sum_{i \le n} \frac{e^{\beta x_i}}{\sum_{j \le n} e^{\beta x_j}} = 1.$$

If we multiply both sides by $p_i(x)$ and subtract $p_i^2(x)$, we get

$$p_i(x) - p_i^2(x) = \sum_{j \ne i} p_i(x) p_j(x).$$

This means that

$$\sum_{i \le n} (\mathbb{E}X_i^2 - \mathbb{E}Y_i^2) \mathbb{E}(p_i - p_i^2)$$

$$= \sum_{i\le n}(\mathbb{E}X_i^2 - \mathbb{E}Y_i^2)\mathbb{E}\sum_{j\ne i} p_i p_j = \sum_{i\ne j}(\mathbb{E}X_i^2 - \mathbb{E}Y_i^2)\mathbb{E}p_i p_j.$$

Switching indices i and j, this can also be written as $\sum_{i\ne j}(\mathbb{E}X_j^2 - \mathbb{E}Y_j^2)\mathbb{E}p_i p_j$, and taking average of the two, we get

$$\sum_{i\le n}(\mathbb{E}X_i^2 - \mathbb{E}Y_i^2)\mathbb{E}(p_i - p_i^2) = \frac{1}{2}\sum_{i\ne j}(\mathbb{E}X_i^2 + \mathbb{E}X_j^2 - \mathbb{E}Y_i^2 - \mathbb{E}Y_j^2)\mathbb{E}p_i p_j.$$

Plugging it into the above expression for the derivative $f'(t)$ and collecting the terms, we get

$$f'(t) = \frac{\beta}{4}\sum_{i\ne j}(\mathbb{E}(X_i - X_j)^2 - \mathbb{E}(Y_i - Y_j)^2)\mathbb{E}p_i p_j \ge 0,$$

since, by the assumption of the theorem, each term is positive. This implies that $f(1) \ge f(0)$ or, in other words, $\mathbb{E}F(X) \ge \mathbb{E}F(Y)$. This finishes the proof. □

Example 10.3.1 (Gaussian contraction inequality). Given a standard Gaussian vector $g = (g_1, \ldots, g_n)$ on $\mathbb{R}^n$, let us consider a Gaussian process

$$X(t) = (t, g) = t_1 g_1 + \ldots + t_n g_n, \tag{10.37}$$

where $t = (t_1, \ldots, t_n) \in T$ for some bounded subset $T \subseteq \mathbb{R}^n$. Let us consider a 1-Lipschitz function $f\colon \mathbb{R}^n \to \mathbb{R}^n$, i.e.

$$\|f(t) - f(t')\| \le \|t - t'\|,$$

and, if we write $f = (f_1, \ldots, f_n)$, consider a process

$$Y(t) = (f(t), g) = f_1(t)g_1 + \ldots + f_n(t)g_n. \tag{10.38}$$

Obviously,

$$\mathbb{E}(Y(t) - Y(t'))^2 = \|f(t) - f(t')\|^2 \le \|t - t'\|^2 = \mathbb{E}(X(t) - X(t'))^2,$$

so Theorem 10.3 implies that

$$\mathbb{E}\sup_{t\in T} Y(t) = \mathbb{E}\sup_{t\in T}(f(t), g) \le \mathbb{E}\sup_{t\in T} X(t) = \mathbb{E}\sup_{t\in T}(t, g), \tag{10.39}$$

which is known as the *Gaussian contraction inequality*. □

Next, we will prove a more general minimax analogue of the first theorem above known as *Gordon's comparison inequality*.

Theorem 10.4 (Gordon's inequality). *Let $(X_{i,j})$, $(Y_{i,j})$ be two Gaussian vectors indexed by $i \le n, j \le m$ such that*

1. $\mathbb{E}X_{i,j}^2 = \mathbb{E}Y_{i,j}^2$ *for all* i, j,
2. $\mathbb{E}X_{i,j}X_{i,\ell} \leq \mathbb{E}Y_{i,j}Y_{i,\ell}$ *for all* i, j, ℓ,
3. $\mathbb{E}X_{i,j}X_{k,\ell} \geq \mathbb{E}Y_{i,j}Y_{k,\ell}$ *for all* i, j, k, ℓ *such that* $i \neq k$.

Then, for any choice of parameters $(\lambda_{i,j})$,

$$\mathbb{P}\Big(\bigcup_{i=1}^{n}\bigcap_{j=1}^{m}\{X_{i,j} \leq \lambda_{i,j}\}\Big) \leq \mathbb{P}\Big(\bigcup_{i=1}^{n}\bigcap_{j=1}^{m}\{Y_{i,j} \leq \lambda_{i,j}\}\Big) \tag{10.40}$$

and

$$\mathbb{E}\min_i \max_j X_{i,j} \geq \mathbb{E}\min_i \max_j Y_{i,j}. \tag{10.41}$$

This reduces to Slepian's inequality when the index i takes only one value, so the condition 3 is empty.

Proof. Let us rewrite the indicator

$$\mathrm{I}\Big(\bigcup_{i=1}^{n}\bigcap_{j=1}^{m}\{x_{i,j} \leq \lambda_{i,j}\}\Big) = 1 - \prod_{i=1}^{n}\Big(1 - \prod_{j=1}^{m}\mathrm{I}(x_{i,j} \leq \lambda_{i,j})\Big).$$

Let us approximate each indicator $\mathrm{I}(x_{i,j} \leq \lambda_{i,j})$ by a smooth nonnegative non-increasing function $\varphi_{i,j}(x_{i,j})$. Define

$$\varphi(x) = 1 - \prod_{i=1}^{n}\Big(1 - \prod_{j=1}^{m}\varphi_{i,j}(x_{i,j})\Big).$$

If we consider the interpolation $f(t) = \mathbb{E}\varphi(\sqrt{t}X + \sqrt{1-t}Y)$ and use the Gaussian interpolation formula (10.14), it is easy to check that the conditions of the theorem on the covariance imply that $f'(t) \leq 0$. Indeed, for $j \neq \ell$,

$$\frac{\partial^2 \varphi}{\partial x_{i,j}\partial x_{i,\ell}} = \prod_{k \neq i}^{n}\Big(1 - \prod_{p=1}^{m}\varphi_{k,p}(x_{k,p})\Big)\frac{\partial^2}{\partial x_{i,j}\partial x_{i,\ell}}\prod_{p=1}^{m}\varphi_{i,p}(x_{i,p}) \geq 0,$$

because the derivatives applied to two factors in the last product will both be non-positive (since all functions $\varphi_{i,j}$ are non-increasing). On the other hand, the difference of the covariances $\mathbb{E}X_{i,j}X_{i,\ell} - \mathbb{E}Y_{i,j}Y_{i,\ell} \leq 0$ is negative in this case, so the corresponding term in (10.14) will be negative. Similarly, for $i \neq k$,

$$\frac{\partial^2 \varphi}{\partial x_{i,j}\partial x_{k,\ell}} = -\prod_{k' \neq i,k}^{n}\Big(1 - \prod_{p=1}^{m}\varphi_{k',p}(x_{k',p})\Big)\frac{\partial}{\partial x_{i,j}}\prod_{p=1}^{m}\varphi_{i,p}(x_{i,p})\frac{\partial}{\partial x_{k,\ell}}\prod_{p=1}^{m}\varphi_{k,p}(x_{k,p})$$

is ≤ 0. By assumption, the difference of the covariances $\mathbb{E}X_{i,j}X_{k,\ell} - \mathbb{E}Y_{i,j}Y_{k,\ell} \geq 0$ is positive in this case, so the corresponding term in (10.14) will again be negative. This proves that $f'(t) \leq 0$ and, therefore,

$$f(1) = \mathbb{E}\varphi(X) \leq f(0) = \mathbb{E}\varphi(Y).$$

Now, letting $\varphi_{i,j}$'s converge to the corresponding indicators, proves that

$$\mathbb{E}\mathrm{I}\Big(\bigcup_{i=1}^{n}\bigcap_{j=1}^{m}\{X_{i,j}\leq\lambda_{i,j}\}\Big)\leq\mathbb{E}\mathrm{I}\Big(\bigcup_{i=1}^{n}\bigcap_{j=1}^{m}\{Y_{i,j}\leq\lambda_{i,j}\}\Big),$$

which is the same as (10.40). If we take all $\lambda_{i,j}=\lambda$, this can be rewritten as

$$\mathbb{P}\Big(\min_{i}\max_{j}X_{i,j}\leq\lambda\Big)\leq\mathbb{P}\Big(\min_{i}\max_{j}Y_{i,j}\leq\lambda\Big).$$

and (10.41) follows by integration by parts as before. □

For the last statement of the previous theorem, one can also remove the assumption about the equalities of variances.

Theorem 10.5 (Gordon's inequality II). *Let $(X_{i,j})$ and $(Y_{i,j})$ be two Gaussian vectors indexed by $i\leq n$, $j\leq m$ such that*

1. $\mathbb{E}(X_{i,j}-X_{i,\ell})^2\geq\mathbb{E}(Y_{i,j}-Y_{i,\ell})^2$ *for all i,j,ℓ,*
2. $\mathbb{E}(X_{i,j}-X_{k,\ell})^2\leq\mathbb{E}(Y_{i,j}-Y_{k,\ell})^3$ *for all i,j,k,ℓ such that $i\neq k$.*

Then,

$$\mathbb{E}\min_{i}\max_{j}X_{i,j}\geq\mathbb{E}\min_{i}\max_{j}Y_{i,j}. \tag{10.42}$$

To prove this inequality, one can use that

$$F(x)=\frac{1}{\beta}\log\sum_{i}\frac{1}{\sum_{j}e^{\beta X_{i,j}}}\to-\min_{i}\max_{j}X_{i,j}\text{ as }\beta\to\infty,$$

and, as in the proof of Theorem 10.3, show that $\mathbb{E}F(X)\leq\mathbb{E}F(Y)$. The calculation is a bit more involved but straightforward. Letting $\beta\to\infty$ proves (10.42).

Exercise 10.3.1. If $g\sim N(0,I)$ is standard Gaussian on $\mathbb{R}^n$, a function $\sigma\colon\mathbb{R}\to\mathbb{R}$ is 1-Lipschitz and a set $T\subseteq\mathbb{R}^n$ is bounded, show that

$$\mathbb{E}\sup_{t\in T}\Big(\sigma(t_1)g_1+\ldots+\sigma(t_n)g_n\Big)\leq\mathbb{E}\sup_{t\in T}\big(t_1g_1+\ldots+t_ng_n\big).$$

Exercise 10.3.2. Prove Theorem 10.5.

10.4 Comparison of bilinear and linear forms

Let us consider the following random bilinear form

$$X(t,u) = \sum_{i=1}^{n}\sum_{j=1}^{m} g_{i,j} t_i u_j, \tag{10.43}$$

where $(g_{i,j})_{i\le n, j\le m}$ are i.i.d. standard Gaussian random variables and parameters

$$t = (t_1,\ldots,t_n) \in \mathbb{R}^n \text{ and } u = (u_1,\ldots,u_m) \in \mathbb{R}^m.$$

In various applications, one is interested in quantities of the form

$$\max_{t\in T}\max_{u\in U} X(t,u),\ \min_{t\in T}\max_{u\in U} X(t,u) \text{ or } \min_{u\in U}\max_{t\in T} X(t,u)$$

for some sets of parameters $T\subseteq \mathbb{R}^n$ and $U\subseteq\mathbb{R}^m$, which can be difficult to calculate or estimate directly. It turns out that one can use comparison inequalities from the previous section to relate these quantities to similar quantities for another, linear or 'almost' linear form. We will give several examples below.

Example 10.4.1. Let us first consider the following random process

$$Y(t,u) = \|u\|\sum_{i=1}^{n} h_i t_i + \|t\|\sum_{j=1}^{m} g_j u_j, \tag{10.44}$$

where $h_1,\ldots,h_n,g_1,\ldots,g_m$ are i.i.d. standard Gaussian random variables. This is not quite a linear form because of the factors $\|t\|$ and $\|u\|$, but the dependence on t and u is simpler and one can often analyze more easily the quantities

$$\max_{t\in T}\max_{u\in U} Y(t,u),\ \min_{t\in T}\max_{u\in U} Y(t,u) \text{ or } \min_{u\in U}\max_{t\in T} Y(t,u).$$

Let us take any two pairs of parameters (t,u) and (t',u') and compute

$$\begin{aligned}\mathbb{E}\big(X(t,u)-X(t',u')\big)^2 &= \mathbb{E}\Big(\sum_{i=1}^{n}\sum_{j=1}^{m} g_{i,j}(t_iu_j - t_i'u_j')\Big)^2 \\ &= \sum_{i=1}^{n}\sum_{j=1}^{m}(t_iu_j-t_i'u_j')^2 = \|t\|^2\|u\|^2 + \|t'\|^2\|u'\|^2 - 2(t,t')(u,u').\end{aligned} \tag{10.45}$$

Similarly, one can compute

$$\begin{aligned}&\mathbb{E}\big(Y(t,u)-Y(t',u')\big)^2 \\ &= \mathbb{E}\Big(\sum_{i=1}^{n} h_i(\|u\|t_i - \|u'\|t_i')\Big)^2 + \mathbb{E}\Big(\sum_{j=1}^{m} g_j(\|t\|u_j - \|t'\|u_j')\Big)^2 \\ &= 2\|t\|^2\|u\|^2 + 2\|t'\|^2\|u'\|^2 - 2\|u\|\|u'\|(t,t') - 2\|t\|\|t'\|(u,u').\end{aligned} \tag{10.46}$$

Subtracting and rearranging the terms, it is easy to see that

$$\mathbb{E}\big(Y(t,u)-Y(t',u')\big)^2-\mathbb{E}\big(X(t,u)-X(t',u')\big)^2= \tag{10.47}$$
$$=\Big(\|t\|\|u\|-\|t'\|\|u'\|\Big)^2+2\Big(\|t\|\|t'\|-(t,t')\Big)\Big(\|u\|\|u'\|-(u,u')\Big).$$

By the Cauchy-Schwarz inequality, $(t,t')\le\|t\|\|t'\|$ and $(u,u')\le\|u\|\|u'\|$, so

$$\mathbb{E}\big(Y(t,u)-Y(t',u')\big)^2\ge\mathbb{E}\big(X(t,u)-X(t',u')\big)^2. \tag{10.48}$$

By Theorem 10.3 in the previous section,

$$\mathbb{E}\max_{t\in T}\max_{u\in U}X(t,u)\le\mathbb{E}\max_{t\in T}\max_{u\in U}Y(t,u). \tag{10.49}$$

□

In order to apply the results for the minimax $\min_{t\in T}\max_{u\in U}$, we need the reverse inequality for $t=t'$. Because of (10.48), we can only hope to get the equality

$$\mathbb{E}\big(Y(t,u)-Y(t,u')\big)^2=\mathbb{E}\big(X(t,u)-X(t,u')\big)^2. \tag{10.50}$$

There are a couple of ways this can be achieved, as we will see in the following examples.

Example 10.4.2. One way to do this is to modify the definition of the bilinear form slightly and consider

$$X^+(t,u)=\sum_{i=1}^{n}\sum_{j=1}^{m}g_{i,j}t_iu_j+z\|t\|\|u\|, \tag{10.51}$$

where z is a standard Gaussian random variable independent of all $g_{i,j}$. Including this additional term will add $(\|t\|\|u\|-\|t'\|\|u'\|)^2$ to (10.45) and, as a result, (10.47) will become

$$\mathbb{E}\big(Y(t,u)-Y(t',u')\big)^2-\mathbb{E}\big(X^+(t,u)-X^+(t',u')\big)^2=$$
$$=2\Big(\|t\|\|t'\|-(t,t')\Big)\Big(\|u\|\|u'\|-(u,u')\Big). \tag{10.52}$$

This is equal to zero when $t=t'$ so, by Theorem 10.5,

$$\mathbb{E}\min_{t\in T}\max_{u\in U}X^+(t,u)\ge\mathbb{E}\min_{t\in T}\max_{u\in U}Y(t,u). \tag{10.53}$$

Notice that the inequality is reversed in this case and together with (10.49), we can write

$$\mathbb{E}\min_{t\in T}\max_{u\in U}Y(t,u)\le\mathbb{E}\min_{t\in T}\max_{u\in U}X^+(t,u)$$

$$\leq \mathbb{E}\max_{t\in T}\max_{u\in U} X^+(t,u) \leq \mathbb{E}\max_{t\in T}\max_{u\in U} Y(t,u). \quad (10.54)$$

Moreover, if we take $t' = t$ in (10.52), we get that

$$\mathbb{E}X^+(t,u)^2 = \mathbb{E}Y(t,u)^2 \text{ for all } (t,u),$$

so we are in a position to apply Theorem 10.4 to compare the probabilities as in (10.40),

$$\mathbb{P}\Big(\bigcup_{t\in T}\bigcap_{u\in U}\{X^+(t,u)\leq\lambda(t,u)\}\Big) \leq \mathbb{P}\Big(\bigcup_{t\in T}\bigcap_{u\in U}\{Y(t,u)\leq\lambda(t,u)\}\Big), \quad (10.55)$$

for arbitrary function $\lambda(t,u)$. Notice that (10.52) is also equal to zero if $u = u'$ so, by the same logic, we can switch the role of the parameters t and u,

$$\begin{aligned}\mathbb{E}\min_{u\in U}\max_{t\in T} Y(t,u) &\leq \mathbb{E}\min_{u\in U}\max_{t\in T} X^+(t,u)\\ &\leq \mathbb{E}\max_{u\in U}\max_{t\in T} X^+(t,u) \leq \mathbb{E}\max_{u\in U}\max_{t\in T} Y(t,u) \quad (10.56)\end{aligned}$$

and

$$\mathbb{P}\Big(\bigcup_{u\in U}\bigcap_{t\in T}\{X^+(t,u)\leq\lambda(t,u)\}\Big) \leq \mathbb{P}\Big(\bigcup_{u\in U}\bigcap_{t\in T}\{Y(t,u)\leq\lambda(t,u)\}\Big). \quad (10.57)$$

This example will be useful to us in Section 10.6. □

Example 10.4.3. There is another special case that is very useful, when

$$\|t\| = a \text{ for all } t\in T,\ \|u\|\leq b \text{ for all } u\in U, \quad (10.58)$$

for some constants $a,b>0$. In other words, T is a subset of the sphere or radius a in $\mathbb{R}^n$ and U is a subset of the ball of radius b in $\mathbb{R}^m$. In this case, we will modify the definition (10.44) slightly and consider a proper random linear form

$$Y^+(t,u) = b\sum_{i=1}^n h_i t_i + a\sum_{j=1}^m g_j u_j. \quad (10.59)$$

As before, by a direct calculation, one can check that

$$\begin{aligned}&\mathbb{E}\big(Y^+(t,u)-Y^+(t',u')\big)^2 - \mathbb{E}\big(X(t,u)-X(t',u')\big)^2 \quad (10.60)\\ &= 2\big(a^2-(t,t')\big)\big(b^2-(u,u')\big). \quad (10.61)\end{aligned}$$

By the Cauchy-Schwarz inequality, this difference is nonnegative. Moreover, it is equal to zero when $t = t'$ by the assumption that $\|t\| = a$. This implies, by Theorem 10.3 and Theorem 10.5,

$$\mathbb{E}\min_{t\in T}\max_{u\in U} Y^+(t,u) \leq \mathbb{E}\min_{t\in T}\max_{u\in U} X(t,u)$$

$$\le \mathbb{E}\max_{t\in T}\max_{u\in U} X(t,u) \le \mathbb{E}\max_{t\in T}\max_{u\in U} Y^+(t,u). \tag{10.62}$$

We will use this example in Section 10.5 below. □

Example 10.4.4. In the setting of the Example 10.4.1, let us suppose that

$$\|t\| = a \text{ for all } t \in T, \tag{10.63}$$

i.e. T is a subset of the sphere or radius a in $\mathbb{R}^n$. If we take $u = u'$ in (10.47), we get the equality

$$\mathbb{E}\big(Y(t,u) - Y(t',u)\big)^2 = \mathbb{E}\big(X(t,u) - X(t',u)\big)^2. \tag{10.64}$$

As in (10.56), this implies

$$\begin{aligned}\mathbb{E}\min_{u\in U}\max_{t\in T} Y(t,u) &\le \mathbb{E}\min_{u\in U}\max_{t\in T} X(t,u)\\ &\le \mathbb{E}\max_{u\in U}\max_{t\in T} X(t,u) \le \mathbb{E}\max_{u\in U}\max_{t\in T} Y(t,u).\end{aligned} \tag{10.65}$$

□

Example 10.4.5. Let us take $T = S^{n-1}$ and $U = S^{m-1}$ to be unit spheres in $\mathbb{R}^n$ and $\mathbb{R}^m$. We will use Example 10.4.3 with $a = b = 1$ and the role of t and u reversed in (10.62). First of all,

$$\min_{\|u\|=1}\max_{\|t\|=1}\Big(\sum_{i=1}^n h_i t_i + \sum_{j=1}^m g_j u_j\Big) = \max_{\|t\|=1}(h,t) - \max_{\|u\|=1}(g,u) = \|h\| - \|g\|,$$

so

$$\mathbb{E}\min_{u\in U}\max_{t\in T} Y^+(t,u) = \mathbb{E}\|h\| - \mathbb{E}\|g\|.$$

This can be computed explicitly, and it is well-known that

$$\frac{n}{\sqrt{n+1}} \le \mathbb{E}\|h\| = \frac{\sqrt{2}\,\Gamma(\frac{n+1}{2})}{\Gamma(\frac{n}{2})} \le \sqrt{n} \tag{10.66}$$

(we leave it as an exercise below). The same holds for $\mathbb{E}\|g\|$ with n replaced by m. On the other hand,

$$\min_{u\in U}\max_{t\in T} X(t,u) = \min_{\|u\|=1}\Big(\sum_{i=1}^n\Big(\sum_{j=1}^m g_{i,j}u_j\Big)^2\Big)^{1/2},$$

which we can rewrite as follows. Let us introduce the notation

$$g_i = (g_{i,1},\ldots,g_{i,m})^T \in \mathbb{R}^m \text{ and } Z = \frac{1}{n}\sum_{i=1}^n g_i g_i^T. \tag{10.67}$$

Vectors g_i are i.i.d. standard Gaussian in $\mathbb{R}^m$ and $m \times m$ matrix Z is their sample covariance matrix. We can then rewrite

$$\sum_{j=1}^{n}\Big(\sum_{j=1}^{m} g_{i,j}u_j\Big)^2 = \sum_{j=1}^{n}(g_i,u)^2 = \sum_{j=1}^{n} u^T g_i g_i^T u = \sum_{j=1}^{n}(g_i g_i^T u, u) = n(Zu,u).$$

Since Z is symmetric and positive semi-definite,

$$\min_{\|u\|=1}(Zu,u) = \lambda_{\min}(Z) \geq 0,$$

where $\lambda_{\min}(Z)$ is the smallest eigenvalue of Z. The first inequality in (10.62) gives

$$\mathbb{E}\|h\| - \mathbb{E}\|g\| \leq \mathbb{E}\sqrt{n\lambda_{\min}(Z)}, \tag{10.68}$$

and, using (10.66), we can write

$$\sqrt{\frac{n}{n+1}} - \sqrt{\frac{m}{n}} \leq \mathbb{E}\sqrt{\lambda_{\min}(Z)}. \tag{10.69}$$

When $n = (1+\varepsilon)m$, i.e. the sample size n is relatively bigger than the dimension m of our space, the left hand side is separated away from zero, so the smallest eigenvalue $\lambda_{\min}(Z)$ is separated away from zero, at least on average. To obtain similar statement in probability, we can use Example 10.4.5. The only difference will be that

$$\min_{u\in U}\max_{t\in T} X^+(t,u) = \sqrt{n\lambda_{\min}(Z)} + z,$$

and using (10.57) with $\lambda(t,u) \equiv \lambda$, we get

$$\mathbb{P}\Big(\sqrt{n\lambda_{\min}(Z)} + z \leq \lambda\Big) \leq \mathbb{P}\Big(\|h\| - \|g\| \leq \lambda\Big). \tag{10.70}$$

If we take $\lambda = t\sqrt{n}$ and flip the inequality, we get

$$\mathbb{P}\Big(\sqrt{\lambda_{\min}(Z)} + \frac{z}{\sqrt{n}} \geq t\Big) \geq \mathbb{P}\Big(\frac{\|h\| - \|g\|}{\sqrt{n}} \geq t\Big). \tag{10.71}$$

By the law of large number, $(\|h\| - \|g\|)/\sqrt{n} \approx 1 - \sqrt{m/n}$, so the second probability will be close to one if n is such that $1 - \sqrt{m/n} > t$, or $n > m/(1-t)^2$. Since $z/\sqrt{n}$ is negligible for large n, this shows that with probability close to one, $\lambda_{\min}(Z) \geq t^2$. This shows that when $n = (1+\varepsilon)m$, $\lambda_{\min}(Z)$ is separated away from zero with high probability, and not only on average.

We can reformulate these results for general Gaussian vectors on $\mathbb{R}^m$ with the covariance C. If $X_i = Ag_i$ for some matrix A such that $C = AA^T$, then $(X_i)_{i\leq n}$ are i.i.d. $N(0,C)$ and

$$Z_C = \frac{1}{n}\sum_{i=1}^{n} X_i X_i^T = AZA^T \tag{10.72}$$

is their sample covariance matrix, where Z was defined in (10.67). Since

$$\lambda_{\min}(Z_C) = \inf_{\|u\|=1}(Z_C u, u) = \inf_{\|u\|=1}(AZA^T u, u) = \inf_{\|u\|=1}(ZA^T u, A^T u)$$
$$\geq \lambda_{\min}(Z) \inf_{\|u\|=1}(A^T u, A^T u) = \lambda_{\min}(Z) \inf_{\|u\|=1}(Cu, u) = \lambda_{\min}(Z)\lambda_{\min}(C),$$

the statements we obtained above for the sample covariance matrix Z can be transferred to similar statements for Z_C, only now scaled by $\lambda_{\min}(C)$. □

Exercise 10.4.1. If $g \sim N(0,I)$ is standard Gaussian on $\mathbb{R}^n$ and $\|g\|$ is its Euclidean norm, show that

$$\frac{n}{\sqrt{n+1}} \leq \mathbb{E}\|g\| = \frac{\sqrt{2}\Gamma(\frac{n+1}{2})}{\Gamma(\frac{n}{2})} \leq \sqrt{n}.$$

10.5 Johnson–Lindenstrauss lemma

In this section, we will give one classical application of the results in the previous section. Let us consider $N \geq 1$ and a set

$$V = \{v_1, \ldots, v_m\} \subseteq \mathbb{R}^N \tag{10.73}$$

of m points in $\mathbb{R}^N$. The dimension N here can be arbitrarily large, and we should think of the number of points m as also being large. The Johnson–Lindenstrauss lemma states that there exists a linear map from $\mathbb{R}^N$ into Euclidean space $\mathbb{R}^n$ of possibly much lower dimension n that preserves the distances between all the points in the set V up to a small relative error. This is called a *low-distortion embedding*. It was discovered in a work on functional analysis, but it found many applications to computational algorithms in various fields as a preprocessing step to reduce the dimensionality of high-dimensional data. Here is the precise statement.

Theorem 10.6 (Johnson–Lindenstrauss lemma). *Given m points $v_1, \ldots, v_m$ in $\mathbb{R}^N$, any $\varepsilon \in (0,1)$, and*

$$n > \frac{4}{\varepsilon^2} \log m, \tag{10.74}$$

there exists a linear map $f\colon \mathbb{R}^N \to \mathbb{R}^n$ such that

$$\frac{\sqrt{n}}{\sqrt{n+1}} - \varepsilon \leq \frac{\|f(v_k) - f(v_\ell)\|}{\|v_k - v_\ell\|} \leq 1 + \varepsilon \tag{10.75}$$

for all $1 \leq k < \ell \leq m$, where $\|\cdot\|$ denotes the Euclidean norm (in $\mathbb{R}^N$ and $\mathbb{R}^n$).

The dimension n in (10.74) that we are allowed to choose depends on the distortion parameter ε and the number of points m, but it does not depend on N. The dependence on m is logarithmic, so the dimension can be relatively small even when the number of points is very large. We will give two proofs of this result, using Gaussian concentration and Gaussian comparison.

Proof (using Gaussian concentration). In our first proof we will assume (10.74) with constant 8 instead of 4 (however, using Gaussian concentration inequality with optimal constant would resolve this issue).

Let us consider a random matrix

$$\mathscr{G} = \begin{bmatrix} g_{11} & g_{12} & g_{13} & \cdots & g_{1N} \\ g_{21} & g_{22} & g_{23} & \cdots & g_{2N} \\ \vdots & \vdots & \vdots & \ddots & \vdots \\ g_{n1} & g_{n2} & g_{n3} & \cdots & g_{nN} \end{bmatrix},$$

where the entries g_{ij} are i.i.d. standard Gaussian random variables, and show that the linear map

$$f(x) = \frac{1}{\sqrt{n}} \mathscr{G} x \tag{10.76}$$

satisfies the low-distortion property (10.75) with positive probability. Let us denote, for $1 \le k < \ell \le m$,

$$a_{k\ell} = \frac{v_k - v_\ell}{\|v_k - v_\ell\|}.$$

Then the equation (10.75) can be rewritten, for f in (10.76), as

$$\frac{\sqrt{n}}{\sqrt{n+1}} - \varepsilon \le \frac{1}{\sqrt{n}} \|\mathscr{G} a_{k\ell}\| \le 1 + \varepsilon$$

for all $1 \le k < \ell \le m$. For a fixed $a = (a_1, \dots, a_N) \in S^{N-1}$, the coordinates of $\mathscr{G} a$,

$$(\mathscr{G} a)_i = g_{i1} a_1 + \ldots + g_{iN} a_N,$$

are independent standard Gaussian random variables, so $\mathscr{G} a$ is a standard Gaussian vector on $\mathbb{R}^n$. Combing Exercise 10.2.2 and Exercise 10.4.1, we get that

$$\mathbb{P}\Big(\frac{\sqrt{n}}{\sqrt{n+1}} - \varepsilon \le \frac{1}{\sqrt{n}} \|\mathscr{G} a\| \le 1 + \varepsilon \Big) \ge 1 - 2e^{-n\varepsilon^2/4}.$$

Since $a_{k\ell} \in S^{N-1}$ for all for $1 \le k < \ell \le m$,

$$\mathbb{P}\Big(\frac{\sqrt{n}}{\sqrt{n+1}} - \varepsilon \le \frac{1}{\sqrt{n}} \|\mathscr{G} a_{k\ell}\| \le 1 + \varepsilon \Big) \ge 1 - 2e^{-n\varepsilon^2/4},$$

and, by the union bound,

$$\mathbb{P}\Big(\forall 1 \le k < \ell \le m, \frac{\sqrt{n}}{\sqrt{n+1}} - \varepsilon \le \frac{1}{\sqrt{n}} \|\mathscr{G} a_{k\ell}\| \le 1 + \varepsilon \Big) \ge 1 - m^2 e^{-n\varepsilon^2/4}.$$

The right hand side is strictly positive if $n > \frac{8}{\varepsilon^2} \log m$. The fact that the probability is positive means that there exists $\mathscr{G}$ such that the low-distortion condition (10.75) is satisfied. □

Proof (using Gaussian comparison). We will now show a proof using Gordon's inequality in the form of the Example 10.4.3. Let us again consider the set

$$A = \Big\{ a_{k\ell} = \frac{v_k - v_\ell}{\|v_k - v_\ell\|} \in S^{N-1} : 1 \le k < \ell \le n \Big\},$$

and let $B = \{x \in \mathbb{R}^n : \|x\| \le 1\}$ be the unit ball in $\mathbb{R}^n$. Given the Gaussian $n \times N$ matrix $\mathscr{G}$ in the previous proof, consider a bilinear form for $a \in A$ and $b \in B$,

$$X(a,b) := \sum_{i=1}^{n} \sum_{j=1}^{N} g_{ij} b_i a_j.$$

Given standard Gaussian vectors $h = (h_i)_{i \le n}$ on $\mathbb{R}^n$ and $g = (g_j)_{j \le N}$ on $\mathbb{R}^N$, we consider a linear form

$$Y(a,b) := \sum_{i=1}^{n} h_i b_i + \sum_{j=1}^{N} g_j a_j.$$

Then, the comparison inequality (10.62) in the Example 10.4.3 gives

$$\begin{aligned} \mathbb{E} \min_{a\in A} \max_{b\in B} Y(a,b) &\le \mathbb{E} \min_{a\in A} \max_{b\in B} X(a,b) \\ &\le \mathbb{E} \max_{a\in A} \max_{b\in B} X(a,b) \le \mathbb{E} \max_{a\in A} \max_{b\in B} Y(a,b). \end{aligned} \tag{10.77}$$

If we denote

$$\Delta := \max_{a\in A} (g,a) = \max_{a\in A} \sum_{j=1}^{N} g_j a_j \tag{10.78}$$

then (10.77) can be rewritten as (also using $-g \overset{d}{=} g$)

$$\mathbb{E}\|h\| - \mathbb{E}\Delta \le \mathbb{E} \min_{a\in A} \|\mathscr{G} a\| \le \mathbb{E} \max_{a\in A} \|\mathscr{G} a\| \le \mathbb{E}\|h\| + \mathbb{E}\Delta.$$

From Exercise 10.4.1 we know that

$$\frac{n}{\sqrt{n+1}} \le \mathbb{E}\|h\| = \frac{\sqrt{2}\Gamma(\frac{n+1}{2})}{\Gamma(\frac{n}{2})} \le \sqrt{n}.$$

To estimate $\mathbb{E}\Delta$, take any $\beta > 0$ and write

$$\begin{aligned} \mathbb{E} \max_{a\in A} (g,a) &\le \mathbb{E} \frac{1}{\beta} \log \sum_{a\in A} e^{\beta(g,a)} \le \frac{1}{\beta} \log \mathbb{E} \sum_{a\in A} e^{\beta(g,a)} \\ &= \frac{1}{\beta} \log \sum_{a\in A} e^{\beta^2/2} = \frac{\beta}{2} + \frac{1}{\beta} \log \operatorname{card}(A). \end{aligned}$$

Optimizing over β gives

$$\mathbb{E}\Delta \le \sqrt{2 \log \operatorname{card} A} \le 2\sqrt{\log m} < \varepsilon\sqrt{n},$$

since $\operatorname{card}(A) \le m^2$ and we assumed that $n > \frac{4}{\varepsilon^2} \log m$. Together, the inequalities above imply

$$\frac{\sqrt{n}}{\sqrt{n+1}} - \varepsilon < \frac{1}{\sqrt{n}} \mathbb{E} \min_{a\in A} \|\mathscr{G} a\| \le \frac{1}{\sqrt{n}} \mathbb{E} \max_{a\in A} \|\mathscr{G} a\| < 1 + \varepsilon.$$

This implies that

$$\Big(\frac{\sqrt{n}}{\sqrt{n+1}} - \varepsilon\Big)^{-1} \mathbb{E} \min_{a\in A} \|\mathscr{G} a\| > 1 > (1+\varepsilon)^{-1} \mathbb{E} \max_{a\in A} \|\mathscr{G} a\|$$

and, therefore,

$$\mathbb{E}\Big[\min_{a\in A}\|\mathcal{G}a\| - (1+\varepsilon)^{-1}\Big(\frac{\sqrt{n}}{\sqrt{n+1}} - \varepsilon\Big)\max_{a\in A}\|\mathcal{G}a\|\Big] > 0.$$

Therefore, there exists $\mathcal{G}$ such that

$$\min_{a\in A}\|\mathcal{G}a\| - (1+\varepsilon)^{-1}\Big(\frac{\sqrt{n}}{\sqrt{n+1}} - \varepsilon\Big)\max_{a\in A}\|\mathcal{G}a\| > 0.$$

If we rescale $\mathcal{G} \to \overline{\mathcal{G}}$ so that $\max_{a\in A}\|\overline{\mathcal{G}}a\| = 1+\varepsilon$, we get that

$$\min_{a\in A}\|\overline{\mathcal{G}}a\| > \frac{\sqrt{n}}{\sqrt{n+1}} - \varepsilon.$$

This shows the existence of a linear map $\overline{\mathcal{G}}$ with the distortion property (10.75). □

Suppose that we also want to control the distortion of the areas of triangles formed by any three points v_i, v_j and v_k. This can be done as follows. First, the areas of all triangles in the same (two-dimensional) plane are distorted by the same factor under a linear transformation. For each three points, we can add another point v_{ijk} in the same plane, which forms an *isosceles right* triangle with v_i, v_j. This means that, instead of m points, we now have $m + \binom{m}{3} \le m^3$ points on the list. We can now find f as in Lemma 10.6 for this new enlarged list of points. Since the dependence on m is logarithmic, we lose a factor of 3 in the bound (10.74) for n. It remains to solve the following exercise to see that this procedure allows us to control the distortion of areas.

Exercise 10.5.1. Suppose that three points v_i, v_j and v_k in the setting of Lemma 10.6 form an isosceles right triangle Δ. Show that

$$\sqrt{1-5\varepsilon} \le \frac{\text{Area}(f(\Delta))}{\text{Area}(\Delta)} \le 1+\varepsilon.$$

Exercise 10.5.2. Show that, for $\delta \in (0,1)$, if the dimension

$$n > \frac{8}{\varepsilon^2}\log\frac{m}{\sqrt{\delta}}$$

then the map $f(x) = \frac{1}{\sqrt{n}}\mathcal{G}x$ in the above proof satisfies (10.75) with probability at least $1-\delta$.

10.6 Intersecting random half-spaces

In this section, we will show another application of Gaussian comparison. Consider the following problem. Let $g^1,\dots,g^m$ be i.i.d. standard Gaussian random vectors on $\mathbb{R}^n$, and $\kappa \in \mathbb{R}$ is a fixed number. Consider a set

$$A = \bigcap_{\ell=1}^{m} \{x \in S^{n-1} : g^\ell \cdot x \geq \kappa\}, \tag{10.79}$$

which is an intersection of a unit sphere with m random half-spaces defined by random directions g^ℓ and some threshold κ. We will think of the dimension n as being large, and we would like to know how many random half-spaces we need to intersect to make this set empty with high probability. In Physics this is known as a "perceptron capacity" problem. For negative κ this is an open problem, but for $\kappa \geq 0$ the answer is known. It turns out that m should be proportional to n, i.e. $m = \alpha n$, and the following holds. We write $x_+ = \max(x,0)$ and let $z \sim N(0,1)$.

Theorem 10.7 (Gardner-Stojnic theorem). *For all* $\kappa \in \mathbb{R}$,

$$\alpha \mathbb{E}(z+\kappa)_+^2 > 1 \implies \mathbb{P}(A = \emptyset) \geq 1 - e^{-cn} \tag{10.80}$$

for some constant $c = c(\alpha,\kappa) > 0$ *and, for all* $\kappa \geq 0$,

$$\alpha \mathbb{E}(z+\kappa)_+^2 < 1 \implies \mathbb{P}(A \neq \emptyset) \geq 1 - e^{-cn} \tag{10.81}$$

for some constant $c = c(\alpha,\kappa) > 0$.

The first statement holds for all κ, but for negative κ this threshold is not the correct one. The second statement shows that the threshold is sharp for nonnegative κ, in the sense that for α below the threshold the set A is not empty with high probability and for α above the threshold the set A is empty with high probability. For $\kappa = 0$, which corresponds to half-spaces through the origin, the threshold is 2, which means that one must intersect more than $2n$ half-spaces to get an empty intersection with the unit sphere with high probability.

Proof (of (10.80)). Assume $\alpha \mathbb{E}(z+\kappa)_+^2 > 1$. Let G be a $m \times n$ matrix whose entries are i.i.d. standard Gaussian random variables, and $\mathbb{1} = (1,\dots,1)^T \in \mathbb{R}^m$. Then A is non-empty if and only if there exists $x \in S^{n-1}$ such that $Gx \geq \kappa \mathbb{1}$ where the inequality of two vectors should be interpreted coordinate-wise. We notice that this is equivalent to

$$\min_x \max_\lambda \lambda^T(\kappa \mathbb{1} - Gx) \leq 0, \tag{10.82}$$

where minimax is taken over $(x,\lambda) \in S^{n-1} \times B_+^m$ and where

$$B_+^m = \{\lambda \in \mathbb{R}^m : \|\lambda\| \leq 1; \lambda_i \geq 0, 1 \leq i \leq m\}.$$

From now on we consider only such pairs $(x,\lambda) \in S^{n-1} \times B^m_+$. Therefore, to show (10.80) it suffices to show that there exists $\delta > 0$ such that

$$\mathbb{P}\Big(\min_x \max_\lambda \lambda^T(\kappa\mathbb{1} - Gx) \geq \delta\Big) \geq 1 - e^{-cn}.$$

We start with the following observation. Let $z \sim N(0,1)$ be a standard Gaussian random variable independent of the entries in G. Then, for any $\delta, \varepsilon > 0$,

$$\begin{aligned}&\mathbb{P}\Big(\min_x \max_\lambda \lambda^T(\kappa\mathbb{1} - Gx) + z \geq \varepsilon\sqrt{n}\Big)\\ &\leq \mathbb{P}\Big(\min_x \max_\lambda \lambda^T(\kappa\mathbb{1} - Gx) \geq \delta\Big) + \mathbb{P}(z \geq \varepsilon\sqrt{n} - \delta).\end{aligned}$$

The second term on the right-hand side can be bounded by $e^{-\varepsilon^2 n/8}$, so we will try to bound the probability on the left-hand side from below. We will use Gordon's inequality in Theorem 10.4 in Section 10.3. Let

$$X(x,\lambda) := -\lambda^T Gx + z,\ Y(x,\lambda) := \lambda^T g + x^T h$$

where $g \in \mathbb{R}^m, h \in \mathbb{R}^n$ are independent standard Gaussian random vectors. It is easy to see that

$$\mathbb{E}X(x_1,\lambda_1)X(x_2,\lambda_2) = \mathbb{E}(-\lambda_1^T Gx_1 + z)(-\lambda_2^T Gx_2 + z) = (\lambda_1 \cdot \lambda_2)(x_1 \cdot x_2) + 1,$$

and

$$\mathbb{E}Y(x_1,\lambda_1)Y(x_2,\lambda_2) = \mathbb{E}(\lambda_1^T g + x_1^T h)(\lambda_2^T g + x_2^T h) = \lambda_1 \cdot \lambda_2 + x_1 \cdot x_2.$$

If $(x_1,\lambda_1) = (x_2,\lambda_2)$ then both variances are equal to 2. Also,

$$\mathbb{E}X(x_1,\lambda_1)X(x_2,\lambda_2) - \mathbb{E}Y(x_1,\lambda_1)Y(x_2,\lambda_2) = (1 - \lambda_1 \cdot \lambda_2)(1 - x_1 \cdot x_2) \geq 0,$$

by the Cauchy-Schwarz inequality. However, if $x_1 = x_2$,

$$\mathbb{E}X(x_1,\lambda_1)X(x_1,\lambda_2) - \mathbb{E}Y(x_1,\lambda_1)Y(x_1,\lambda_2) = (1 - \lambda_1 \cdot \lambda_2)(1 - \|x_1\|^2) = 0 \leq 0.$$

Therefore, the assumptions in Theorem 10.4 are satisfied and

$$\begin{aligned}&\mathbb{P}\Big(\min_x \max_\lambda \big(X(x,\lambda) + \kappa\lambda^T\mathbb{1}\big) \geq \varepsilon\sqrt{n}\Big)\\ &\geq \mathbb{P}\Big(\min_x \max_\lambda \big(Y(x,\lambda) + \kappa\lambda^T\mathbb{1}\big) \geq \varepsilon\sqrt{n}\Big).\end{aligned}$$

The minimax on the right hand side can be computed explicitly,

$$\min_x \max_\lambda \big(Y(x,\lambda) + \kappa\lambda^T\mathbb{1}\big) = \max_{\lambda \in S^{m-1}_+} \lambda^T(g + \kappa\mathbb{1}) + \min_{x \in S^{n-1}} x^T h$$

$$= \|(g+\kappa\mathbb{1})_+\| - \|h\|,$$

where $(g+\kappa\mathbb{1})_+ = ((g_i+\kappa)_+)_{i\le m}$. By the assumption $\alpha\mathbb{E}(z+\kappa)_+^2 > 1$, we can choose $\varepsilon > 0$ small enough such that $\alpha\mathbb{E}(z+\kappa)_+^2 \ge (1+3\varepsilon)^2$, or

$$(m\mathbb{E}(z+\kappa)_+^2)^{1/2} - \sqrt{n} - 2\varepsilon\sqrt{n} \ge \varepsilon\sqrt{n}.$$

Therefore,

$$\begin{aligned}
&\mathbb{P}\Big(\min_x \max_\lambda \big(X(x,\lambda) + \kappa\lambda^T\mathbb{1}\big) \ge \varepsilon\sqrt{n}\Big) \\
&\ge \mathbb{P}\Big(\|(g+\kappa\mathbb{1})_+\| - \|h\| \ge \varepsilon\sqrt{n}\Big) \\
&\ge \mathbb{P}\Big(\|(g+\kappa\mathbb{1})_+\| - \|h\| \ge (m\mathbb{E}(z+\kappa)_+^2)^{1/2} - \sqrt{n} - 2\varepsilon\sqrt{n}\Big) \\
&\ge 1 - \mathbb{P}\Big(\|h\| \ge \sqrt{n} + \varepsilon\sqrt{n}\Big) \\
&\quad - \mathbb{P}\Big(\|(g+\kappa\mathbb{1})_+\| \le (m\mathbb{E}(z+\kappa)_+^2)^{1/2} - \varepsilon\sqrt{n}\Big).
\end{aligned}$$

The first probability is exponentially small by Gaussian concentration inequality for the norm $\|h\|$. In the last term, we can rewrite the event as

$$\sum_{i=1}^m (g_i+\kappa)_+^2 \le m\mathbb{E}(z+\kappa)_+^2 - C_\varepsilon n,$$

where $C_\varepsilon = 2\varepsilon(\alpha\mathbb{E}(z+\kappa)_+^2)^{1/2} - \varepsilon^2 \ge 2\varepsilon - \varepsilon^2 \ge \varepsilon$ for $\varepsilon \in (0,1)$. One can then rewrite this as

$$\sum_{i=1}^m \Big[\mathbb{E}(g_i+\kappa)_+^2 - (g_i+\kappa)_+^2\Big] \ge C_\varepsilon n,$$

and use Markov's inequality as in Section 2.1 to show that this probability is exponentially small. We will leave this as an exercise. □

Proof (of (10.81)). Now we assume that $\alpha\mathbb{E}(z+\kappa)_+^2 < 1$ and $\kappa \ge 0$. Because $\kappa \ge 0$, the existence of a point $x \in S^{n-1}$ such that $Gx \ge \mathbb{1}\kappa$ is equivalent to the existence of a point $x \in B^n$ in the unit ball such that $Gx \ge \mathbb{1}\kappa$. Analogously to (10.82), this is equivalent to

$$\min_x \max_\lambda \lambda^T(\kappa\mathbb{1} - Gx) \le 0, \tag{10.83}$$

where the minimax is now taken over $(x,\lambda) \in B^n \times B_+^m$. Because both B^n and B_+^m are convex and compact, by Sion's minimax theorem,

$$\min_x \max_\lambda \lambda^T(\kappa\mathbb{1} - Gx) = \max_\lambda \min_x \lambda^T(\kappa\mathbb{1} - Gx) = -\min_\lambda \max_x \lambda^T(Gx - \kappa\mathbb{1}).$$

Therefore, our goal is to show that

$$\mathbb{P}\Big(\min_{\lambda}\max_{x}\lambda^T(Gx-\kappa\mathbb{1})\geq 0\Big)\geq 1-e^{-cn}$$

for some $c>0$. In this direction, the calculation will be based on a comparison inequality in the Example 10.4.2 in Section 10.4.

As above, let $z\sim N(0,1)$ be a standard Gaussian random variable independent of the entries in G. Then, for any $\varepsilon>0$,

$$\begin{aligned}&\mathbb{P}\Big(\min_{\lambda}\max_{x}\Big[\lambda^T Gx-\kappa\lambda^T\mathbb{1}+z\|\lambda\|\|x\|-\varepsilon\sqrt{n}\|\lambda\|\|x\|\Big]\geq 0\Big)\\ &\leq\mathbb{P}\Big(\min_{\lambda}\max_{x}\Big[\lambda^T Gx-\kappa\lambda^T\mathbb{1}\Big]\geq 0\Big)+\mathbb{P}(z\geq\varepsilon\sqrt{n}).\end{aligned}$$

The second term on the right-hand side can be bounded by $e^{-\varepsilon^2 n/2}$, so we will try to bound the probability on the left-hand side from below. We will again use Gordon's inequality in Theorem 10.4, only now

$$X(x,\lambda):=\lambda^T Gx+z\|\lambda\|\|x\|,\ Y(x,\lambda):=\|x\|\lambda^T g+\|\lambda\|x^T h$$

where $g\in\mathbb{R}^m, h\in\mathbb{R}^n$ are independent standard Gaussian random vectors. One can check that that

$$\mathbb{E}X(x_1,\lambda_1)X(x_2,\lambda_2)=(\lambda_1\cdot\lambda_2)(x_1\cdot x_2)+\|x_1\|\|x_2\|\|\lambda_1\|\|\lambda_2\|,$$

and

$$\mathbb{E}Y(x_1,\lambda_1)Y(x_2,\lambda_2)=\|x_1\|\|x_2\|(\lambda_1\cdot\lambda_2)+\|\lambda_1\|\|\lambda_2\|(x_1\cdot x_2).$$

If $(x_1,\lambda_1)=(x_2,\lambda_2)$ then the variances are again equal. Also,

$$\begin{aligned}&\mathbb{E}X(x_1,\lambda_1)X(x_2,\lambda_2)-\mathbb{E}Y(x_1,\lambda_1)Y(x_2,\lambda_2)\\ &=(\|\lambda_1\|\|\lambda_2\|-\lambda_1\cdot\lambda_2)(\|x_1\|\|x_2\|-x_1\cdot x_2)\geq 0,\end{aligned}$$

by the Cauchy-Schwarz inequality. However, if $\lambda_1=\lambda_2$,

$$\mathbb{E}X(x_1,\lambda_1)X(x_2,\lambda_1)-\mathbb{E}Y(x_1,\lambda_1)Y(x_2,\lambda_1)=0\leq 0.$$

Therefore, the assumptions in Theorem 10.4 are satisfied and

$$\begin{aligned}&\mathbb{P}\Big(\min_{\lambda}\max_{x}\Big[X(x,\lambda)-\kappa\lambda^T\mathbb{1}-\varepsilon\sqrt{n}\|\lambda\|\|x\|\Big]\geq 0\Big)\\ &\geq\mathbb{P}\Big(\min_{\lambda}\max_{x}\Big[Y(x,\lambda)-\kappa\lambda^T\mathbb{1}-\varepsilon\sqrt{n}\|\lambda\|\|x\|\Big]\geq 0\Big).\end{aligned}$$

The minimax on the right hand side can be computed explicitly. First, we maximize over $x\in B^n$,

$$\max_{x}\Big(\|x\|\lambda^T g+\|\lambda\|x^T h-\varepsilon\sqrt{n}\|x\|\|\lambda\|\Big)$$

$$\begin{aligned}
&= \max_{r\in[0,1],v\in S^{n-1}} r\Big(\lambda^T g + \|\lambda\| v^T h - \varepsilon\sqrt{n}\|\lambda\|\Big) \\
&= \max_{r\in[0,1]} r\Big(\lambda^T g + \|\lambda\|\|h\| - \varepsilon\sqrt{n}\|\lambda\|\Big) \\
&= \max\Big(0, \lambda^T g + \|\lambda\|\|h\| - \varepsilon\sqrt{n}\|\lambda\|\Big) \\
&\geq \lambda^T g - \varepsilon\sqrt{n}\|\lambda\| + \|\lambda\|\|h\|.
\end{aligned}$$

Therefore, the above probability is bounded from below by

$$\mathbb{P}\Big(\min_{\lambda}\Big[\lambda^T g - \varepsilon\sqrt{n}\|\lambda\| + \|\lambda\|\|h\| - \kappa\lambda^T \mathbb{1}\Big] \geq 0\Big).$$

If we represent $\lambda \in B_+^m$ as $\lambda = rv$ for $r \in [0,1]$ and

$$v \in S_+^{m-1} = \Big\{\lambda \in S^{m-1} : \lambda_i \geq 0, 1 \leq i \leq m\Big\},$$

the minimum over $\lambda \in B_+^m$ above can be computed as follows,

$$\begin{aligned}
&\min_{r\in[0,1],v\in S_+^{m-1}} r\Big[v^T(g - \kappa\mathbb{1}) - \varepsilon\sqrt{n} + \|h\|\Big] \\
&= \min_{r\in[0,1]} r\Big(-\|(\kappa\mathbb{1} - g)_+\| - \varepsilon\sqrt{n} + \|h\|\Big) \\
&= \min\Big(0, -\|(\kappa\mathbb{1} - g)_+\| - \varepsilon\sqrt{n} + \|h\|\Big).
\end{aligned}$$

Therefore, the above probability equals

$$\begin{aligned}
&\mathbb{P}\Big(-\|(\kappa\mathbb{1} - g)_+\| - \varepsilon\sqrt{n} + \|h\| \geq 0\Big) \\
&= \mathbb{P}\Big(-\|(\kappa\mathbb{1} + g)_+\| + \|h\| \geq \varepsilon\sqrt{n}\Big),
\end{aligned}$$

using symmetry $-g \stackrel{d}{=} g$. Since $\alpha\mathbb{E}(z+\kappa)_+^2 < 1$, we can find $\varepsilon > 0$ such that

$$-(m\mathbb{E}(z+\kappa)_+^2)^{1/2} + \sqrt{n} - 2\varepsilon\sqrt{n} > \varepsilon\sqrt{n}.$$

As a result, the probability above can be bounded from below by

$$\begin{aligned}
1 &- \mathbb{P}\Big(\|h\| \leq \sqrt{n} - \varepsilon\sqrt{n}\Big) \\
&- \mathbb{P}\Big(\|(g + \kappa\mathbb{1})_+\| \geq (m\mathbb{E}(z+\kappa)_+^2)^{1/2} + \varepsilon\sqrt{n}\Big).
\end{aligned}$$

The rest of the argument is the same as in the first part above, based on Gaussian concentration and the exercise below. □

Exercise 10.6.1. Show that

$$\mathbb{P}\Big(\sum_{i=1}^{m}\Big[\mathbb{E}(g_i+\kappa)_+^2-(g_i+\kappa)_+^2\Big]\geq C_\varepsilon n\Big)\leq e^{-cn}$$

for some constant that depends on κ, α and C_ε.